AF358698

Numerical and Analytical Methods in Electromagnetics

Numerical and Analytical Methods in Electromagnetics

Editor

Hristos T. Anastassiu

MDPI • Basel • Beijing • Wuhan • Barcelona • Belgrade • Manchester • Tokyo • Cluj • Tianjin

Editor
Hristos T. Anastassiu
Department of Informatics Engineering
Technological and Educational Institute of Central Macedonia at Serres
Greece

Editorial Office
MDPI
St. Alban-Anlage 66
4052 Basel, Switzerland

This is a reprint of articles from the Special Issue published online in the open access journal *Applied Sciences* (ISSN 2076-3417) (available at: https://www.mdpi.com/journal/applsci/special_issues/Numerical_Analytical_Methods).

For citation purposes, cite each article independently as indicated on the article page online and as indicated below:

LastName, A.A.; LastName, B.B.; LastName, C.C. Article Title. *Journal Name* **Year**, *Volume Number*, Page Range.

ISBN 978-3-0365-0064-5 (Hbk)
ISBN 978-3-0365-0065-2 (PDF)

Contents

About the Editor

Hristos T. Anastassiu, M.Sc.(1), M.Sc.(2), Ph.D., was born in Edessa, Greece, on July 11, 1966. He obtained the Diploma Degree in Electrical Engineering from the Aristotle University of Thessaloniki, Greece, in 1989. He was awarded the National Fellowship for academic excellence in every year of his studies. From 1989 to 1992 he was a Graduate Research Assistant at the ElectroScience Laboratory, the Ohio State University, U.S.A., where he obtained the M.Sc. degree in Electrical Engineering. From 1992 to 1997 he was a Graduate Research Assistant at the Radiation Laboratory, University of Michigan, U.S.A., where he obtained the Ph.D. degree in Electrical Engineering. In December 1995 he obtained the M.Sc. degree in Mathematics from the same university. He received the second place award in the student paper contest of the 1993 IEEE-AP (Antennas and Propagation) International Symposium held in Ann Arbor, MI. He served in the Hellenic Army (Artillery) from 1997 to 1998. Between 1999 and 2004 he was a Research Scientist at the Institute of Communication and Computer Systems (ICCS) of the National Technical University of Athens (NTUA). From June 2004 to March 2011 he was affiliated with the Hellenic Aerospace Industry (HAI), Tanagra, Greece. In September 2005 he was elected Adjunct Assistant Professor at the Hellenic Air Force Academy, Dekelia, Greece, where he taught courses in Electromagnetics and Communications until 2010. In March 2011 he was elected Associate Professor at the Department of Informatics and Communications of the Technological and Educational Institute (now International Hellenic University) of Serres, Greece, where he was promoted to Professor in June 2015. He is a Senior Member of IEEE, an Associate Member of $\Sigma\Xi$ (Sigma Xi), and a Member of the Technical Chamber of Greece. He has been chairman of the SET-085/RTG 49 and SET-138 RTG 75 RFT NATO/RTO research groups. He has participated in several research projects, and as of November 2020, he has been author or co-author of more than 140 scientific publications in international journals and conferences, which have gathered more than 780 citations.

Editorial

Special Issue "Numerical and Analytical Methods in Electromagnetics"

Hristos T. Anastassiu

Department of Informatics, Computer and Communications Engineering, International Hellenic University, Serres Campus, End of Magnisias Str., GR-62124 Serres, Greece; hristosa@teiser.gr

Received: 9 October 2020; Accepted: 13 October 2020; Published: 16 October 2020

1. Introduction

Like all branches of physics and engineering, electromagnetics relies on mathematical methods for modeling, simulation, and design procedures in all of its aspects (radiation, propagation, scattering, imaging, etc.). Originally, rigorous analytical techniques were the only machinery available to produce any useful results. Essentially, the aim was the solution of partial differential equations (such as the Laplace, Poisson, Helmholtz, and wave equations) since the electric and magnetic fields are the unknown quantities in such expressions, although exact analytical methods (e.g., the Wiener–Hopf technique) were limited to canonical geometries, which are unfortunately rare in nature. Hence, in the 1960s and 1970s, emphasis was placed on asymptotic techniques, which produced approximations of the fields for very high frequencies when closed-form solutions were not feasible. Typical examples of such techniques were the geometrical and physical optics (GO and PO, respectively), improved by the geometrical, physical, and uniform theories of diffraction (GTD, PTD and UTD, respectively). Later, when computers demonstrated explosive progress, numerical techniques were utilized to develop approximate results of controllable accuracy for arbitrary geometries. Either differential or integral equations were discretized, leading to standard techniques, such as the method of moments (MoM), finite element method (FEM), the finite difference time domain method (FDTD), finite integration technique (FIT), and the method of auxiliary sources (MAS). Researchers soon realized that several practical problems required extremely high computational resources, in terms of memory and CPU time, to handle, typically, millions of unknowns. Therefore, "fast" variants of the latter techniques were developed to suppress the computational cost, such as the adaptive integral method (AIM); the fast multipole method (FMM); its parallel version, called the multi-level fast multipole algorithm (MLFMA); and its time domain counterpart, i.e., the plane wave time domain (PWTD) method. The lists above are by no means exhaustive; there is a plethora of additional algorithms, having evolved particularly over the last few years, designed to reduce the complexity and simultaneously improve the accuracy of calculations. In this Special Issue, the most recent advances thereof were presented to illustrate the state-of-the-art mathematical techniques in electromagnetics.

2. The Contents of This Special Issue

A wide variety of practical electromagnetic problems were addressed in this Special Issue and further solved via appropriate mathematical methods. In [1], Wei Gao et al. used partial differential equation techniques to solve the nonlinear Schrödinger equations applied to wave propagation in optical fibers with nonlinear impacts. Two powerful analytical methods, namely the $(m+G'/G)$ improved expansion method and the $\exp(-\phi\,(\xi))$ expansion method were utilized to construct novel solutions of the governing equations.

In [2], the application topic is geophysics; Yanju Ji et al. propose an efficient approach of the GRounded Electrical-source Airborne Transient Electromagnetics (GREATEM), which is a widespread detection method among researchers in the field. Maxwell's equations are transformed via the relationship between the diffusion field and fictitious wave field. The fractional order Cole–Cole model is introduced into the fictitious wave field and the final solution is obtained by using the finite difference time domain (FDTD) method. Finally, an integral transformation is applied to obtain the calculation results in the actual diffusion field form.

Moreover, electromagnetic properties of composites filled with carbon nano tubes (CNTs) are modeled by A. Plyushch et al. in [3]. The total conductivity of the composite is governed by the inter-tube tunneling equation. In this framework, the direction for the conductivity computation is selected and the nanotubes near the initial and final boarders are collected. The Dijkstra algorithm is used to trace the paths of minimal resistance between the initial and final tubes, and, finally, conductivity is computed in a highly accurate way.

Electromagnetic wave amplification by non-linear wave mixing is targeted in [4] by Ö. E. Aşırım and M. Kuzuoğlu. Suitable numerical analysis is performed that provides evidence for the high-gain amplification of a low-power stimulus wave, via intense pump waves of ultra-short duration, inside a several-micrometers-long micro-resonator, by maximizing the electric energy density of the pump wave in the resonator. In order to perform the optimization of the stimulus wave magnitude at a given wave frequency, an efficient optimization procedure, namely the quasi-Newton Broyden–Fletcher–Goldfarb–Shanno (BFGS) algorithm is implemented.

Although analytical solutions are rare in real-world problems, efficient approximations are still applicable to important devices. Such a case is a birdcage radio frequency coil used in nuclear magnetic resonance (NMR) imaging applications, as demonstrated in [5] by Young Cheol Kim et al. A novel analytical solution for the characteristic properties of the coil is derived via equivalent circuit modeling and T-matrix theory, facilitating and efficient design strategy.

Scattering analysis is another important aspect of applied electromagnetics which is in need of powerful mathematical tools. In [6], V.G. Iatropoulos et al. describe how the method of auxiliary sources (MAS) can be optimized for cylindrical scattering geometries containing curvilinear wedges. Instead of retaining a conformal auxiliary surface, which is customary in MAS, auxiliary sources are locally positioned close to the wedge tips with variable density. Numerical results clearly show the reduction of the calculation error and the improvement in the accuracy of the radar cross section.

An enjoyable application of numerical techniques to robotic system design is described in [7]: a coil gun found in soccer ball launchers is analyzed and designed by V. Gies et al. A coupled electromagnetic, electrical and mechanical model is used to simulate the performance of reluctance coil guns. Four different mechatronic coupled models thereof are proposed, and for each one of these the electromagnetic behavior is investigated on the basis of a finite element (FEM) software tool, whereas commercial sofware was used for modeling the electrical and mechanical parts.

Comparison of simplified analytical models based on the principles of superposition and reflections and the finite element method (FEM) was performed by R. Deltuva and R. Lukočius in [8], where the modeling of electric power lines is facilitated. The target of the analysis is an actual high-voltage AC, double-circuit 400 kV overhead power transmission line that runs from the city of Elk, Poland, to the city of Alytus, Lithuania.

Integral equation methods could by no means be absent from this Special Issue. Indeed, an integral formulation was used in [9] by Tung Le-Duc and G. Meunier to model thin surfaces coupled with an external circuit. A hybrid integral formulation is proposed to allow for the modeling of an inhomogeneous structure constituted by conductors and thin magnetic and conducting shells. The resulting integral equations are discretized via a Galerkin procedure and are further transformed to a linear system of equations, finally solved by standard linear algebra techniques.

The impressive application range of computational electromagnetics is clearly demonstrated in [10]. I. B. Yeboah et al. address a problem in biomedical engineering, namely fibroadenoma, which is one of the commonest benign female breast diseases. A particular form of treatment involves nanomedicine, which is based on the use of nanomaterials—metallic and ceramic (iron-oxide) nanoparticles (NPs)—for theranostic purposes in living organisms. The authors characterize the material properties and quantify the photothermal heat generation of specific NPs by experimental measurements, obtain their optical absorption coefficient via experimentally guided Mie scattering theory and integrate it into a computational—finite element method (FEM)—model to predict the in-vivo thermal damage of an NP-embedded tumor located in a multi-tissue breast model during irradiation by a near-infrared (NIR) 810 nm laser.

Finally, the relationship of mathematical methods in electromagnetics with other research disciplines is underlined in multiphysics problems, such as the rail launcher addressed in [11]. The in-house integral equation code named "EN4EM" (Equivalent Network for Electromagnetic Modeling), developed by V. Consolo et al., is able to take into account all relevant electromechanical quantities and phenomena (i.e., eddy currents, velocity skin effect, sliding contacts etc.).

Acknowledgments: The Guest Editor would like to express his sincere appreciation to all authors, reviewers and the editorial board of *Applied Sciences* for their extraordinary work and excellent cooperation during the preparation of this Special Issue. Combined efforts by all were essential for the final success of this project. Special thanks are reserved for Jennifer Li, SI Managing Editor from the MDPI Branch Office, Beijing.

Conflicts of Interest: The author declares no conflict of interest.

References

1. Gao, W.; Ismael, H.F.; Husien, A.M.; Bulut, H.; Baskonus, H.M. Optical soliton solutions of the cubic-quartic nonlinear Schrödinger and resonant nonlinear Schrödinger equation with the parabolic law. *Appl. Sci.* **2019**, *10*, 219. [CrossRef]
2. Ji, Y.; Meng, X.; Huang, W.; Wu, Y.; Li, G. 3D numerical modeling of induced-polarization grounded electrical-source airborne transient electromagnetic response based on the fictitious wave field methods. *Appl. Sci.* **2020**, *10*, 1027. [CrossRef]
3. Plyushch, A.; Lyakhov, D.A.; Šimėnas, M.; Bychanok, D.; Macutkevič, J.; Michels, D.; Banys, J.; Lamberti, P.; Kuzhir, P. Percolation and transport properties in the mechanically deformed composites filled with carbon nanotubes. *Appl. Sci.* **2020**, *10*, 1315. [CrossRef]
4. Aşırım, Ö.E.; Kuzuoğlu, M. Super-gain optical parametric amplification in dielectric micro-resonators via BFGS algorithm-based non-linear programming. *Appl. Sci.* **2020**, *10*, 1770. [CrossRef]
5. Kim, Y.C.; Kim, H.D.; Yun, B.-J.; Ahmad, S.F. A Simple analytical solution for the designing of the birdcage RF coil used in NMR imaging applications. *Appl. Sci.* **2020**, *10*, 2242. [CrossRef]
6. Iatropoulos, V.G.; Anastasiadou, M.-T.; Anastassiu, H.T. Electromagnetic scattering from surfaces with curved wedges using the method of auxiliary sources (MAS). *Appl. Sci.* **2020**, *10*, 2309. [CrossRef]
7. Gies, V.; Soriano, T.; Marzetti, S.; Barchasz, V.; Barthélemy, H.; Glotin, H.; Hugel, V. Optimisation of energy transfer in reluctance coil guns: Application to soccer ball launchers. *Appl. Sci.* **2020**, *10*, 3137. [CrossRef]
8. Deltuva, R.; Lukočius, R. Distribution of magnetic field in 400 kV double-circuit transmission lines. *Appl. Sci.* **2020**, *10*, 3266. [CrossRef]
9. Le-Duc, T.; Meunier, G. 3-D integral formulation for thin electromagnetic shells coupled with an external circuit. *Appl. Sci.* **2020**, *10*, 4284. [CrossRef]
10. Yeboah, I.B.; Hatekah, S.W.K.; Konku-Asase, Y.; Yaya, A.; Kan-Dapaah, K. Destruction of fibroadenomas using photothermal heating of Fe_3O_4 nanoparticles: Experiments and models. *Appl. Sci.* **2020**, *10*, 5844. [CrossRef]
11. Consolo, V.; Musolino, A.; Rizzo, R.; Sani, L. Numerical 3D simulation of a full system air core compulsator-electromagnetic rail launcher. *Appl. Sci.* **2020**, *10*, 5903. [CrossRef]

Publisher's Note: MDPI stays neutral with regard to jurisdictional claims in published maps and institutional affiliations.

applied
sciences

Article

Optical Soliton Solutions of the Cubic-Quartic Nonlinear Schrödinger and Resonant Nonlinear Schrödinger Equation with the Parabolic Law

Wei Gao [1,*]**, Hajar Farhan Ismael** [2,3]**, Ahmad M. Husien** [4]**, Hasan Bulut** [3] **and Haci Mehmet Baskonus** [5]

[1] School of Information Science and Technology, Yunnan Normal University, Kunming 650500, China
[2] Department of Mathematics, Faculty of Science, University of Zakho, Zakho 42002, Iraq; hajar.ismael@uoz.edu.krd
[3] Department of Mathematics, Faculty of Science, Firat University, Elazig 23000, Turkey; hbulut@firat.edu.tr
[4] Department of Mathematics, College of Science, University of Duhok, Duhok 42001, Iraq; ahmad.husien@uod.ac
[5] Department of Mathematics and Science Education, Faculty of Education, Harran University, Sanliurfa 63100, Turkey; hmbaskonus@gmail.com
* Correspondence: gaowei@ynnu.edu.cn

Received: 23 October 2019; Accepted: 20 December 2019; Published: 27 December 2019

Featured Application: The optical soliton solutions obtained in this research paper may be of concern and useful in many fields of science, such as mathematical physics, applied physics, nonlinear science, and engineering.

Abstract: In this paper, the cubic-quartic nonlinear Schrödinger and resonant nonlinear Schrödinger equation in parabolic law media are investigated to obtain the dark, singular, bright-singular combo and periodic soliton solutions. Two powerful methods, the $\left(m + \frac{G'}{G}\right)$ improved expansion method and the $\exp(-\varphi(\xi))$ expansion method are utilized to construct some novel solutions of the governing equations. The obtained optical soliton solutions are presented graphically to clarify their physical parameters. Moreover, to verify the existence solutions, the constraint conditions are utilized.

Keywords: cubic-quartic Schrödinger equation; cubic-quartic resonant Schrödinger equation; parabolic law

1. Introduction

In the current century, many entropy problems have been expressed by using mathematical models that are nonlinear partial differential equations. New results in the last few years have shown that the relation between non-standard entropies and nonlinear partial differential equations can be applied on new nonlinear wave equations inspired by quantum mechanics. Nonlinear models of the celebrated Klein–Gordon and Dirac equations have been found to admit accurate time dependent soliton-like solutions with the shapes of the so-called q-plane waves. Such q-plane waves are generalizations of the complex exponential plane wave solutions of the linear Klein–Gordon and Dirac equations [1]. Wave progressing of soliton forming and its application in the differential equation has been noticeable in the last few years. The physical phenomena of nonlinear partial differential equations (NLPDEs) may connect to many areas of sciences, for example plasma physics, optical fibers, nonlinear optics, fluid mechanics, chemistry, biology, geochemistry, and engineering sciences. The nonlinear Schrödinger equations describe wave propagation in optical fibers with nonlinear impacts [2–4].

Various numeric and analytic techniques have been used to seek solutions for nonlinear differential equations such as the homotopy perturbation scheme [5], the Adams–Bashforth–Moulton method [6], the shooting technique with fourth-order Runge–Kutta scheme [7–10], the group preserving method [11], the finite forward difference method [12,13], the Adomian decomposition method [14,15], the sine-Gordon expansion method [16–18], the modified auxiliary expansion method [19], the modified $\exp(-\varphi(\zeta))$ expansion function method [20,21], the improved Bernoulli sub-equation method [22,23], the Riccati–Bernoulli sub-ODE method [24], the modified exponential function method [25], the improved $\tan(\phi(\zeta)/2)$ [26,27], the Darboux transformation method [28,29], the double $\left(\frac{G'}{G},\frac{1}{G'}\right)$ expansion method [30,31], the $\left(\frac{1}{G'}\right)$ expansion method [32,33], the decomposition Sumudu-like-integral transform method [34], and the inverse scattering method [35].

In recent years, many researchers have carried out investigations on the governing models in optical fibers. The nonlinear Schrödinger equation, involving cubic and quartic-order dispersion terms, has been investigated to seek the exact optical soliton solutions via the undetermined coefficients method [36], the modified Kudryashov approach [37], the complete discrimination system method [38], the generalized tanh function method [39], the sin-cosine method, as well as the Bernoulli equation approach [40], the semi-inverse variation method [41], a simple equation method [3], and the extended sinh-Gordon expansion method [42].

Now, optical solitons are the exciting research area of nonlinear optics studies, and this research field has led to tremendous advances in their extensive applications. It is identified that the dynamics of nonlinear optical solitons and Madelung fluids are based on the generalized nonlinear Schrödinger dispersive equation and resonant nonlinear Schrödinger dispersive equation. In the research of chirped solitons in Hall current impacts in the field of quantum mechanics, a specific resonant term must be given [43].

Dispersion and nonlinearity are the two key elements for the propagation of solitons over intercontinental ranges. Normally, group velocity dispersion (GVD) leveling with self-phase modulation in a sensitive way allows such solitons to maintain long distance travel. Now, it could occur that GVD is minuscule and therefore completely overlooked, so in this condition, the dispersion impact is rewarded for by third-order (3OD) and fourth-order (4OD) dispersion impacts. This is generally referred to as solitons that are cubic-quartic (CQ). This term was implemented in 2017 for the first time. This model was later extensively researched through different points of view such as the semi-inverse variation principle [41], Lie symmetry [44], conservation rules [45], and the system of undetermined coefficients [37]. Consider the nonlinear Schrödinger and resonant nonlinear Schrödinger equations in the appearance of 3OD and 4OD without both GVD and disturbance. The equations are as follows:

$$iu_t + i\alpha u_{xxx} + \beta u_{xxxx} + cF\left(|u|^2\right)u = 0, \tag{1}$$

$$iu_t + i\alpha u_{xxx} + \beta u_{xxxx} + cF\left(|u|^2\right)u + c_3\left(\frac{|u|_{xx}}{|u|}\right)u = 0. \tag{2}$$

In Equations (1) and (2), $u(x,t)$ is the complex valued wave function and x (space) and t (time) are independent variables. The coefficients α and β are real constants, while c_3 is the Bohm potential that occurs in Madelung fluids. The Bohm potential term of disturbance generates quantum behavior, so that quantum characteristics are closely related to their special characteristics. Therefore, we have the chirped NLSE's disturbance expression giving us the introduction of the theory of hidden variables. Therefore, it will be more crucial to retrieve accurate solutions for the development of quantum mechanics from disturbed chiral (resonant) NLSE [46]. Furthermore, the functional F is a real valued algebraic function that represents the source of nonlinearity and $F\left(|u|^2\right)u : C \rightarrow C$. In more detail, the function $F\left(|u|^2\right)u$ is p-times continuously differentiable, so that:

$$F\left(|u|^2\right) u \in \bigcup_{m,n=1}^{\infty} C^p\left((-n,n) \times (-m,m) : R^2\right).$$

Suppose that $F(u) = c_1 u + c_2 u^2$, so Equations (1) and (2) can be rewritten as:

$$iu_t + i\alpha u_{xxx} + \beta u_{xxxx} + \left(c_1|u|^2 + c_2|u|^4\right) u = 0, \tag{3}$$

$$iu_t + i\alpha u_{xxx} + \beta u_{xxxx} + \left(c_1|u|^2 + c_2|u|^4\right) u + c_3 \left(\frac{|u|_{xx}}{|u|}\right) u = 0. \tag{4}$$

Equation (3) was investigated by making $c_2 = 0$ in [47] via the Kudryashov approach. The conservation laws to obtain the conserved densities for Schrödinger's nonlinear cubic-quarter equation have been analyzed in Kerr and power-law media [45]. The undetermined coefficients method has been employed to construct bright soliton and singular soliton solutions of Equation (1), when nonlinearity has been taken into consideration in the instances of the Kerr law and power law [37]. In this study, we use two methods to investigate soliton solutions of the cubic-quartic nonlinear Schrödinger equation and cubic-quartic resonant nonlinear Schrödinger equation with the parabolic law, namely Equations (3) and (4).

2. Instructions for the Methods

Assume a nonlinear partial differential equation (NLPDE) as follows:

$$P(U, U_x, U_t, U_{xx}, U_{tt}, U_{tx}, \ldots) = 0, \tag{5}$$

and define the traveling wave transformation as follows,

$$U(x, y, t) = \phi(\zeta), \quad \zeta = x - vt. \tag{6}$$

Putting Equation (6) into Equation (5), the outcome is:

$$N(\phi, \phi', \phi'', \ldots) = 0. \tag{7}$$

For the $m + \frac{G'(\zeta)}{G(\zeta)}$ expansion method, we take the trial solution for Equation (7) as follows:

$$\phi(\zeta) = \sum_{i=-n}^{n} a_i (m + F)^i = a_{-n}(m + F)^{-n} + \ldots + m\,a_0 + a_1(m + F) + \ldots + a_n(m + F)^n, \tag{8}$$

where a_i, $i = 0, 1, \ldots, n$ and m are nonzero constants. According to the principles of balance, we find the value of n. In this manuscript, we define F to be a function as:

$$F = \frac{G'(\zeta)}{G(\zeta)}, \tag{9}$$

where $G(\zeta)$ satisfy $G'' + (\lambda + 2m)G' + \mu G = 0$.

Putting Equation (8) into Equation (7) by using Equation (9), then collecting all terms with the same order of $(m + F)^n$, we get the system of algebraic equations for v, a_n, $n = 0, 1, \ldots, n$, λ, and μ. As a result, solving the obtained system, we get the explicit and exact solutions of Equation (5).

For the $(\exp - \varphi(\xi))$ expansion method, we use the trial solution as follows:

$$\phi(\xi) = \sum_{i=0}^{n} b_i (\exp(-\varphi(\xi)))^i, \quad i = 1, 2, \ldots, n \tag{10}$$

where b_i are non-zero constants. The auxiliary ODE $\varphi\left(\xi\right)$ is defined as follows:

$$\varphi'\left(\xi\right) = \exp\left(-\varphi\left(\xi\right)\right) + \mu\exp\left(\varphi\left(\xi\right)\right) + \lambda. \tag{11}$$

Solving Equation (11), we have:

Case 1. When $\Delta > 0$ and $\mu \neq 0$, we get the hyperbolic function solution:

$$\varphi\left(\xi\right) = \ln\left(\frac{-\lambda - \sqrt{\Delta}\tanh\left(\frac{1}{2}\sqrt{\Delta}\left(\xi + c\right)\right)}{2\mu}\right). \tag{12}$$

Case 2. When $\Delta < 0$ and $\mu \neq 0$, we get the trigonometric function solution:

$$\varphi\left(\xi\right) = \ln\left(\frac{-\lambda + \sqrt{-\Delta}\tan\left(\frac{1}{2}\sqrt{-\Delta}\left(\xi + c\right)\right)}{2\mu}\right). \tag{13}$$

Case 3. When $\Delta > 0$, $\mu = 0$, and $\lambda \neq 0$, we get hyperbolic function solution

$$\varphi\left(\xi\right) = -\ln\left(\frac{\lambda}{-1 + \cosh\left(\lambda\left(\xi + c\right)\right) + \sinh\left(\lambda\left(\xi + c\right)\right)}\right). \tag{14}$$

Case 4. When $\Delta = 0$, $\mu \neq 0$ and $\lambda \neq 0$, we get the rational function solution:

$$\varphi\left(\xi\right) = \ln\left(\frac{-2 - 2\lambda\left(\xi + c\right)}{\lambda^2\left(\xi + c\right)}\right). \tag{15}$$

Case 5. When $\Delta = 0$, $\mu = 0$, and $\lambda = 0$, we get:

$$\varphi\left(\xi\right) = \ln\left(\xi + c\right), \tag{16}$$

where c is the non-zero constant of integration and $\Delta = \lambda^2 - 4\mu$.

3. Application to the $\left(m + \frac{G'}{G}\right)$ Expansion Method

In this section, we use the $\left(m + \frac{G'}{G}\right)$ expansion method for the cubic-quartic nonlinear Schrödinger and cubic-quartic resonant nonlinear Schrödinger equations.

3.1. The Cubic-Quartic Nonlinear Schrödinger Equation

To solve Equation (3), by the $\left(m + \frac{G'}{G}\right)$ expansion method, we use the following transformation:

$$u\left(x, t\right) = U(\xi)e^{i\theta}, \qquad \xi = x - vt, \qquad \theta = -\kappa x + \omega t. \tag{17}$$

In the above equation, $\theta\left(x, t\right)$ symbolize the phase component of the soliton, κ represent the soliton frequency, while ω denote the wave number, and v symbolize the velocity of the soliton. Substitute wave transformation into Equation (3), and separate the outcome equation into real and imaginary parts. We can write the real part as follows:

$$-\left(\alpha\kappa^3 - \beta\kappa^4 + \omega\right)U + c_1 U^3 + c_2 U^5 + 3\alpha\kappa U'' - 6\beta\kappa^2 U'' + \beta U^{(4)} = 0, \tag{18}$$

and the imaginary part can be written as:

$$\left(3\alpha\kappa^2 - 4\beta\kappa^3 + v\right)U' - \left(\alpha - 4\beta\kappa\right)U^{(3)} = 0. \tag{19}$$

From Equation (19) $U' \neq 0$ and $U''' \neq 0$, then:

$$v = 4\beta\kappa^3 - 3\alpha\kappa^2, \quad \alpha = 4\beta\kappa. \tag{20}$$

Hence, Equation (18) can be rewritten as:

$$\left(3\beta\kappa^4 + \omega\right)U - c_1 U^3 - c_2 U^5 - 12\beta\kappa^2 U'' + 6\beta\kappa^2 U'' - \beta U^{(4)} = 0. \tag{21}$$

Multiplying both sides of Equation (21) by U' and taking its integration with respect to ξ, we get:

$$\beta\left(-12\left(U''\right)^2 + 24U''' U'\right) + 6c_1 U^4 + 4c_2 U^6 + 72\beta\kappa^2 \left(U'\right)^2 - \left(36\beta\kappa^4 + 12\omega\right)U^2 = 0. \tag{22}$$

Finding the balance, we gain $n = 1$. Replacing this value of balance into Equation (8), we get:

$$U\left(\xi\right) = a_{-1}(m + F)^{-1} + a_0 + a_1\left(m + F\right). \tag{23}$$

By substituting Equation (23) into Equation (3) by using Equation (9), we get the following solutions:

Case 1. When $a_0 = \frac{\lambda a_1}{2}$, $\kappa = \mp \frac{\sqrt{(2m+\lambda)^2 - 4\mu}}{\sqrt{6}}$, $c_1 = \frac{8\beta\left((2m+\lambda)^2 - 4\mu\right)}{a_1^2}$, $c_2 = -\frac{24\beta}{a_1^4}$, $a_{-1} = 0$, and $\Delta = (\lambda + 2m)^2 - 4\mu$, we get an exponential function solution as follows:

$$u\left(x,t\right) = e^{i\left(\sqrt{\frac{\Delta}{6}}x + \frac{5}{12}\beta\Delta^2 t\right)}$$
$$\left(\frac{\lambda a_1}{2} + a_1\left(m + \frac{1}{2}\left(-2m + \left(1 - \frac{2A_1}{A_1 + A_2 e^{\sqrt{\Delta}\left(x - \frac{2}{3}\sqrt{\frac{2}{3}}\beta(\Delta)^{3/2}t\right)}}\right)\sqrt{\Delta} - \lambda\right)\right)\right), \tag{24}$$

which is a dark solution, as shown in Figure 1, A_1 and A_2 are non-zero numbers, and $\Delta > 0$. Figure 1 shows that Equation (24) is a dark soliton under the suitable values of parameters.

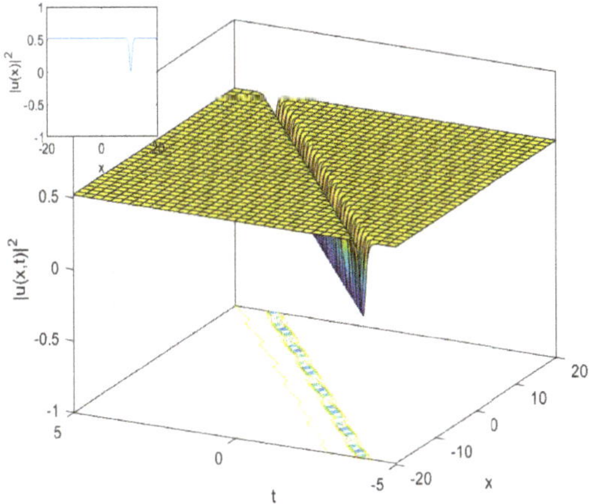

Figure 1. 3D surface of Equation (24), which is a dark optical soliton solution plotted when $A_1 = 1, A_2 = 0.3, \beta = 0.2, a_1 = 0.4, \lambda = 1, m = 1, \mu = -1$, and $t = 2$ for 2D.

Case 2. When $a_0 = -\frac{\lambda a_{-1}}{2m(m+\lambda)-2\mu}$, $a_1 = 0$, $a_2 = \frac{12\omega}{5\left((2m+\lambda)^2-4\mu\right)^2}$, $\kappa = \frac{\sqrt{(2m+\lambda)^2-4\mu}}{\sqrt{6}}$, $c_1 = \frac{96(-m(m+\lambda)+\mu)^2\omega}{5\left((2m+\lambda)^2-4\mu\right)a_{-1}^2}$, $c_2 = -\frac{288(-m(m+\lambda)+\mu)^4\omega}{5\left((2m+\lambda)^2-4\mu\right)^2 a_{-1}^4}$, and $\Delta = (\lambda+2m)^2 - 4\mu$, we obtain an exponential function solution:

$$u(x,t) = \frac{a_{-1}\,e^{i\left(-\frac{x\sqrt{(2m+\lambda)^2-4\mu}}{\sqrt{6}}+t\omega\right)}}{m+\frac{1}{2}\left(-2m+\left(1-\frac{2A_1}{A_1+A_2 e^{\sqrt{\Delta}\left(x+\frac{8\sqrt{\frac{2}{3}}\omega}{5\sqrt{\Delta}}T\right)}}\right)\sqrt{\Delta}-\lambda\right)} + \frac{\lambda a_{-1}\,e^{i\left(-\frac{x\sqrt{(2m+\lambda)^2-4\mu}}{\sqrt{6}}+t\omega\right)}}{2m(m+\lambda)-2\mu},\qquad(25)$$

which is a soliton solution, as shown in Figure 2, A_1 and A_2 are non-zero numbers, and $\Delta > 0$. With the suitable values, Figure 2 presents that Equation (25) is a singular soliton.

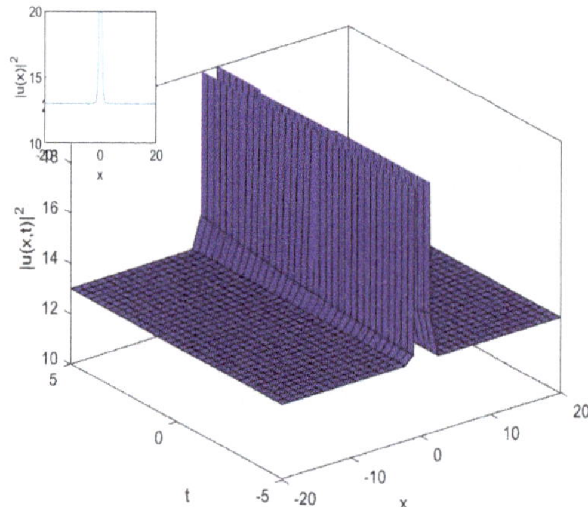

Figure 2. 3D surface of Equation (25), which is a singular soliton solution plotted when $A_1 = 2$, $A_2 = 3$, $\beta = 6$, $a_{-1} = 6$, $\lambda = 1$, $m = 1$, $\mu = -1$, and $t = 2$ for 2D.

Case 3. When $a_{-1} = -\frac{i\sqrt{3}c_1(m(m+\lambda)-\mu)}{\sqrt{c_2\left((2m+\lambda)^2-4\mu\right)}}$, $a_0 = \frac{i\sqrt{3}c_1\lambda}{2\sqrt{c_2\left((2m+\lambda)^2-4\mu\right)}}$, $a_1 = 0$, $\omega = -\frac{5c_1^2}{32c_2}$, $\kappa = \mp\frac{\sqrt{(2m+\lambda)^2-4\mu}}{\sqrt{6}}$, $\gamma = -\frac{3c_1^2}{8c_2\left((2m+\lambda)^2-4\mu\right)^2}$, and $\Delta = (\lambda+2m)^2 - 4\mu$, we have an exponential function solution:

$$u(x,t) = e^{i\left(-\frac{5c_1^2}{32c_2}t+\frac{\sqrt{\Delta}}{\sqrt{6}}x\right)}\left(\frac{i\sqrt{3}\sqrt{c_1}\lambda}{2\sqrt{c_2\Delta}} - \frac{i\sqrt{3}\sqrt{c_1}\,(m(m+\lambda)-\mu)}{\left(m+\frac{1}{2}\left(-2m+\left(1-\frac{2A_1}{A_1+A_2 e^{\sqrt{\Delta}\left(x+\frac{c_1^2 t}{2\sqrt{6}c_2\sqrt{\Delta}}\right)}}\right)\sqrt{\Delta}-\lambda\right)\right)\sqrt{c_2\Delta}}\right),\qquad(26)$$

which is a soliton solution, as shown in Figure 3, A_1 and A_2 are non-zero numbers, and $\Delta > 0$. Considering some values of parameters, Figure 3 shows singular soliton solution.

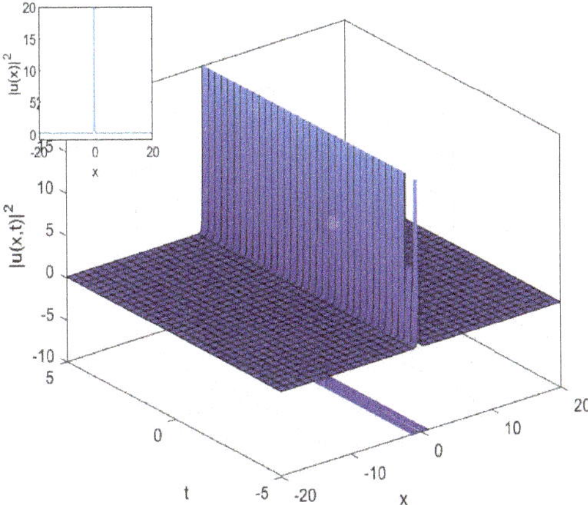

Figure 3. 3D surface of Equation (26), which is a singular soliton solution plotted when $A_1 = 0.3, A_2 = 2, c_1 = 0.3, c_2 = 2, \lambda = 1, m = 1, \mu = -1$, and $t = 2$ for 2D.

3.2. The Cubic-Quartic Resonant Nonlinear Schrödinger Equation

To solve Equation (4), by the $\left(m + \frac{G'}{G}\right)$ expansion method, we consider wave transformation Equation (17). Replacing Equation (17) into Equation (4) and separating the outcome equation into real and imaginary parts, we can write the real part as follows:

$$\left(\kappa^3 \left(\alpha - \beta\kappa\right) + \omega\right) U - c_1 U^3 - c_2 U^5 - \left(c_3 + 3\kappa \left(\alpha - 2\beta\kappa\right)\right) U'' - \beta U^{(4)} = 0, \tag{27}$$

and the imaginary part can be written as:

$$\left(3\alpha\kappa^2 - 4\beta\kappa^3 + v\right) U' - \left(\alpha - 4\beta\kappa\right) U''' = 0. \tag{28}$$

From Equation (28) $U' \neq 0$ and $U''' \neq 0$, then:

$$v = 4\beta\kappa^3 - 3\alpha\kappa^2, \qquad \alpha = 4\beta\kappa. \tag{29}$$

Hence, Equation (27) can be rewritten as:

$$\left(3\beta\kappa^4 + \omega\right) U - c_1 U^3 - c_2 U^5 - \left(c_3 + 6\beta\kappa^2\right) U'' - \beta U^{(4)} = 0. \tag{30}$$

Multiplying both sides of Equation (30) by U' and integrating it once with respect to ξ, we get:

$$\left(36\beta\kappa^4 + 12\omega\right) U^2 - 6c_1 U^4 - 4c_2 U^6 - \left(12c_3 + 72\beta\kappa^2\right) \left(U'\right)^2 + \beta \left(12\left(U''\right)^2 - 24U'U'''\right) = 0. \tag{31}$$

Finding the balance, we gain $n = 1$. Putting this value into Equation (8), we get the same result of Equation (23). Substituting Equation (23) with Equation (9) into Equation (4), we get the following solutions:

Case 1. When $a_{-1} = -\frac{2(m(m+\lambda)-\mu)a_0}{\lambda}$, $a_1 = 0$, $\omega = \frac{1}{2}\beta\left(-6\kappa^4 + \left((2m+\lambda)^2 - 4\mu\right)^2\right)$, $c_1 = \frac{2\beta\lambda^2\left((2m+\lambda)^2 - 4\mu\right)}{a_0^2}$, $c_2 = -\frac{3\beta\lambda^4}{2a_0^4}$, and $c_3 = \beta\left(-6\kappa^2 + (2m+\lambda)^2 - 4\mu\right)$, we obtain the following solutions:

Solution 1. In the case $\Delta > 0$, we have an exponential function solution:

$$u(x,t) = e^{i\left(-\kappa x + \frac{1}{2}\beta\left(-6\kappa^4 + \left((2m+\lambda)^2 - 4\mu\right)^2\right)t\right)}$$
$$\left(a_0 - \frac{2\left(m(m+\lambda) - \mu\right)a_0}{\left(m + \frac{1}{2}\left(-2m + \left(1 - \frac{2A_1}{A_1 + A_2 e^{\sqrt{\Delta}(x + 8\beta\kappa^3 t)}}\right)\sqrt{\Delta} - \lambda\right)\right)\lambda}\right). \tag{32}$$

Considering some values of parameters, Figure 4 shows singular soliton solution.

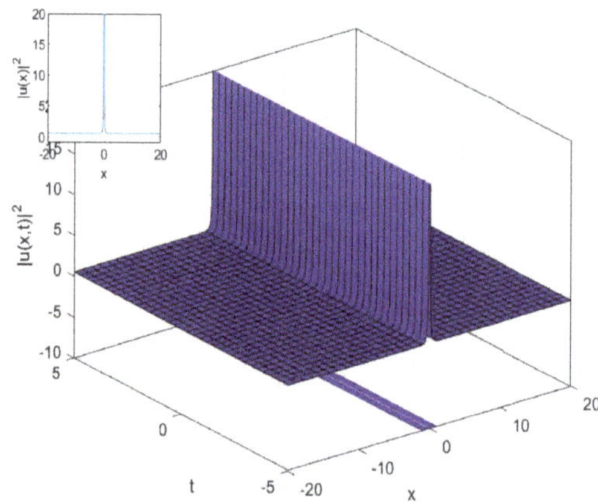

Figure 4. 3D figure of Equation (32), which is a singular soliton solution plotted when $A_1 = 1$, $A_2 = 3$, $\lambda = 1$, $m = 1$, $\mu = -1$, $\beta = 0.2$, $a_0 = 0.2$, $\kappa = 0.01$, and $t = 2$ for 2D.

Solution 2. In the case $\Delta < 0$, we have a trigonometric function solution:

$$u(x,t) = e^{-i\kappa x + \frac{1}{2}i\beta\left(-6\kappa^4 + \left((2m+\lambda)^2 - 4\mu\right)^2 t\right)}$$
$$\left(a_0 + \frac{4a_0\left(m^2 + m\lambda - \mu\right)\left(A_2\cos(\alpha) + A_1\sin(\alpha)\right)}{\lambda\left(\left(-A_1\sqrt{-\Delta} + A_2\lambda\right)\cos(\alpha) + \left(A_2\sqrt{-\Delta} + A_1\lambda\right)\sin(\alpha)\right)}\right), \tag{33}$$

which is $\alpha = \frac{1}{2}\sqrt{-\Delta}\left(x + 8\beta\kappa^3 t\right)$.

Periodic singular solution is plotted in Figure 5.

Case 2. When $a_{-1} = 0$, $a_1 = \frac{2a_0}{\lambda}$, $\omega = \frac{1}{2}\beta\left(-6\kappa^4 + \left((2m+\lambda)^2 - 4\mu\right)^2\right)$, $c_1 = \frac{2\beta\lambda^2\left((2m+\lambda)^2 - 4\mu\right)}{a_0^2}$, $c_2 = -\frac{3\beta\lambda^4}{2a_0^4}$, and $c_3 = \beta\left(-6\kappa^2 + (2m+\lambda)^2 - 4\mu\right)$, we obtain the following solutions:

Solution 1. In the case $\Delta > 0$, we get dark solution, as shown in Figure 6 :

$$u\left(x,y\right)=e^{i\left(-\kappa x+\frac{1}{2}\beta\left(-6\kappa^4+\left(\left(2m+\lambda\right)^2-4\mu\right)^2\right)t\right)}$$

$$\left(a_0+\frac{2\left(m+\frac{1}{2}\left(-2m+\left(1-\frac{2A_1}{A_1+A_2e^{\sqrt{\Delta}\left(x+8\beta\kappa^3 t\right)}}\right)\sqrt{\Delta}-\lambda\right)\right)a_0}{\lambda}\right). \tag{34}$$

Figure 6 shows the dark structure this solution.

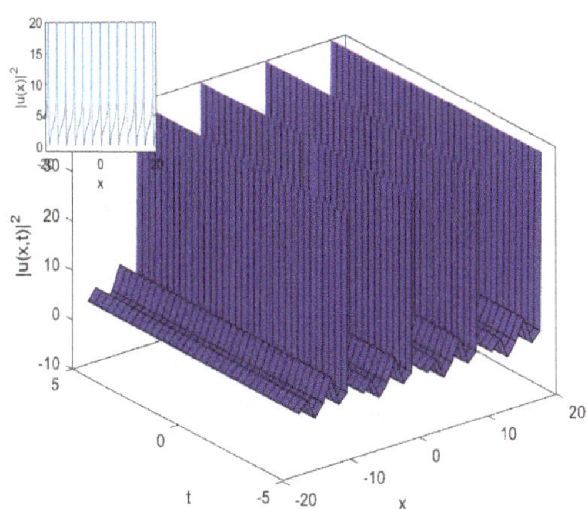

Figure 5. 3D surface of Equation (33), which is a periodic singular soliton solution plotted when $A_1=1, A_2=2, \lambda=1, m=\frac{1}{2}, \mu=2, \beta=0.1, a_0=2, \kappa=0.1$, and $t=2$ for 2D.

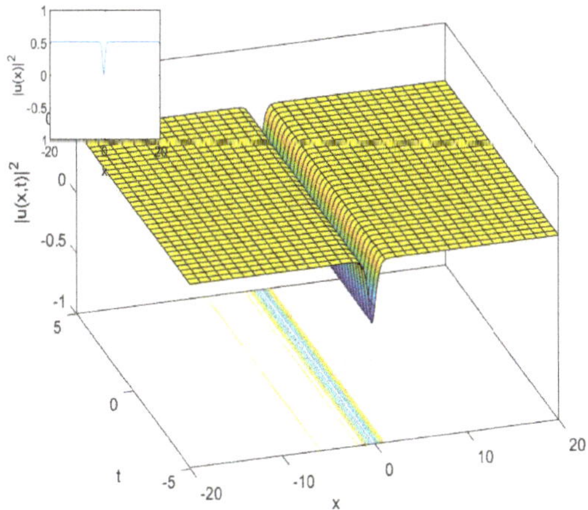

Figure 6. 3D surface of Equation (34), which is a dark soliton solution plotted when $A_1=1, A_2=3, \lambda=1, m=1, \mu=-1, \beta=0.2, a_0=0.2, \kappa=0.01$, and $t=2$ for 2D.

Solution 2. In the case $\Delta<0$, we have a trigonometric function solution:

$$u(x,t) = e^{-ix\kappa + \frac{1}{2}i\beta\left(-6\kappa^4 + \left((2m+\lambda)^2 - 4\mu\right)^2\right)t}$$
$$\left(\frac{\sqrt{-\Delta}\left(A_1 \cos\left(\frac{1}{2}\sqrt{-\Delta}\left(x + 8\beta\kappa^3 t\right)\right) - A_2 a_0 \sin\left(\frac{1}{2}\sqrt{-\Delta}\left(x + 8\beta\kappa^3 t\right)\right)\right)}{\lambda\left(A_2 \cos\left(\frac{1}{2}\sqrt{-\Delta}\left(x + 8\beta\kappa^3 t\right)\right) + A_1 \sin\left(\frac{1}{2}\sqrt{-\Delta}\left(x + 8\beta\kappa^3 t\right)\right)\right)}\right). \tag{35}$$

Periodic singular solution is plotted in Figure 7.

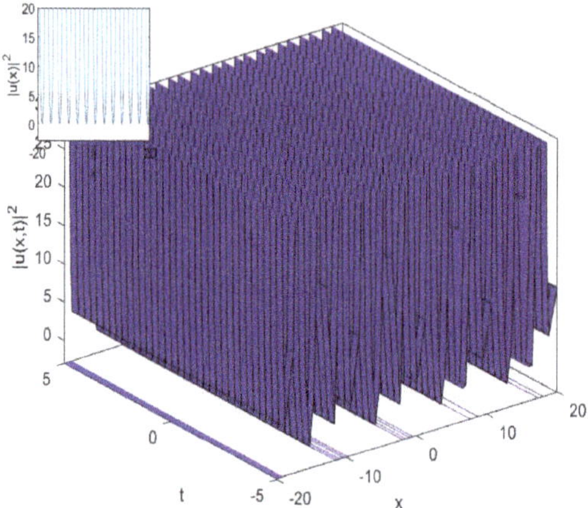

Figure 7. 3D figure of Equation (35), which is a periodic singular soliton solution plotted when $A_1 = 1, A_2 = 2, \lambda = 1, m = \frac{1}{2}, \mu = 2, \beta = 0.1, a_0 = 0.2, \kappa = 0.1$, and $t = 2$ for 2D.

4. Application to the Exp $(-\varphi(\xi))$ Expansion Method

In this section, we apply the $\exp(-\varphi(\xi))$ expansion method to the cubic-quartic nonlinear Schrödinger and resonant nonlinear Schrödinger equations.

4.1. The Cubic-Quartic Nonlinear Schrödinger Equation

To apply this method on the cubic-quartic nonlinear Schrödinger equation, Equation (3), we utilize the same wave transformation of Equation (17). As a result, we get Equation (22). Finding the balance, we gain $n = 1$. By inserting the value of the balance into Equation (10), we get:

$$U(\xi) = b_0 + b_1 e^{-\varphi(\xi)}. \tag{36}$$

Substituting Equation (40) into Equation (22) and setting each summation of the coefficients of the exponential identities of the same power to be zero, we discuss the following cases of the solutions.

Case 1. When $b_0 = \frac{\lambda b_1}{2}, c_1 = \frac{8a_2(\lambda^2 - 4\mu)}{b_1^2}, c_2 = -\frac{24a_2}{b_1^4}, \kappa = -\frac{\sqrt{\lambda^2 - 4\mu}}{\sqrt{6}}$, and $\omega = \frac{5}{12}a_2(\lambda^2 - 4\mu)^2$, we get the following solutions:

Solution 1. In the case $\lambda^2 - 4\mu > 0$ and $\mu \neq 0$, we have a hyperbolic function solution:

$$u\left(x,t\right)=e^{\,i\left(\frac{\sqrt{\lambda^2-4\mu}}{\sqrt{6}}x+\frac{5}{12}a_2\left(\lambda^2-4\mu\right)^2 t\right)}$$

$$\left(\frac{\lambda b_1}{2}+\frac{2\mu b_1}{-\lambda-\sqrt{\lambda^2-4\mu}\tanh\left[\frac{1}{2}\left(c+x-\frac{2}{3}\sqrt{\frac{2}{3}}a_2(\lambda^2-4\mu)^{3/2}t\right)\sqrt{\lambda^2-4\mu}\right]}\right). \tag{37}$$

This is a dark soliton solution, as shown in Figure 8.

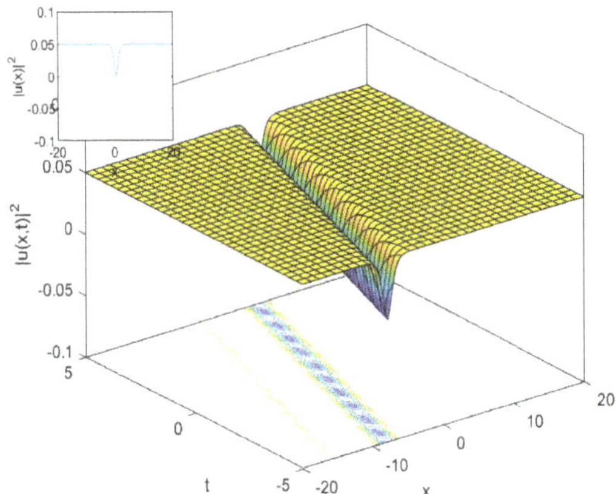

Figure 8. 3D surface of Equation (37), which is a bright singular combo soliton solution plotted when $b_1=0.2, \beta=0.2, c=1, \lambda=3, \mu=1$, and $t=2$ for 2D.

Solution 2. When $\lambda^2-4\mu>0$, $\mu=0$, and $\lambda\neq0$, we have hyperbolic function solutions:

$$u\left(x,t\right)=e^{\,i\left(\frac{5}{12}\beta\lambda^4 t+\frac{\sqrt{\lambda^2}}{\sqrt{6}}x\right)}$$

$$\left(\frac{\lambda b_1}{2}+\frac{\lambda b_1}{-1+\cosh\left(\lambda\left(c+x-\frac{2}{3}\sqrt{\frac{2}{3}}\beta(\lambda^2)^{3/2}t\right)\right)+\sinh\left(\lambda\left(c+x-\frac{2}{3}\sqrt{\frac{2}{3}}\beta(\lambda^2)^{3/2}t\right)\right)}\right). \tag{38}$$

This is a bright singular combo soliton solution, as shown in Figure 9.

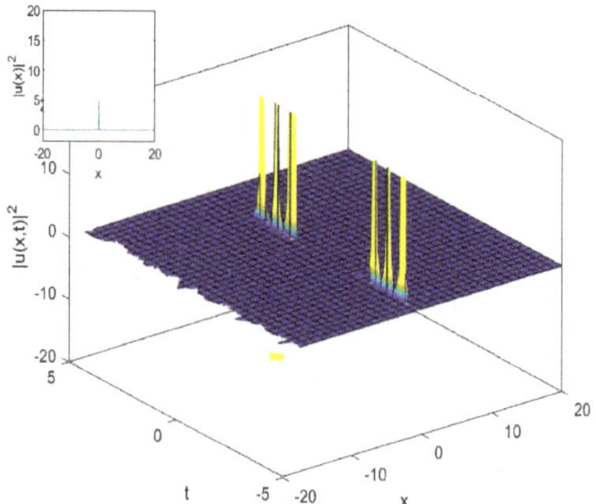

Figure 9. 3D surface of Equation (38), which is a bright singular combo soliton solution plotted when $b_1 = 0.04, \beta = 0.2, c = 0.2, \lambda = 1, \mu = 0$, and $t = 2$ for 2D.

Case 2. When $b_0 = \dfrac{\sqrt{3}\sqrt{c_1}\lambda}{2\sqrt{-c_2(\lambda^2-4\mu)}}, b_1 = \dfrac{\sqrt{3}\sqrt{c_1}}{\sqrt{-c_2(\lambda^2-4\mu)}}, \kappa = -\dfrac{\sqrt{\lambda^2-4\mu}}{\sqrt{6}}, \omega = -\dfrac{5c_1^2}{32c_2}$, and $\beta = -\dfrac{3c_1^2}{8c_2(\lambda^2-4\mu)^2}$, we get the following solutions:

Solution 1. When $\lambda^2 - 4\mu > 0$ and $\mu \neq 0$, we get a dark solution, as shown in Figure 10:

$$
u\left(x,t\right) = e^{i\left(-\frac{5c_1^2}{32c_2}t + \frac{\sqrt{\lambda^2-4\mu}}{\sqrt{6}}x\right)}
$$
$$
\left(\frac{\sqrt{3}\sqrt{c_1}\left(\lambda^2 - 4\mu + \lambda\sqrt{\lambda^2-4\mu}\tanh\left(\frac{c_1^2 t}{4\sqrt{6}c_2} + \frac{1}{2}(c+x)\sqrt{\lambda^2-4\mu}\right)\right)}{2\sqrt{-c_2(\lambda^2-4\mu)}\left(\lambda + \sqrt{\lambda^2-4\mu}\tanh\left(\frac{c_1^2 t}{4\sqrt{6}c_2} + \frac{1}{2}(c+x)\sqrt{\lambda^2-4\mu}\right)\right)}\right). \tag{39}
$$

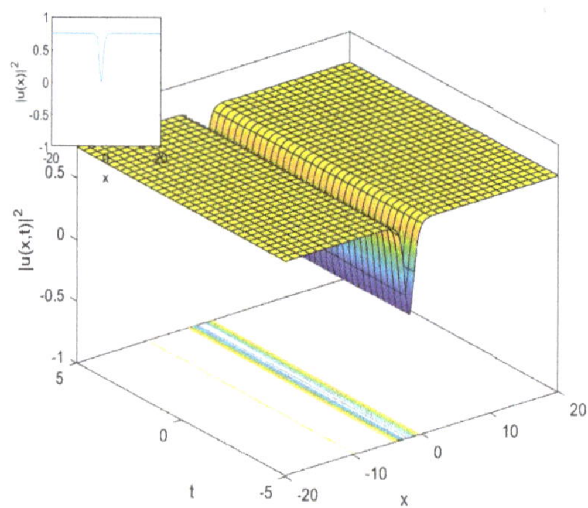

Figure 10. 3D surface of Equation (39), which is a dark soliton solution plotted when $b_1 = 0.2, \beta = 0.2, c_1 = 0.1, c_2 = -0.1, c = 1, \lambda = 3, \mu = 1$, and $t = 2$ for 2D.

Solution 2. When $\lambda^2 - 4\mu > 0$ and $\mu = 0$, we get hyperbolic function solution:

$$u\left(x,t\right) = e^{i\left(-\frac{5c_1^2}{32c_2}t + \frac{\sqrt{\lambda^2-4\mu}}{\sqrt{6}}x\right)}\left(\frac{\sqrt{3}\sqrt{c_1}\lambda\coth\left(\frac{1}{24}\lambda\left(12\left(c+x\right)+\frac{\sqrt{6}c_1^2}{c_2\sqrt{\lambda^2}}t\right)\right)}{2\sqrt{-c_2\lambda^2}}\right). \tag{40}$$

This is a singular soliton solution, as shown in Figure 11.

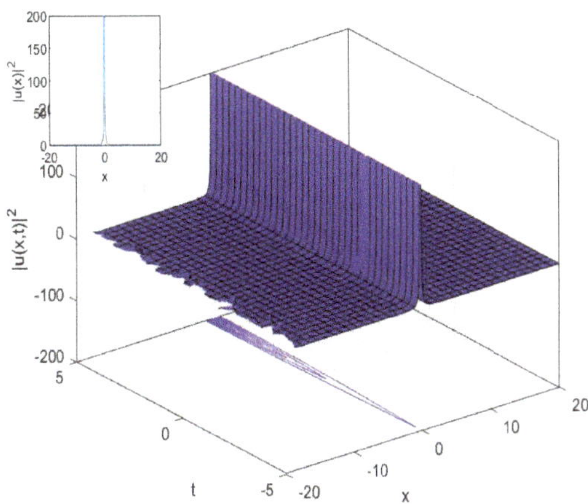

Figure 11. 3D figure of Equation (40), which is a singular soliton solution plotted when $b_1 = 4$, $\beta = 0.2$, $c_1 = 0.1, c_2 = -0.1, c = 0.2, \lambda = 1, \mu = 0$, and $t = 2$ for 2D.

4.2. The Cubic-Quartic Resonant Nonlinear Schrödinger Equation

To apply the $\exp\left(-\varphi\left(\xi\right)\right)$ expansion method on the cubic-quartic resonant nonlinear Schrödinger equation, Equation (4), we utilize the same wave transformation of Equation (17). As a result, we get Equation (31). Finding the balance, we gain $n = 1$. Via inserting the value of the balance into Equation (10), we get the same result of Equation (36). Substituting Equation (36) into Equation (31) and setting each summation of the coefficients of the exponential identities of the same power to be zero, we discuss the following cases of solutions.

Case 1. When $b_1 = \frac{2b_0}{\lambda}$, $c_1 = \frac{2\beta\lambda^2\left(\lambda^2-4\mu\right)}{b_0^2}$, $c_2 = -\frac{3\beta\lambda^4}{2b_0^4}$, $\kappa = \frac{\sqrt{-c_3+\beta\left(\lambda^2-4\mu\right)}}{\sqrt{6}\sqrt{\beta}}$, and $\omega = \frac{-c_3^2+2c_3\beta\left(\lambda^2-4\mu\right)+5\beta^2\left(\lambda^2-4\mu\right)^2}{12\beta}$, we get the following solutions:

Solution 1. When $\lambda^2 - 4\mu > 0$ and $\mu \neq 0$, we get hyperbolic function solution:

$$u\left(x,t\right) = e^{i\left(-\frac{\sqrt{-c_3+\beta\left(\lambda^2-4\mu\right)}}{\sqrt{6}\sqrt{\beta}}x + \frac{\left(-c_3^2+2c_3\beta\left(\lambda^2-4\mu\right)+5\beta^2\left(\lambda^2-4\mu\right)^2\right)}{12\beta}t\right)}$$

$$\left(b_0 + \frac{4\mu b_0}{\lambda\left(-\lambda - \sqrt{\lambda^2-4\mu}\tanh\left[\frac{1}{2}\left(c+x+\frac{2\sqrt{\frac{2}{3}}t\left(-c_3+\beta\left(\lambda^2-4\mu\right)\right)^{3/2}}{3\sqrt{\beta}}\right)\sqrt{\lambda^2-4\mu}\right]\right)}\right). \tag{41}$$

This is a dark soliton solution, as shown in Figure 12.

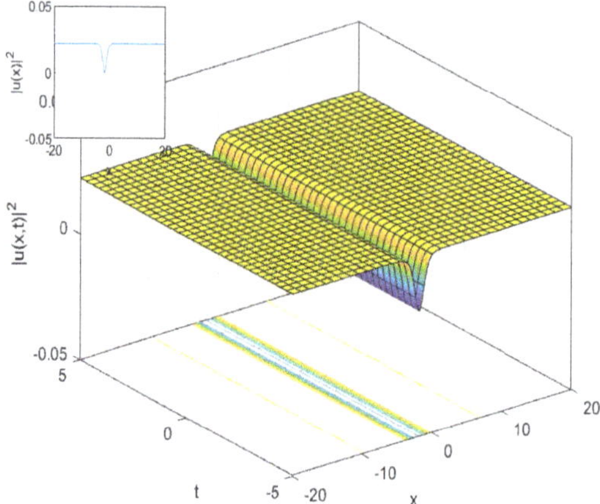

Figure 12. 3D surface of Equation (41), which is a dark soliton solution plotted when $b_0 = 0.2$, $\beta = 0.2$, $c_3 = 1$, $c = 1$, $\lambda = 3$, $\mu = 1$, and $t = 2$ for 2D.

Solution 2. When $\lambda^2 - 4\mu < 0$ and $\mu \neq 0$, we get trigonometric function solution:

$$u\left(x,t\right) = e^{i\left(-\frac{x\sqrt{-c_3+\beta\left(\lambda^2-4\mu\right)}}{\sqrt{6}\sqrt{\beta}} + \frac{t\left(-c_3^2+2c_3\beta\left(\lambda^2-4\mu\right)+5\beta^2\left(\lambda^2-4\mu\right)^2\right)}{12\beta}\right)} \left(b_0 - \frac{4\mu b_0}{\lambda^2 - \lambda\sqrt{-\lambda^2+4\mu}\tan\left[\frac{1}{2}\left(c+x+\frac{2\sqrt{\frac{2}{3}}\left(-c_3+\beta\left(\lambda^2-4\mu\right)\right)^{3/2}}{3\sqrt{\beta}}t\right)\sqrt{-\lambda^2+4\mu}\right]}\right). \tag{42}$$

This is a periodic singular soliton solution, as shown in Figure 13.

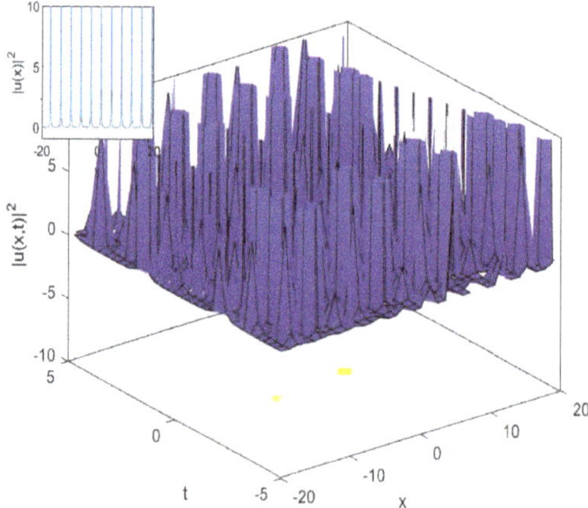

Figure 13. 3D surface of Equation (42), which is a periodic singular soliton solution plotted when $b_0 = 0.4$, $\beta = 0.1$, $c_3 = -5$, $c = 1$, $\lambda = 1$, $\mu = 1$, and $t = 2$ for 2D.

Solution 3. When $\lambda^2 - 4\mu > 0$ and $\mu = 0$, we get hyperbolic function solution:

$$u(x,t) = e^{i\left(-\frac{x\sqrt{-c_3+\beta\lambda^2}}{\sqrt{6}\sqrt{\beta}} + \frac{t\left(-c_3^2+2c_3\beta\lambda^2+5\beta^2\lambda^4\right)}{12\beta}\right)}$$
$$\left(b_0 + \frac{2b_0}{\cosh\left[\lambda\left(c+x+\frac{2\sqrt{\frac{2}{3}}t(-c_3+\beta\lambda^2)^{3/2}}{3\sqrt{\beta}}\right)\right] + \sinh\left[\lambda\left(c+x+\frac{2\sqrt{\frac{2}{3}}t(-c_3+\beta\lambda^2)^{3/2}}{3\sqrt{\beta}}\right)\right]}\right). \tag{43}$$

This is a bright singular combo soliton solution, as shown in Figure 14.

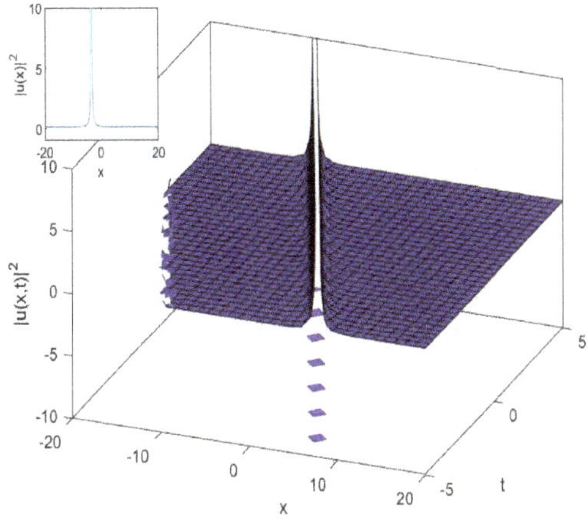

Figure 14. 3D surface of Equation (43), which is a singular soliton solution plotted when $b_0 = 0.4$, $\beta = 0.2, c_3 = -1, c = 0.2, \lambda = 3, \mu = 0$, and $t = 2$ for 2D.

Case 2. When $b_0 = \frac{\sqrt{2}\sqrt{\beta\lambda^2(\lambda^2-4\mu)}}{\sqrt{c1}}$, $b_1 = \frac{2\sqrt{2}\sqrt{\beta\lambda^2(\lambda^2-4\mu)}}{\sqrt{c_1}\lambda}$, $c_2 = -\frac{3c_1^2}{8\beta(\lambda^2-4\mu)^2}$, $c_3 = \beta\left(-6\kappa^2 + \lambda^2 - 4\mu\right)$, and $\omega = \frac{1}{2}\beta\left(-6\kappa^4 + \left(\lambda^2 - 4\mu\right)^2\right)$, we get the following solutions:

Solution 1. When $\lambda^2 - 4\mu > 0$ and $\mu \neq 0$, we get hyperbolic function solution:

$$u(x,t) = \sqrt{2\beta\lambda^2(\lambda^2-4\mu)}\, e^{-i\kappa x + \frac{1}{2}i\beta\left(-6\kappa^4+\left(\lambda^2-4\mu\right)^2\right)t}$$
$$\frac{\left(1 - \frac{4\mu}{\lambda^2+\lambda\sqrt{\lambda^2-4\mu}\tanh\left[\frac{1}{2}(c+x+8\beta\kappa^3 t)\sqrt{\lambda^2-4\mu}\right]}\right)}{\sqrt{c_1}}. \tag{44}$$

This is a dark soliton solution, as shown in Figure 15.

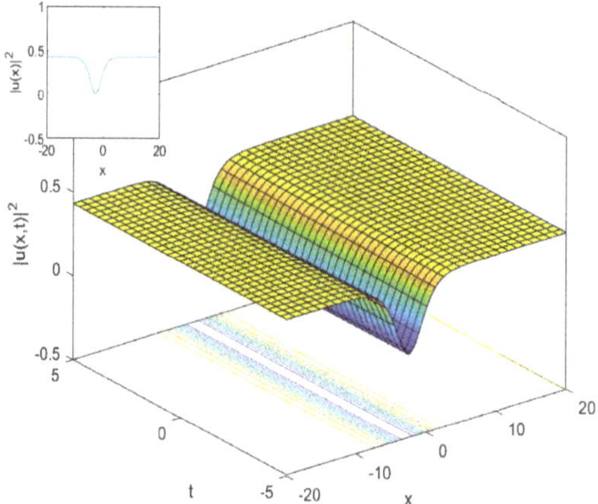

Figure 15. 3D surface of Equation (44), which is a dark soliton solution plotted when $\beta = 3, c_1 = 5$, $c = 0.03, \lambda = 1, \mu = 0.1, \kappa = 0.1$, and $t = 2$ for 2D.

Solution 2. When $\lambda^2 - 4\mu < 0$ and $\mu \neq 0$, we get trigonometric function solution:

$$u(x,t) = \sqrt{2\beta\lambda^2(\lambda^2 - 4\mu)}\, e^{-i\kappa x + \frac{1}{2}i\beta\left(-6\kappa^4 + \left(\lambda^2 - 4\mu\right)^2\right)t}$$
$$\frac{\left(1 - \dfrac{4\mu}{\lambda^2 - \lambda\sqrt{-\lambda^2 + 4\mu}\,\tan\left[\frac{1}{2}(c + x + 8t\beta\kappa^3)\sqrt{-\lambda^2 + 4\mu}\right]}\right)}{\sqrt{c_1}}. \tag{45}$$

This is a periodic singular soliton solution, as shown in Figure 16.

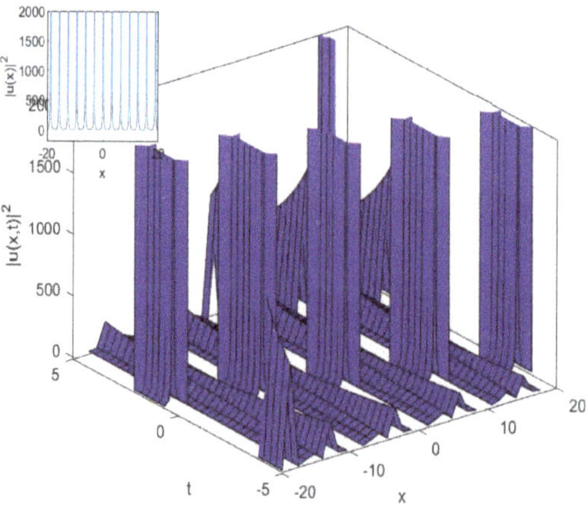

Figure 16. 3D surface of Equation (45), which is a periodic singular soliton solution plotted when $\beta = -3, c_1 = 5, c = 0.03, \lambda = 0.1, \mu = 1, \kappa = 0.1$, and $t = 2$ for 2D.

Solution 3. When $\lambda^2 - 4\mu > 0$ and $\mu = 0$, we get the hyperbolic function solution:

$$u\left(x,t\right)=\frac{\sqrt{2}e^{-i\kappa x+\frac{1}{2}i\beta\left(-6\kappa^4+\lambda^4\right)t}\sqrt{\beta\lambda^4}\coth\left(\frac{1}{2}\left(c+x+8\beta\kappa^3t\right)\lambda\right)}{\sqrt{c_1}},\qquad(46)$$

which is a periodic singular solution, as shown in Figure 17.

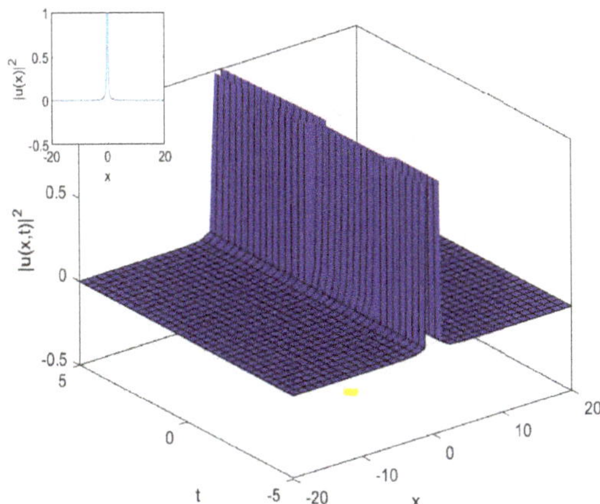

Figure 17. 3D surface of Equation (46), which is a singular soliton solution plotted when $\beta=3,c_1=5,$ $c=0.03,\lambda=0.1,\mu=0,\kappa=0.1,$ and $t=2$ for 2D.

5. Conclusions

In this research, the new dark, singular, bright singular combo soliton, and periodic singular solutions of the cubic-quantic nonlinear Schrödinger equation and the cubic-quantic resonant nonlinear Schrödinger equation were shown. Figures 1, 6, 8, 10, 12 and 15 are dark soliton solutions, Figures 2–4, 11, 14 and 17 are singular soliton solutions, Figures 5, 7, 13 and 16 are periodic singular solutions, and Figure 9 is bright singular combo soliton solution. The $(m+G'/G)$ expansion and $\exp(-\varphi\left(\zeta\right))$ expansion methods were utilized to study these two models with the parabolic law. The new solutions verified the main equations after we substituted them into Equations (3) and (4) for the existence of the equation.

Conte and Musette introduced that wave transformation, which we considered in this paper, protects the Painleve conditions and its properties [48]. Therefore, it can be seen that all results verified their physical properties and presented their estimated wave behaviors. Therefore, one can observe that the wave transformation considered in this paper in Equation (17) satisfies these conditions. We substituted all solutions to the main equations Equations (3) and (4), and they verified it; the constraint conditions Equations (20) and (29) were also used to verify this existence. The optical soliton solutions obtained in this research paper may be of concern and useful in many fields of science, such as mathematical physics, applied physics, nonlinear science, and engineering.

Author Contributions: All authors contributed equally.

Funding: This research received no external funding.

Conflicts of Interest: The authors declare no conflict of interest.

Abbreviations

The following abbreviations are used in this manuscript:

NLPDE   Nonlinear partial differential equation
ODE   Ordinary differential equation

References

1. Plastino, A. Entropic aspects of nonlinear partial differential equations: Classical and quantum mechanical perspectives. *Entropy* **2017**, *19*, 166. [CrossRef]
2. Ilhan, O.A.; Esen, A.; Bulut, H.; Baskonus, H.M. Singular solitons in the pseudo-parabolic model arising in nonlinear surface waves. *Results Phys.* **2019**, *12*, 1712–1715. [CrossRef]
3. Arshad, M.; Seadawy, A.R.; Lu, D. Exact bright–dark solitary wave solutions of the higher-order cubic–quintic nonlinear Schrödinger equation and its stability. *Optik* **2017**, *138*, 40–49. [CrossRef]
4. Arshad, M.; Seadawy, A.R.; Lu, D. Elliptic function and solitary wave solutions of the higher-order nonlinear Schrödinger dynamical equation with fourth-order dispersion and cubic-quintic nonlinearity and its stability. *Eur. Phys. J. Plus* **2017**, *132*, 371. [CrossRef]
5. Yousif, M.A.; Mahmood, B.A.; Ali, K.K.; Ismael, H.F. Numerical simulation using the homotopy perturbation method for a thin liquid film over an unsteady stretching sheet. *Int. J. Pure Appl. Math.* **2016**, *107*, 289–300. [CrossRef]
6. Baskonus, H.M.; Bulut, H. On the numerical solutions of some fractional ordinary differential equations by fractional Adams-Bashforth-Moulton method. *Open Math.* **2015**, *13*, 547–556. [CrossRef]
7. Ismael, H.F. Carreau-Casson fluids flow and heat transfer over stretching plate with internal heat source/sink and radiation. *Int. J. Adv. Appl. Sci.* **2017**, *4*, 11–15. [CrossRef]
8. Ali, K.K.; Ismael, H.F.; Mahmood, B.A.; Yousif, M.A. MHD Casson fluid with heat transfer in a liquid film over unsteady stretching plate. *Int. J. Adv. Appl. Sci.* **2017**, *4*, 55–58. [CrossRef]
9. Ismael, H.F.; Arifin, N.M. Flow and heat transfer in a maxwell liquid sheet over a stretching surface with thermal radiation and viscous dissipation. *JP J. Heat Mass Transf.* **2018**, *15*, 847–866. [CrossRef]
10. Zeeshan, A.; Ismael, H.F.; Yousif, M.A.; Mahmood, T.; Rahman, S.U. Simultaneous Effects of Slip and Wall Stretching/Shrinking on Radiative Flow of Magneto Nanofluid Through Porous Medium. *J. Magn.* **2018**, *23*, 491–498. [CrossRef]
11. Gao, W.; Partohaghighi, M.; Baskonus, H.M.; Ghavi, S. Regarding the group preserving scheme and method of line to the numerical simulations of Klein–Gordon model. *Results Phys.* **2019**, *15*, 102555. [CrossRef]
12. Yokus, A.; Baskonus, H.M.; Sulaiman, T.A.; Bulut, H. Numerical simulation and solutions of the two-component second order KdV evolutionarysystem. *Numer. Methods Partial Differ. Equ.* **2018**, *34*, 211–227. [CrossRef]
13. Sulaiman, T.A.; Bulut, H.; Yokus, A.; Baskonus, H.M. On the exact and numerical solutions to the coupled Boussinesq equation arising in ocean engineering. *Indian J. Phys.* **2019**, *93*, 647–656. [CrossRef]
14. Bulut, H.; Ergüt, M.; Asil, V.; Bokor, R.H. Numerical solution of a viscous incompressible flow problem through an orifice by Adomian decomposition method. *Appl. Math. Comput.* **2004**, *153*, 733–741. [CrossRef]
15. Ismael, H.F.; Ali, K.K. MHD casson flow over an unsteady stretching sheet. *Adv. Appl. Fluid Mech.* **2017**, *20*, 533–541. [CrossRef]
16. Baskonus, H.M.; Bulut, H.; Sulaiman, T.A. New Complex Hyperbolic Structures to the Lonngren-Wave Equation by Using Sine-Gordon Expansion Method. *Appl. Math. Nonlinear Sci.* **2019**, *4*, 141–150. [CrossRef]
17. Cattani, C.; Sulaiman, T.A.; Baskonus, H.M.; Bulut, H. On the soliton solutions to the Nizhnik-Novikov-Veselov and the Drinfel'd-Sokolov systems. *Opt. Quantum Electron.* **2018**, *50*, 138. [CrossRef]
18. Eskitaşçıoğlu, E.İ.; Aktaş, M.B.; Baskonus, H.M. New Complex and Hyperbolic Forms for Ablowitz–Kaup–Newell–Segur Wave Equation with Fourth Order. *Appl. Math. Nonlinear Sci.* **2019**, *4*, 105–112. [CrossRef]
19. Gao, W.; Ismael F.H.; Bulut H.; Baskonus, H.M. Instability modulation for the (2+1)-dimension paraxial wave equation and its new optical soliton solutions in Kerr media. *Phys. Scr.* **2019**. [CrossRef]
20. Houwe, A.; Hammouch, Z.; Bienvenue, D.; Nestor, S.; Betchewe, G. Nonlinear Schrödingers equations with cubic nonlinearity: M-derivative soliton solutions by $\exp(-\Phi(\xi))$-Expansion method. *Preprints* **2019**. [CrossRef]
21. Cattani, C.; Sulaiman, T.A.; Baskonus, H.M.; Bulut, H. Solitons in an inhomogeneous Murnaghan's rod. *Eur. Phys. J. Plus* **2018**, *133*, 228. [CrossRef]

22. Baskonus, H.M.; Yel, G.; Bulut, H. Novel wave surfaces to the fractional Zakharov-Kuznetsov-Benjamin-Bona-Mahony equation. *Proc. Aip Conf. Proc.* **2017**, *1863*, 560084.

23. Baskonus, H.M.; Bulut, H. Exponential prototype structures for (2+1)-dimensional Boiti-Leon-Pempinelli systems in mathematical physics. *Waves Random Complex Media* **2016**, *26*, 189–196. [CrossRef]

24. Abdelrahman, M.A.E.; Sohaly, M.A. The Riccati-Bernoulli Sub-ODE Technique for Solving the Deterministic (Stochastic) Generalized-Zakharov System. *Int. J. Math. Syst. Sci.* **2018**, *1*. [CrossRef]

25. Gao, W.; Ismael, H.F.; Mohammed, S.A.; Baskonus, H.M.; Bulut H. Complex and real optical soliton properties of the paraxial nonlinear Schrödinger equation in Kerr media with M-fractional. *Front. Phys.* **2019**, *7*, 197. doi:10.3389/fphy.2019.00197 [CrossRef]

26. Hammouch, Z.; Mekkaoui, T.; Agarwal, P. Optical solitons for the Calogero-Bogoyavlenskii-Schiff equation in (2 + 1) dimensions with time-fractional conformable derivative. *Eur. Phys. J. Plus* **2018**, *133*, 248. [CrossRef]

27. Manafian, J.; Aghdaei, M.F. Abundant soliton solutions for the coupled Schrödinger-Boussinesq system via an analytical method. *Eur. Phys. J. Plus* **2016**, *131*, 97. [CrossRef]

28. Guo, L.; Zhang, Y.; Xu, S.; Wu, Z.; He, J. The higher order rogue wave solutions of the Gerdjikov-Ivanov equation. *Phys. Scr.* **2014**, *89*, 035501. [CrossRef]

29. Ling, L.; Feng, B.F.; Zhu, Z. General soliton solutions to a coupled Fokas–Lenells equation. *Nonlinear Anal. Real World Appl.* **2018**, *40*, 185–214. [CrossRef]

30. Miah, M.M.; Ali, H.M.S.; Akbar, M.A.; Seadawy, A.R. New applications of the two variable $G'/G, 1/G$-expansion method for closed form traveling wave solutions of integro-differential equations. *J. Ocean Eng. Sci.* **2019**. [CrossRef]

31. Miah, M.M.; Ali, H.M.S.; Akbar, M.A.; Wazwaz, A.M. Some applications of the $(G'/G, 1/G)$-expansion method to find new exact solutions of NLEEs. *Eur. Phys. J. Plus* **2017**, *132*, 252. [CrossRef]

32. Yokuş, A. Comparison of Caputo and conformable derivatives for time-fractional Korteweg–de Vries equation via the finite difference method. *Int. J. Mod. Phys. B* **2018**, *32*, 1850365. [CrossRef]

33. Yokuş, A.; Kaya, D. Conservation laws and a new expansion method for sixth order Boussinesq equation. *Proc. AIP Conf. Proc.* **2015**, *1676*, 020062.

34. Yang, X.; Yang, Y.; Cattani, C.; Zhu, C.M. A new technique for solving the 1-D Burgers equation. *Therm. Sci.* **2017**, *21*, 129–136. [CrossRef]

35. Vakhnenko, V.O.; Parkes, E.J.; Morrison, A.J. A Bäcklund transformation and the inverse scattering transform method for the generalised Vakhnenko equation. *Chaos Solitons Fractals* **2003**, *17*, 683–692. [CrossRef]

36. Jawad, A.J.M.; Abu-AlShaeer, M.J.; Biswas, A.; Zhou, Q.; Moshokoa, S.; Belic, M. Optical solitons to Lakshmanan-Porsezian-Daniel model for three nonlinear forms. *Optik* **2018**, *160*, 197–202. [CrossRef]

37. Biswas, A.; Triki, H.; Zhou, Q.; Moshokoa, S.P.; Ullah, M.Z.; Belic, M. Cubic–quartic optical solitons in Kerr and power law media. *Optik* **2017**, *144*, 357–362. [CrossRef]

38. Xie, Y.; Yang, Z.; Li, L. New exact solutions to the high dispersive cubic–quintic nonlinear Schrödinger equation. *Phys. Lett. Sect. A Gen. At. Solid State Phys.* **2018**, *382*, 2506–2514. [CrossRef]

39. Li, H.M.; Xu, Y.S.; Lin, J. New optical solitons in high-order dispersive cubic-quintic nonlinear Schrödinger equation. *Commun. Theor. Phys.* **2004**, *41*, 829.

40. Nawaz, B.; Ali, K.; Abbas, S.O.; Rizvi, S.T.R.; Zhou, Q. Optical solitons for non-Kerr law nonlinear Schrödinger equation with third and fourth order dispersions. *Chin. J. Phys.* **2019**, *60*, 133–140. [CrossRef]

41. Biswas, A.; Arshed, S. Application of semi-inverse variational principle to cubic-quartic optical solitons with kerr and power law nonlinearity. *Optik* **2018**, *172*, 847–850. [CrossRef]

42. Seadawy, A.R.; Kumar, D.; Chakrabarty, A.K. Dispersive optical soliton solutions for the hyperbolic and cubic-quintic nonlinear Schrödinger equations via the extended sinh-Gordon equation expansion method. *Eur. Phys. J. Plus* **2018**, *133*, 182. [CrossRef]

43. El-Dessoky, M.M.; Islam, S. Chirped Solitons in Generalized Resonant Dispersive Nonlinear Schrödinger's equation. *Comput. Sci.* **2019**, *14*, 737--752.

44. Bansal, A.; Biswas, A.; Zhou, Q.; Babatin, M.M. Lie symmetry analysis for cubic–quartic nonlinear Schrödinger's equation. *Optik* **2018**, *169*, 12–15. [CrossRef]

45. Biswas, A.; Kara, A.H.; Ullah, M.Z.; Zhou, Q.; Triki, H.; Belic, M. Conservation laws for cubic–quartic optical solitons in Kerr and power law media. *Optik* **2017**, *145*, 650–654. [CrossRef]

46. Younis, M.; Cheemaa, N.; Mahmood, S.A.; Rizvi, S.T.R. On optical solitons: The chiral nonlinear Schrödinger equation with perturbation and Bohm potential. *Opt. Quantum Electron.* **2016**, *48*, 542. [CrossRef]

Appl. Sci. **2020**, *10*, 219

47. Hosseini, K.; Samadani, F.; Kumar, D.; Faridi, M. New optical solitons of cubic-quartic nonlinear Schrödinger equation. *Optik* **2018**, *157*, 1101–1105. [CrossRef]
48. Conte, R.; Musette, M. Elliptic general analytic solutions. *Stud. Appl. Math.* **2009**, *123*, 63–81. [CrossRef]

Article

3D Numerical Modeling of Induced-Polarization Grounded Electrical-Source Airborne Transient Electromagnetic Response Based on the Fictitious Wave Field Methods

Yanju Ji [1], Xiangdong Meng [2], Weimin Huang [3], Yanqi Wu [2] and Gang Li [2,*]

[1] Key Laboratory of Geophysical Exploration Equipment, Ministry of Education, Jilin University, Changchun 130026, China; jiyj@jlu.edu.cn
[2] College of Instrumentation and Electrical Engineering, Jilin University, Changchun 130026, China; mengxd16@sina.com (X.M.); wuyq@16mails.jlu.edu.cn (Y.W.)
[3] Faculty of Engineering and Applied Science, Memorial University, St. John's, NL A1B 3X5, Canada; weimin@mun.ca
* Correspondence: ligang2013@jlu.edu.cn; Tel.: +86-13756683304

Received: 29 December 2019; Accepted: 31 January 2020; Published: 4 February 2020

Abstract: The grounded electrical-source airborne transient electromagnetic (GREATEM) system is widely used in mineral exploration. Meanwhile, the induced polarization (IP) effect, which indicates the polarizability of the earth, is often found. In this paper, the Maxwell equations in the frequency domain are transformed into fictitious wave domain, where Maxwell equations are solved by the time domain finite difference method. Then, an integral transformation method is used to convert the calculation results back to the time domain. A three-dimensional (3D) numerical simulation in a polarizable medium is presented. The accuracy of this method is proven by comparing it with the analytical solution and the existing method, and the calculation efficiency is increased five-fold. The simulation results show that the GREATEM system has a higher response amplitude in the conductive region, while IP effects cannot be identified in the conductive area. The GREATEM system has a higher response amplitude in the low-resistance region, but IP effects cannot be identified in the low-resistance area, and the detection of IP effects is more suitable for the high-resistance area. Therefore, it is necessary to improve the detection ability of the GREATEM system in the low-resistance area.

Keywords: wave field transformation; finite difference method; Cole–Cole model

1. Introduction

Electromagnetic technology has a wide range of applications in various fields. Nasim (2019) proposed a metamaterial platform for solving integral equations using a monochromatic electromagnetic field [1]. The application of electromagnetic technology in medical diagnosis equipment realizes the measurement of blood glucose and the detection of cancer stage [2,3]. In addition, it is necessary to consider electromagnetic fields in plasma electronics [4,5].

Here, we apply electromagnetic technology to geophysics. Grounded electrical-source airborne transient electromagnetic (GREATEM) is a useful detection geophysical method where the transmitter is set on the ground and the receiver is installed on the aircraft [6]. Combining the ground and airborne electromagnetic systems, GREATEM has the advantages of large detection depth, high resolution, wide range, and fast speed. In particular, it has unique superiority in mountains, forest-covered areas, swamps, and other special areas [7,8]. Ji (2013) used an airship as the carrier and the two-dimensional finite difference method to calculate the time-domain GREATEM response with a long wire source [9].

Ren (2017) investigated a three-dimensional time-domain airborne electromagnetic model based on the finite volume method, and then used the strategy of separating the primary field from the secondary field, which significantly saved on calculation time [10]. For airborne transient electromagnetic measurement, a three-dimensional forward simulation method which considered attitude change was proposed, and the electromagnetic response of a shallow surface was analyzed in Reference [11]. Later, the spectral element method (SE) was used to carry out three-dimensional modeling of GREATEM, effectively improving the modeling accuracy [12].

To interpret GREATEM, researchers often make the assumption that the earth is just conductive or resistive. However, there is an induced polarization (IP)effect in the earth, which is usually reflected as a reverse signal. The existence of the IP effect has a great impact on the electromagnetic response [13,14]. The IP effect is related to conductivity and frequency, and its dispersion characteristics can be expressed by the Debye model and the Cole–Cole model [15,16]. At present, numerical simulations of the polarization effect are mainly conducted in the frequency domain. Most of the early studies simulated the IP effect by converting the electromagnetic field from the frequency domain to the time domain [17,18]. Unfortunately, when broadband frequency content is considered, the methods are not so efficient [19]. In order to simulate the IP effect in the time domain, Kang (2015) proposed a method to generate a three-dimensional pseudo charge distribution using airborne time-domain electromagnetic data and effectively extracted the polarization signal [20]. Wu (2017) used the FDTD (finite difference time-domain) method to simulate the IP effect with a small loop source [21]. Commer (2017) expanded the finite difference time-domain scheme, modeled and analyzed the IP effect, and developed a simple calculation of the Cole–Cole model [22].

However, three-dimensional (3D) numerical simulations of IP effects based on FDTD require a high computation load, when including the air layer. In addition, it is difficult to calculate the source in the FDTD method; thus, the upward continuation method is often used to avoid directly loading the emission source. Mittet (2010) put forward the theory of wave field transformation and successfully applied it to marine CSEM (controllable source electromagnetic method), increasing the calculation efficiency by nearly 10-fold [23]. Mittet (2018) used the least squares fitting method to calculate the IP response of sea water in the fictitious wave field [24]. Ji (2017) improved the emission source, applied the wave field transformation method to the ground transient electromagnetic simulation, and realized a three-dimensional transient electromagnetic numerical simulation including the air layer [25]. However, it did not consider the IP effect of the earth. Therefore, here, we perform the wave field transformation in GREATEM, and we realize a three-dimensional numerical simulation in polarizable media using the Cole–Cole model. In addition, the computing efficiency is also improved.

The wave field transformation method is proposed for the grounded electrical-source airborne transient electromagnetic system in this paper. The Maxwell equations in the real diffusion field are transformed into the fictitious wave field by using the transformation relationship between the diffusion field and fictitious wave field. The fractional order Cole–Cole model is introduced into the fictitious wave field, and the Maxwell equations with the Cole–Cole model are solved using the finite difference time-domain method. Finally, an integral transformation method is used to transform the calculation results back to the real diffusion field. In the case of including polarized media, the two-field numerical simulation of the induction field and polarization field is realized. In addition, the computing efficiency is also improved.

2. Wave Field Transformation Theory Based on Fractional-Order Cole–Cole Model

2.1. Transformation from Real Diffusion Field to Fictitious Wave Field

The Maxwell equation in the frequency domain of the real diffusion field is written as follows:

$$-\nabla \times H(x,\omega) + \sigma(\omega)E(x,\omega) = -J(x,\omega), \tag{1}$$

$$\nabla \times E(x,\omega) - i\omega\mu H(x,\omega) = -K(x,\omega), \tag{2}$$

where E and H are the electric and magnetic fields in the real diffusion field, respectively, J denotes the current density, and K denotes the magnetic current density. ω is the angular frequency, μ is permeability tensor, and σ is conductivity, while x is the direction.

The IP effect can be described by the Cole–Cole model [16].

$$\sigma(\omega) = \sigma_\infty \left(1 - \frac{\eta}{1 + (-i\omega\tau)^c}\right), \tag{3}$$

where σ_∞ is the conductivity corresponding to infinite frequency, τ is the characteristic time constant, and c and η are the frequency dependence and chargeability, respectively.

Applying Equation (3) to the frequency-domain Maxwell equation (Equation (1)), we can get

$$-\nabla \times H(x,\omega) + \sigma_\infty E(x,\omega) - \frac{\sigma_\infty \eta}{1 + (-i\omega\tau)^c} E(x,\omega) = -J(x,\omega). \tag{4}$$

A fictitious dielectric constant is defined by the conductivity tensor (Mittet, 2010).

$$\sigma_\infty = 2\omega_0 \varepsilon_\infty', \tag{5}$$

where $\omega_0 = 2\pi \times f_0$ is the scaling parameter, $f_0 = 1Hz$, and ε_∞' is the fictitious dielectric permittivity tensor. Substituting Equation (5) into Equation (4), we can get

$$-\nabla \times H(x,\omega) + 2\omega_0 \varepsilon_\infty' E(x,\omega) - \frac{\eta}{1 + (-i\omega\tau)^c} 2\omega_0 \varepsilon_\infty' E(x,\omega) = -J(x,\omega). \tag{6}$$

Multiplying both sides of Equation (6) by $\sqrt{\frac{-i\omega}{2\omega_0}}$ gives

$$-\nabla \times \left[\sqrt{\frac{-i\omega}{2\omega_0}} H(x,\omega)\right] + \left[\sqrt{\frac{-i\omega}{2\omega_0}} 2\omega_0 \varepsilon_\infty' E(x,\omega)\right] - \left[\frac{\eta}{1+(-i\omega\tau)^c} \sqrt{\frac{-i\omega}{2\omega_0}} 2\omega_0 \varepsilon_\infty' E(x,\omega)\right] = -\sqrt{\frac{-i\omega}{2\omega_0}} J(x,\omega). \tag{7}$$

Equation (2) can be rewritten as

$$\nabla \times E(x,\omega) - \sqrt{2i\omega\omega_0}\mu \sqrt{\frac{-i\omega}{2\omega_0}} H(x,\omega) = K(x,\omega). \tag{8}$$

The relationship between the real diffusion field and fictitious wave field was given by Reference [23].

$$-i\omega' = \sqrt{-2i\omega\omega_0}, \tag{9}$$

$$E'(x,\omega') = E(x,\omega), \tag{10}$$

$$H'(x,\omega') = \sqrt{\frac{-i\omega}{2\omega_0}} H(x,\omega), \tag{11}$$

$$J'(x,\omega') = \sqrt{\frac{-i\omega}{2\omega_0}} J(x,\omega), \tag{12}$$

$$K'(x,\omega') = K(x,\omega), \tag{13}$$

where $\omega_0 = 2\pi$ is the scaling parameter, ω' is the angular frequency in the fictitious wave field, J' and K' are the current and magnetic current densities in the fictitious wave field, and E' and H' are the electric and magnetic fields in the fictitious wave domain.

Using Equations (9)–(13), Equations (7) and (8) can be written in terms of fictitious wave fields.

$$-\nabla \times H'(x,\omega') - i\omega' \varepsilon_\infty' E'(x,\omega') + \frac{\eta i\omega' \varepsilon_\infty'}{1 + \left(\frac{\tau}{2\omega_0}\right)^c (i\omega')^{2c}} E'(x,\omega') = -J'(x,\omega'), \tag{14}$$

$$\nabla \times E'(x, \omega') + i\omega'\mu H'(x, \omega') = K'(x, \omega').\tag{15}$$

Because the appropriate kernel used to decompose the spectra is closer to a Cole–Cole function with an exponent c of 0.5 [26], then

$$-\nabla \times \mathbf{H}'(\mathbf{x}, \omega') - i\omega'\varepsilon_\infty'\mathbf{E}'(\mathbf{x}, \omega') + \frac{\eta i\omega'\varepsilon_\infty'}{1 + \sqrt{\frac{\tau}{2\omega_0}i\omega'}}\mathbf{E}'(\mathbf{x}, \omega') = -\mathbf{J}'(\mathbf{x}, \omega'),\tag{16}$$

$$\nabla \times E'(x, \omega') + i\omega'\mu H'(x, \omega') = K'(x, \omega').\tag{17}$$

Assuming $A'(\mathbf{x}, \omega') = \frac{\eta i\omega'\varepsilon_\infty'}{1 + \sqrt{\frac{\tau}{2\omega_0}i\omega'}}\mathbf{E}'(\mathbf{x}, \omega')$, the time-domain expressions (Equations (16) and (17)) can be obtained through inverse Fourier transform.

$$-\nabla \times \mathbf{H}'(\mathbf{x}, t') - \varepsilon_\infty'(\mathbf{x})\partial_{t'}\mathbf{E}'(\mathbf{x}, t') + A'(\mathbf{x}, t') = -\mathbf{J}'(\mathbf{x}, t'),\tag{18}$$

$$\nabla \times E'(x, t') + \mu\partial_{t'}H'(x, t') = K'(x, t').\tag{19}$$

By combining Equations (18) and (19), the electric and magnetic field x-components can be found as

$$E_{x_{i+1/2,jk}}^{n+1} = E_{x_{i+1/2,jk}}^n + \Delta t\frac{2\omega_0}{\sigma_\infty}(\partial_y^- H_{z_{1+1/2,j+1/2,k}}^{n+1/2} - \partial_z^- H_{y_{i+1/2,jk+1/2}}^{n+1/2}) + \Delta t\frac{2\omega_0}{\sigma_\infty}A'(x, t') - \Delta t\frac{2\omega_0}{\sigma_\infty}J_x,\tag{20}$$

$$H_{x_{i,j+1/2,k+1/2}}^{n+1/2} = H_{x_{i,j+1/2,k+1/2}}^{n-1/2} - \frac{\Delta t}{\mu}(\partial_y^+ E_{z_{i,jk+1/2}}^n - \partial_z^+ E_{y_{i,j+1/2,k}}^n),\tag{21}$$

where

$$A'(\mathbf{x}, t') = \sqrt{\frac{2\omega_0}{\tau}}e^{-\sqrt{\frac{2\omega_0}{\tau}}t'}\int_0^{t'} e^{\sqrt{\frac{2\omega_0}{\tau}}t''}\eta\varepsilon_\infty'[\partial_{t''}\mathbf{E}'(\mathbf{x}, t'')]dt''.\tag{22}$$

Equation (22) can be solved by iteration, as shown in Equation (23).

$$A'(\mathbf{x}, t^{n+1}) = A'(\mathbf{x}, t^n)e^{\sqrt{\frac{2\omega_0}{\tau}}\Delta t} - \eta\varepsilon_\infty'\mathbf{E}'(\mathbf{x}, t^N)(1 - e^{\sqrt{\frac{2\omega_0}{\tau}}\Delta t}),\tag{23}$$

where $t^n = n\Delta t$, and the forward staggered time step is $t^N = (n + \frac{1}{2})\Delta t$.

2.2. Electrical Source Loading in Fictitious Wave Field

As shown in Figure 1, Tx is the long wire source and Rx is the receiver. The fictitious emission source is only related to x- and y-components of the electric field, and the z-component is zero.

Here, the first-order Gaussian pulse is selected as the fictitious emitter, as shown in Equation (24).

$$J'(t') = -2\beta(t' - t_0)\sqrt{\frac{\beta}{\pi}}e^{-\beta(t'-t_0)^2},\tag{24}$$

where $\beta = \pi f_{max}^2$, f_{max} is the maximum frequency in the fictitious wave field.

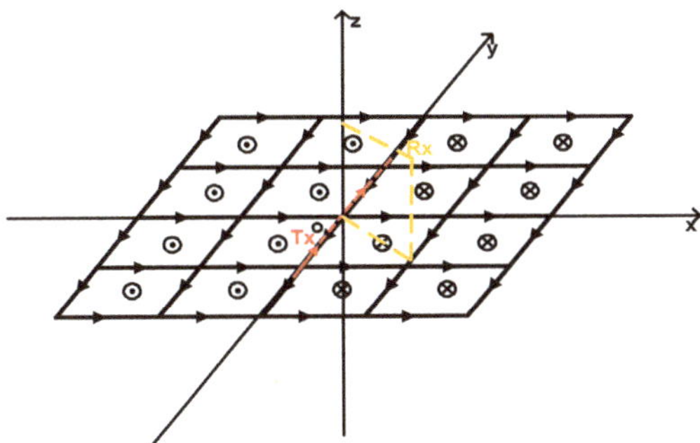

Figure 1. Loading mode of fictitious wave field long wire source.

2.3. *The Transformation from Fictitious Wave Field to Real Diffusion Field*

It should be noted that the fictitious wave field is only used for the convenience of calculation, and it does not exist in reality. Therefore, we need to transform it back to the real diffusion field. The transformations are as follows [23]:

$$J(\omega) = \int_0^T J'(t')e^{-\sqrt{\omega\omega_0}t'} e^{i\sqrt{\omega\omega_0}t'} dt', \tag{25}$$

$$E(\omega) = \sqrt{\frac{-i\omega}{2\omega_0}} \int_0^T E'(t')e^{-\sqrt{\omega\omega_0}t'} e^{i\sqrt{\omega\omega_0}t'} dt', \tag{26}$$

$$H(\omega) = \int_0^T H'(t')e^{-\sqrt{\omega\omega_0}t'} e^{i\sqrt{\omega\omega_0}t'} dt', \tag{27}$$

$$K(\omega) = \sqrt{\frac{-i\omega}{2\omega_0}} \int_0^T K'(t')e^{-\sqrt{\omega\omega_0}t'} e^{i\sqrt{\omega\omega_0}t'} dt', \tag{28}$$

where T is the total calculation time in the fictitious wave field, and t' is the sample time.

2.4. *Loading Complex Frequency Shifted Perfectly Matched Layer (CFS-PML) in Fictitious Wave Field*

Roden (2000) proposed CFS-PML (complex frequency shifted perfectly matched layer) boundary conditions, which significantly saved on memory [27]. The CFS-PML boundary conditions are introduced into the fictitious wave field to improve the absorption effect of the boundary on the low-frequency wave [28].

In x-components, Maxwell's curl equations are as follows:

$$\nabla_s \times H'(\omega') + i\omega'\varepsilon'E'(\omega') = 0, \tag{29}$$

$$\nabla_s = \frac{1}{\hat{s}_x}\frac{\partial}{\partial x}\mathbf{i} + \frac{1}{\hat{s}_y}\frac{\partial}{\partial y}\mathbf{j} + \frac{1}{\hat{s}_z}\frac{\partial}{\partial z}\mathbf{k}. \tag{30}$$

It should be noted that i is the imaginary unit in Maxwell equations, and i, j, and k are the three direction vectors of the coordinate system. $\hat{s}_x$, $\hat{s}_y$, and $\hat{s}_z$ are the coordinate stretching factors, $\hat{s}_{(x'y'z)} = k_{(x,y,z)} + \frac{\sigma_{(x,y,z)}}{\alpha_{(x,y,z)}+i\omega\varepsilon'}$ $\alpha_{(x,y,z)} > 0$, $\sigma_{(x,y,z)} > 0$, and $k_{(x,y,z)} \geq 1$. By setting $\alpha_{(x,y,z)} = \sigma_{(x,y,z)} = 0$, $k_{(x,y,z)} = 1$, we can get the original stretched coordinate metrics, which is the PML boundary condition.

Taking E'_x as an example, and using Equations (29) and (30), we can get

$$i\omega'\varepsilon'E'_x(\omega') = -\left(\frac{1}{\hat{s}_y}\frac{\partial H'_z(\omega')}{\partial y} - \frac{1}{\hat{s}_z}\frac{\partial H'_y(\omega')}{\partial z}\right). \tag{31}$$

Applying Laplace transform to (31) to the time domain, the iterative solution can be obtained.

$$E'^{n+1}_{x_{i+1/2,j,k}} = E'^{n}_{x_{i+1/2,j,k}} + \Delta t' \frac{2\omega_0}{\sigma} \left[\partial_y^{-1} H'^{n+1/2}_{z_{i+1/2,j+1/2,k}} - \partial_z^{-1} H'^{n+1/2}_{y_{i+1/2,j,k+1/2}}\right] + \Delta t' \frac{2\omega_0}{\sigma}\left(\Psi^{n+1/2}_{e_{xy_{i+1/2,j,k}}} - \Psi^{n+1/2}_{e_{xz_{i+1/2,j,k}}}\right), \tag{32}$$

$$\Psi^{n+1/2}_{e_{xy_{i+1/2,j,k}}} = e^{-\left(\frac{\sigma_y}{k_y}+\alpha_y\right)\frac{\Delta t'}{\varepsilon'}} \Psi^{n-1/2}_{e_{xy_{i+1/2,j,k}}} + \frac{\sigma_y}{(\sigma_y k_y + k_y^2 \alpha_y)}\left(e^{-\left(\frac{\sigma_y}{k_y}+\alpha_i\right)\frac{\Delta t'}{\varepsilon'}} - 1\right)\left(H'^{n+1/2}_{z_{i+1/2,j+1/2,k}} - H'^{n+1/2}_{z_{i+1/2,j-1/2,k}}\right)/\Delta y, \tag{33}$$

$$\Psi^{n+1/2}_{e_{xz_{i+1/2,j,k}}} = e^{-\left(\frac{\sigma_z}{k_z}+\alpha_z\right)\frac{\Delta t'}{\varepsilon'}} \Psi^{n-1/2}_{e_{xz_{i+1/2,j,k}}} + \frac{\sigma_z}{(\sigma_z k_z + k_z^2 \alpha_z)}\left(e^{-\left(\frac{\sigma_z}{k_z}+\alpha_z\right)\frac{\Delta t'}{\varepsilon'}} - 1\right)\left(H'^{n+1/2}_{y_{i+1/2,j+1/2,k}} - H'^{n+1/2}_{y_{i+1/2,j-1/2,k}}\right)/\Delta z. \tag{34}$$

On the boundary parameters, σ_i and k_i, which depend on the relative positions of nodes and target areas, are not fixed, whereas $\Psi^{n+1/2}_{e_{xy_{i+1/2,j,k}}}$ and $\Psi^{n+1/2}_{e_{xz_{i+1/2,j,k}}}$ are two discrete variables.

$$\sigma_i = \sigma_{\max}\frac{|k-k_0|^m}{d^m}, (i = x, y, z), \tag{35}$$

$$k_i = 1 + (k_{\max} - 1)\frac{|k-k_0|^m}{d^m}, (i = x, y, z), \tag{36}$$

where k_0 indicates the intersection of the boundary and the target area boundary, d is the boundary thickness, and m is the polynomial parameter.

3. Accuracy and Efficiency Verification of the Calculation Method

3.1. Half Space Model with IP Effect

In the field measurement with GREATEM, only the change of the underground secondary field is considered. In the process of numerical simulation, people usually divide the real space into two parts (air and ground), without considering the interface between the ground and air. Therefore, in the research process of this paper, the ground and the interface are not considered. In order to verify the effectiveness of this method, we design a half-space model incorporating the IP effect, as shown in Figure 2. Let the conductivity of the air layer $\sigma_{\text{air}}= 10^{-6}$ S/m, conductivity at the infinite frequency of the half-space model $\sigma_\infty = 0.1$ S/m, chargeability $\eta= 0.6$, characteristic time constant $\tau= 1$ s, and frequency dependence c $= 0.5$. Tx represents the transmitter, Rx is the receiver, the horizontal distance between Tx and Rx (offset distance) is 300 m, and the flight altitude is h $= 0$ m. In Figure 3, the calculation results are compared with Reference [29]. We can see that the solution of the fictitious domain finite difference method (FW-FDTD) is in good agreement with that of the transformed frequency domain method. The average relative error of both methods is less than 10%.

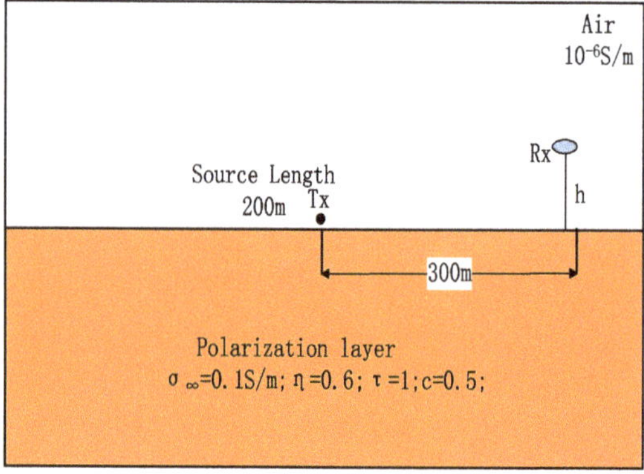

Figure 2. The half-space model with induced polarization (IP) effect.

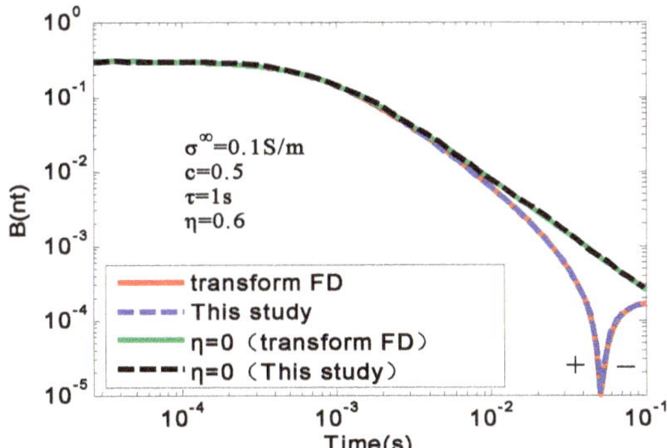

Figure 3. The response curve of the half-space model.

When the earth contains the IP effects, the polarization parameters have great influence on the induced-polarization (IP) effects. In order to analyze the influence of polarization parameters on IP effects, we take the uniform half space model (Figure 2) as an example. Set the air layer conductivity 10^{-6} S/m, a 200-m-long wire source is selected as the transmitter, and the offset distance is r = 300 m. Furthermore, chargeability η= 0.6, conductivity at infinite frequency σ_∞= 0.1 S/m, frequency dependence c = 0.5, and flight altitude h = 25 m. The characteristic time constant is selected as τ= 1 s, τ= 0.1 s, and τ= 0.01 s. The IP response under different characteristic time constants is calculated as shown in Figure 4. Other parameters remain unchanged. The characteristic time constant is set to τ= 0.01 s, and η= 0.5. The conductivity at infinite frequency is σ_∞= 0.01 S/m, σ_∞= 0.05 S/m, and σ_∞= 0.1 S/m. The IP response under different conductivity is calculated as shown in Figure 5. Furthermore, σ_∞= 0.01 S/m, τ= 0.01 s, and the chargeability is η= 0.3, η= 0.5, and η= 0.7. The response curves under different chargeability are calculated as shown in Figure 6.

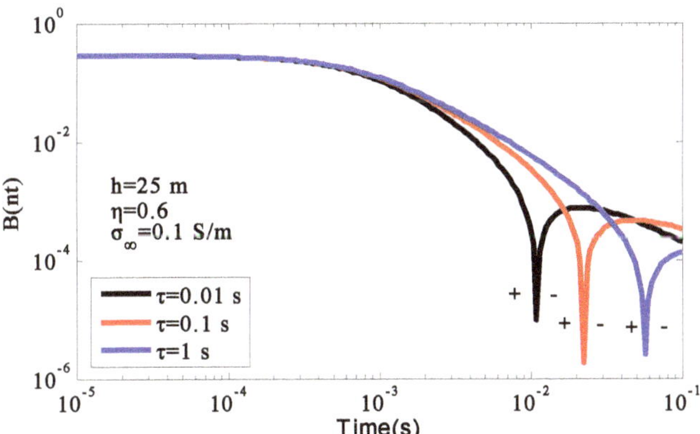

Figure 4. Influence of characteristic time constant on IP effect.

From Figures 4–6, it can be seen that characteristic time constant, chargeability, and conductivity at infinite frequency affect the occurrence time of negative response, and all responses are monotonic. Compared with the conductivity at infinite frequency, the effect of characteristic time constant and chargeability on the IP response is lower, mainly in the late stage, and the change trend of the early induction field is basically the same. With the decrease in characteristic time constant, the time of minus response is gradually advanced. With the decrease in chargeability, the time of minus response is delayed. The conductivity at infinite frequency has a great influence on the IP effect. In addition to the minus response at the later stage, it also has a great influence on the early induction field. In the early stage, the response amplitude of the induction field increases gradually with the increase in conductivity at infinite frequency. In the later period, a smaller conductivity at infinite frequency results in the minus response appearing earlier.

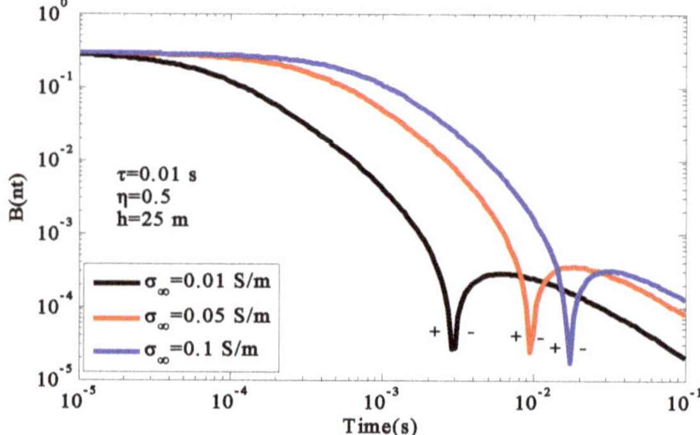

Figure 5. Influence of conductivity at infinite frequency on IP effect.

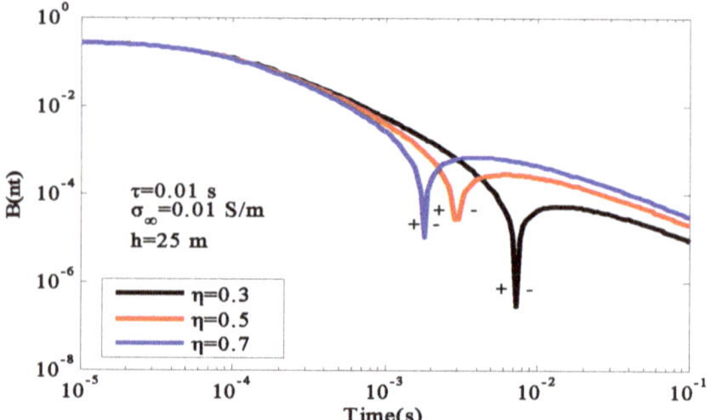

Figure 6. Influence of chargeability on IP effect.

3.2. Three-Dimensional Model

In addition, we calculated a three-dimensional model with a chargeable anomaly and compared the result with Reference [30]. The dimensions of the cell mesh were $70 \times 70 \times 70$, and the minimum grid size was 25 m. The modeling volume was $3500 \times 3500 \times 3500$ m^3. The size of the chargeable anomaly was $100 \times 100 \times 80$ m^3, while the distance from the top to the bottom of the chargeable anomaly was 40 m. The parameters of the Cole–Cole model were $\sigma_\infty = 0.1$ S/m, $\tau = 0.1$ s, $\eta = 0.3$, and c = 0.5. The background conductivity was $\sigma_\infty = 0.001$ S/m. A vertical point dipole emitter, which was located 30 m above the surface, was used, and the receiver and transmitter were at the same location. The calculation results are shown in Figure 7, from which it can been seen that the two IP effect curves were consistent. The relative error was less than 10%, which further verifies the correctness of the method.

In terms of computation load, the iterations in the fictitious wave field are as follows:

$$N_t = \frac{\sqrt{3}R_{\max}}{\Delta x} \times \sqrt{\frac{\sigma_{\max}}{\sigma_{\min}}}, \tag{37}$$

where N_t represents the number of iteration steps, $R_{\max}$ is the maximum transceiver distance, $\sigma_{\max}$ and $\sigma_{\min}$ are the maximum and minimum conductivity, respectively, and Δx is the minimum grid size. Using Equation (37), the number of iteration steps was 76,681. It should be noted that only one transmitter rather than an array was selected in this work. Compared with Reference [30], the calculation efficiency was improved by about five-fold.

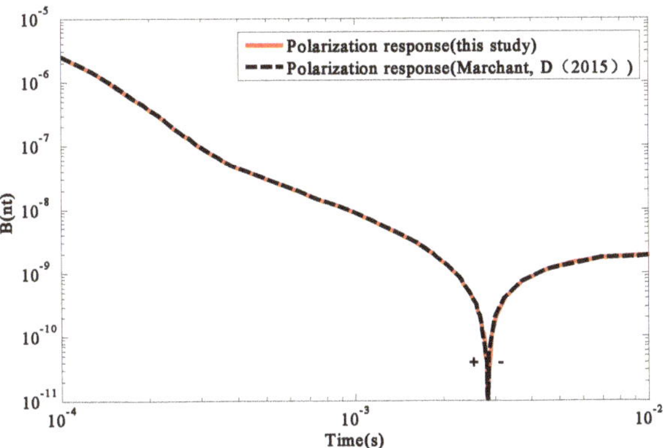

Figure 7. Comparison of this study with that of Reference [29].

4. Calculation Results of the Three-Dimensional Chargeable Body Model

The phenomenon of the inverse sign (IP effect) often appears in GREATEM. Most researchers focused on the influence of the IP parameters. However, in addition to the IP parameters, the system parameters of GREATEM also have an impact on the IP effect. In this section, we mainly analyze the effect of the system parameters, and we provide a theoretical basis for the field measurements.

4.1. Forward Modeling Results of 3D Chargeable Body

In order to analyze the influence of system parameters on the IP effect, we build a three-dimensional model with the consideration of a chargeable body, as shown in Figure 8. The ground conductivity was $\sigma_\infty = 0.001$ S/m, the air layer conductivity was $\sigma_{air} = 10^{-6}$ S/m, Tx was the long wire source, the length was $L_{Tx} = 400$ m, Rx represents the receiving point, the height from the ground was h $= 25$ m, and the offset distance was r $= 200$ m. The dimensions of the chargeable body were $400 \times 400 \times 300$ m^3, which was 100 m below the surface. The model parameters of the chargeable body were selected as follows: chargeability $\eta = 0.5$, frequency dependence c $= 0.5$, and characteristic time constant $\tau = 0.01$ s. The conductivities at infinite frequency were $\sigma_\infty = 0.01$ S/m, $\sigma_\infty = 0.5$ S/m, and $\sigma_\infty = 0.1$ S/m. The calculation results are shown in Figure 9. The snapshots of the induced current system in the fictitious wave field are shown in Figure 10a,b.

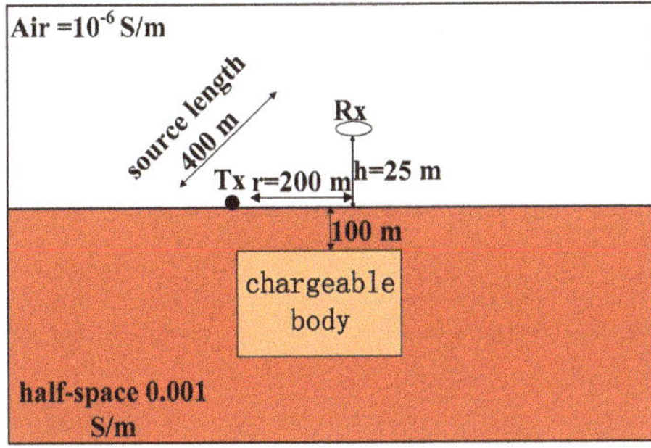

Figure 8. Three-dimensional (3D) model with IP effect.

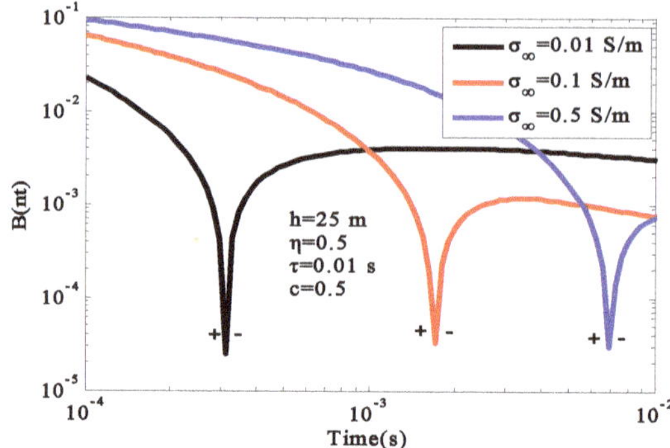

Figure 9. Response curves under different conductivities.

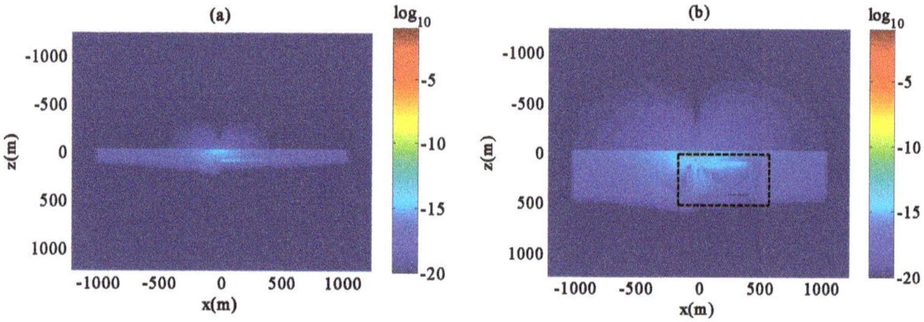

Figure 10. The snapshots of the induced current system in the fictitious wave field (**a**) early time; (**b**) late time.

It can be seen from Figure 9 that, with the increase in conductivity at the infinite frequency, the attenuation of the early induction field gradually decreased, the attenuation of the late IP response gradually quickened, and the occurrence time of the inverse sign gradually delayed. This is because the influence of the induction field of the low-resistance body was greater than that of the high-resistance body. Therefore, the second field decayed slowly, and the negative response appeared later. Thus, the polarization phenomenon was more easily observed on the high-resistance body. The chargeable body position can be clearly observed from Figure 10, and the method of wave field transformation could effectively identify the polarizable body. Since we set the air layer as a finite high-resistance body, the conductivity was not zero, and the propagation of induced current in the air layer was not zero. In the fictitious wave field, due to the influence of the chargeable body, the propagation mode of the electromagnetic field changed, and it no longer propagated in the form of symmetry.

After setting $\sigma_\infty = 0.1$ S/m, as well as the ground conductivity $\sigma_{\text{ground}} = 0.001$ S/m, $\sigma_{\text{ground}} = 0.01$ S/m, and $\sigma_{\text{ground}} = 0.1$ S/m, which are associated with high-, medium-, and low-resistance areas, the IP effects were calculated as shown in Figure 11.

With the increase in ground conductivity, the amplitude of the early induction field and the late polarization field increased. In the high- and middle-resistance regions, the negative responses occurred at 1.7 ms and 3.3 ms. On the other hand, in the low-resistance region, the negative response could not be observed. When the negative value reached the minimum, the curve decayed approximately exponentially. The GREATEM had better resolution in the low-resistance region, but it is more suitable for the high-resistance region when measuring the IP effect.

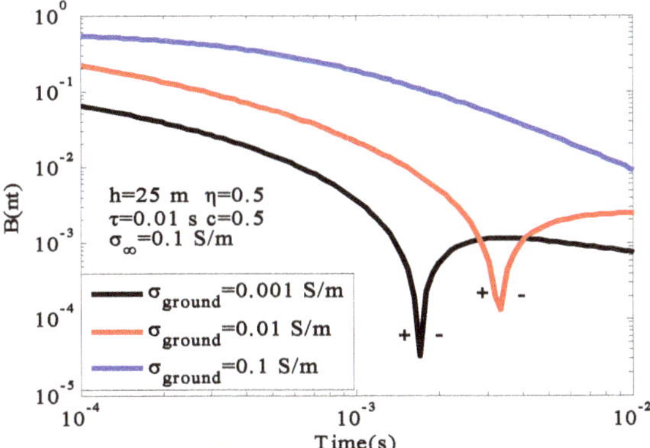

Figure 11. IP effects under different background conductivities.

4.2. The Influence of System Parameters on IP Response

In the field measurements of GREATEM, due to the influence of airflow and environment, the flight height was not a constant. Therefore, it was necessary to analyze the influence of flight altitude on the IP effect. We take the 3D model (Figure 8) as an example. The air layer conductivity was set to $\sigma_{air} = 10^{-6}$ S/m and the earth conductivity was set to 0.001 S/m, while the size of the chargeable body was $400 \times 400 \times 300$ m. The model parameters of the chargeable body were as follows: chargeability $\eta = 0.5$, frequency dependence $c = 0.5$, characteristic time constant $\tau = 0.01$ s, conductivity at the infinite frequency $\sigma_\infty = 0.1$ S/m, offset distance r = 200 m, and flight altitude h = 25 m, h = 50 m, h = 75 m, and h = 100 m. The induced polarization (IP) effect at different flight altitudes was calculated as depicted in Figure 12. In order to understand the influence of the parameters of GREATEM on the IP response, other parameters remained unchanged, while the flight height was set to h = 25 m, and the emission source length was $L_{Tx} = 50$ m, $L_{Tx} = 100$ m, and $L_{Tx} = 200$ m. The responses of different emission source lengths are shown in Figure 13. Taking $L_{Tx} = 400$ m and the offset distance r = 100 m, r = 200 m, and r = 400 m, the responses of different offsets are shown in Figure 14.

Based on Figure 12, flight altitude mainly affected the amplitude of the response. The effect on the time of sign reversal was not obvious, and the sign reversal phenomenon was mainly concentrated in the range 1.5–2 ms. With the increase in flight altitude, the received electromagnetic response decreased gradually. It can be seen from Figure 13 that the length of the emission source mainly affected the amplitude of the response. The length of emission source increased gradually, and the amplitude of response increased gradually, which also shows that a longer emission source resulted in a higher resolution of the polarizable body. In Figure 14, with the decrease in offset distance, the time of the sign reversal appeared earlier, and the polarization response became easier to observe at a short offset distance.

Next, we set the air layer conductivity to $\sigma = 10^{-6}$ S/m and earth conductivity to 0.001 S/m. The size of the chargeable body was $400 \times 400 \times 100$ m^3. The model parameters of the chargeable body were chosen as $\eta = 0.5$, $c = 0.5$, $\tau = 0.01$ s, $\sigma_\infty = 0.1$ S/m, offset distance r = 200 m, emission source length $L_{Tx} = 400$ m, and flight altitude h = 25 m. The chargeable body was buried 25 m and 50 m below the surface. The response curves of the chargeable body at different depths are shown in Figure 15. Moreover, we changed the size of the chargeable body to $300 \times 300 \times 200$ m^3 and the chargeable body depth to 100 m, while keeping other parameters the same, and setting the distance from the left boundary of the chargeable body to the emission source as 0 m, 50 m, and 100 m. The response curves at different source positions are shown in Figure 16.

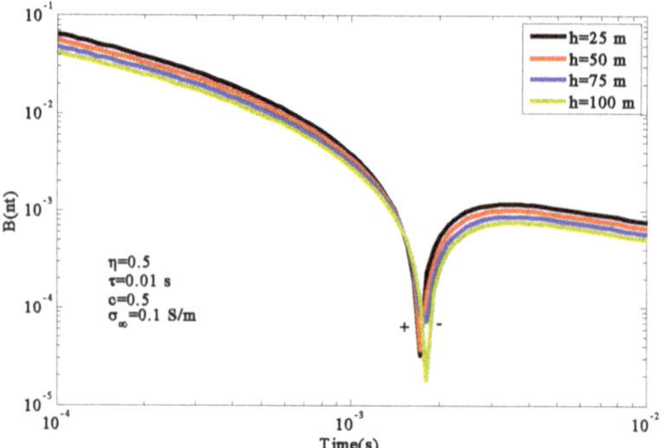

Figure 12. IP effects at different flight altitudes.

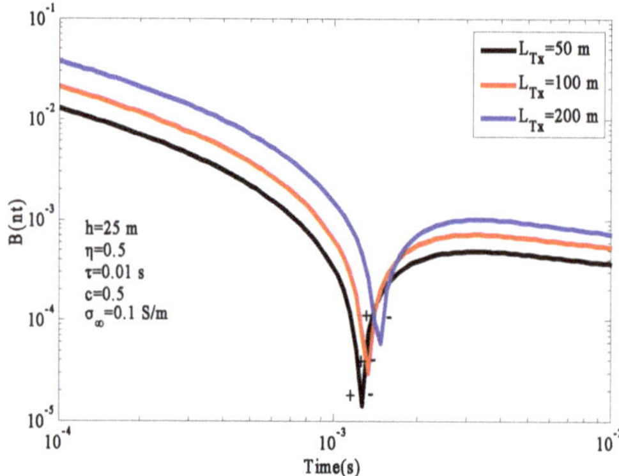

Figure 13. The response curves under different emission source lengths.

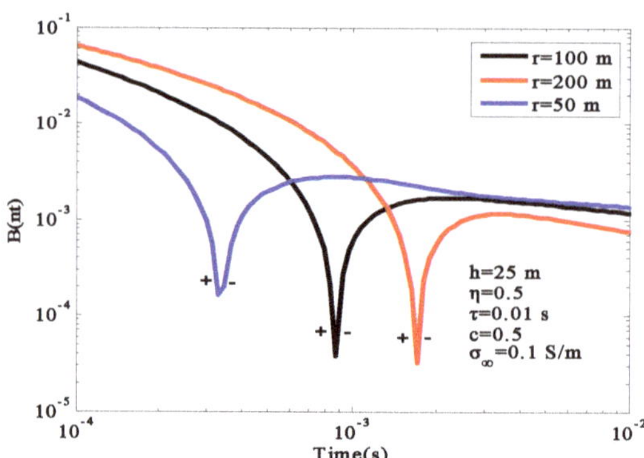

Figure 14. The response curves under different offset distances.

From Figures 15 and 16, it can be seen that all the amplitudes decreased, while the depth of chargeable body increased. However, the time of the reversal sign did not change. In the early stage, as the distance between the emitter source and the chargeable body decreased, the amplitude of the induced field increased, the attenuation of the curve was delayed. In the later stage, the time of sign reversal appeared later when the distance between the emitter and the chargeable body decreased.

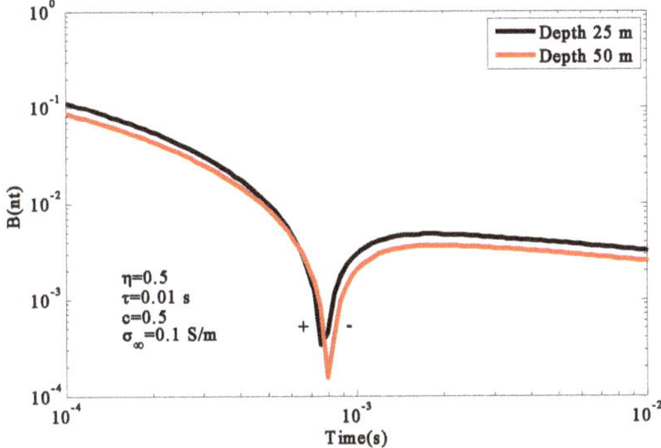

Figure 15. The influence of chargeable body depth on response curve.

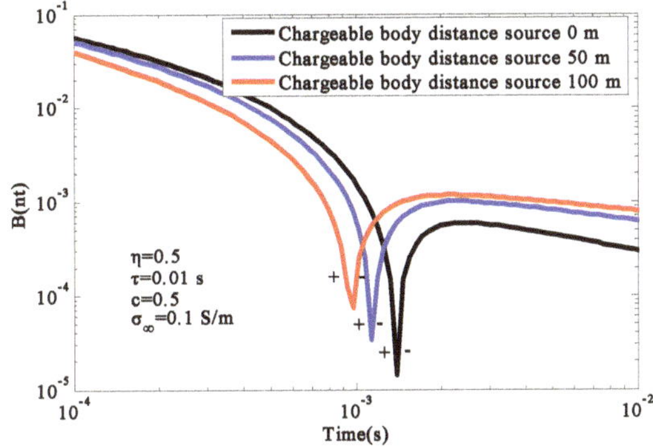

Figure 16. The influence of different emitter locations on IP effect.

5. Conclusions

Three-dimensional numerical simulations are the basis of data interpretation and imaging. In this paper, a three-dimensional numerical simulation of underground polarized media was realized, which achieved simulation results closer to the actual mineral resources. In the actual measurement process with GREATEM, the geodetic model could be established using this method, and the optimal configuration structure of the equipment was obtained using the numerical simulation method. This not only provides the basis for the exploration of underground mineral resources, but also provides theoretical guidance for the design of detection instruments. Wave field transformation was proposed to calculate the response of the GREATEM system using the Cole–Cole model. The electromagnetic wave was absorbed by the boundary condition of CFS-PML in the fictitious wave field. Compared with the analytical solution and the transformed FD method, the relative error was less than 10%, satisfying the standard calculation requirements. Compared with the existing method, the calculation time was reduced five-fold. According to the simulations, the flight altitude, electrical source length, electrical source position, and the offset distance all affected the electromagnetic response monotonically. In the homogeneous half-space model, when the characteristic time constant ($\tau > 1$ s) and conductivity at the infinite frequency ($\sigma_\infty > 1$ S/m) were too large, the negative response was likely overwhelmed by the noise; thus, the negative response could not be observed. Therefore, the occurrence time of the negative response could be enhanced by reducing the offset distance. Low-altitude flights can facilitate the recognition ability of the chargeable body, but they are often limited by geological conditions. The response of the GREATEM system had good recognition ability in the low-resistance area; however, the chargeable body could hardly be detected. Therefore, it is necessary to improve the detection accuracy of the GREATEM in low-resistance areas.

This paper realized a three-dimensional numerical simulation of polarized media based on the Cole–Cole model, and improved the calculation efficiency. However, in the actual measurement process, the underground medium is diverse. It is not comprehensive to only use the Cole–Cole model to describe the characteristics of polarized medium; the existence of the GEMTIP model in particular is a major problem in the field of geophysics. In future work, we need to consider the GEMTIP model in the numerical simulation, so as to improve the universality of this method.

Author Contributions: Conceptualization, X.M. and Y.J.; methodology, X.M.; software, X.M.; validation, X.M., Y.J. and G.L.; formal analysis, Y.W.; investigation, Y.W.; resources, W.H.; data curation, X.M.; writing—original draft preparation, X.M.; writing—review and editing, W.H.; supervision, G.L.; funding acquisition, Y.J. All authors have read and agreed to the published version of the manuscript.

Funding: This work was supported by the National Natural Science Foundation of China (Grant Number 41674109).

Acknowledgments: We thank the editors and reviewers for comment. We thank the National Natural Science Foundation of China (Grant Number 41674109) for support.

Conflicts of Interest: The authors declare no conflicts of interest.

References

1. Estakhri, N.M.; Edwards, B.; Engheta, N. Inverse-designed metastructures that solve equations. *Science* **2019**, *363*, 1333–1338. [CrossRef]
2. Luigi, L.S.; Lucio, V. Electromagnetic Nanoparticles for Sensing and Medical Diagnostic Applications. *Materials* **2018**, *11*, 603.
3. la Spada, L. Metasurfaces for Advanced Sensing and Diagnostics. *Sensors* **2019**, *19*, 355. [CrossRef] [PubMed]
4. Lee, I.H.; Yoo, D.; Avouris, P.; Low, T.; Oh, S.H. Graphene acoustic plasmon resonator for ultrasensitive infrared spectroscopy. *Nat. Nanotechnol.* **2019**, *14*, 313–319. [CrossRef]
5. Greybush, N.J.; Pacheco-Peña, V.; Engheta, N.; Murray, C.B.; Kagan, C.R. Plasmonic Optical and Chiroptical Response of Self-Assembled Au Nanorod Equilateral Trimers. *ACS Nano* **2019**, *13*, 1617–1624. [CrossRef] [PubMed]
6. Mogi, T.; Kusunoki, K.I.; Kaieda, H.; Ito, H.; Jomori, A.; Jomori, N.; Yuuki, Y. Grounded electrical-source airborne transient electromagnetic (GREATEM) survey of Mount Bandai, north-eastern Japan. *Explor. Geophys.* **2009**, *40*, 1–7. [CrossRef]
7. Ito, H.; Mogi, T.; Jomori, A.; Yuuki, Y.; Kiho, K.; Kaieda, H.; Suzuki, K.; Tsukuda, K.; Allah, S.A. Further investigations of underground resistivity structures in coastal areas using grounded-source airborne electromagnetics. *Earth Planets Space* **2011**, *63*, e9–e12. [CrossRef]
8. Allah, S.A.; Mogi, T.; Ito, H.; Jymori, A.; Yukki, Y.; Fomenko, E.; Kiho, K.; Kaieda, H.; Suzuki, K.; Tsukuda, K. Three-dimensional resistivity characterization of a coastal area: Application of Grounded Electrical-Source Airborne Transient Electromagnetic (GREATEM) survey data from Kujukuri Beach, Japan. *J. Appl. Geophys.* **2013**, *99*, 1–11. [CrossRef]
9. Yan-Ju, J.; Yuan, W.; Jiang, X.; Feng-Dao, Z.; Su-Yi, L.; Yi-Ping, Z.; Jun, L. Development and application of the grounded long wire source airborne electromagnetic exploration system based on an unmanned airship. *Chin. J. Geophys.* **2013**, *56*, 3640–3650.
10. Ren, X.; Yin, C.; Liu, Y.; Cai, J.; Wang, C.; Ben, F. Efficient Modeling of Time-domain AEM using Finite-volume Method. *J. Environ. Eng. Geophys.* **2017**, *22*, 267–278. [CrossRef]
11. Qi, Y.; Huang, L.; Wang, X.; Fang, G.; Yu, G. Airborne Transient Electromagnetic Modeling and Inversion Under Full Attitude Change. *IEEE Geosci. Remote Sens. Lett.* **2017**, *14*, 1575–1579. [CrossRef]
12. Yin, C.; Huang, X.; Liu, Y.; Cai, J. 3-D Modeling for Airborne EM using the Spectral-element Method. *J. Environ. Eng. Geophys.* **2017**, *22*, 13–23. [CrossRef]
13. Fitterman, D.V.; Stewart, M.T. Transient electromagnetic sounding for groundwater. *Geophysics* **1986**, *51*, 995–1005. [CrossRef]

14. Schamper, C.; Auken, E.; Sørensen, K. Coil response inversion for very early time modelling of helicopter-borne time-domain electromagnetic data and mapping of near-surface geological layers. *Geophys. Prospect.* **2014**, *62*, 658–674. [CrossRef]

15. Debye, P.; Falkenhagen, H. Dispersion of the conductivity and dielectric constants of strong electrolytes. *Physikalische Zeitschrift* **1928**, *29*, 121.

16. Cole, K.S.; Cole, R.H. Dispersion and absorption in dielectrics. Alternating current characteristics. *J. Chem. Phys.* **1941**, *9*, 341–351. [CrossRef]

17. Yin, C.C.; Liu, B. The research on the 3D TDEM modeling and IP effect. *Chin. J. Geophys.* **1994**, *37*, 486–492.

18. Yin, C.C.; Miao, J.J.; Liu, Y.H.; Qiu, C.K.; Cai, J. The effect of induced polarization on time-domain airborne EM diffusion. *Chin. J. Geophys.* **2016**, *59*, 4710–4719.

19. Marchant, D.; Haber, E.; Oldenburg, D.W. Three-dimensional modeling of IP effects in time-domain electromagnetic data. *Geophysics* **2014**, *79*, E303–E314. [CrossRef]

20. Kang S, W.; Oldenburg, D. Recovering IP information in airborne-time domain electromagnetic data. *ASEG Ext. Abstr.* **2015**, *2015*, 1–4. [CrossRef]

21. Wu, Y. 3-D Forward modeling of Induced Polarization Effects of Transient Electromagnetic Method. In Proceedings of the AGU Fall Meeting, New Orleans, LA, USA, 11–15 December 2017.

22. Commer, M.; Petrov, P.V.; Newman, G.A. FDTD modeling of induced polarization phenomena in transient electromagnetics. *Geophys. J. Int.* **2017**, *209*, 387–405. [CrossRef]

23. Mittet, R. High-order finite-difference simulations of marine CSEM surveys using a correspondence principle for wave and diffusion fields. *Geophysics* **2010**, *75*, 33–50. [CrossRef]

24. Mittet, R. Electromagnetic modeling of induced polarization with the fictitious wave-domain method. In *SEG Technical Program Expanded Abstracts*; Swiss Federal Institute of Technology: Lausanne, Switzerland, 2018.

25. Ji, Y.; Hu, Y.; Imamura, N. Three-Dimensional Transient Electromagnetic Modeling Based on Fictitious Wave Domain Methods. *Pure Appl. Geophys.* **2017**, *174*, 2077–2088. [CrossRef]

26. Revil, A.; Mao, D.; Shao, Z.; Sleevi, M.F.; Wang, D. Induced polarization response of porous media with metallic particles—Part 6: The case of metals and semimetals. *Geophysics* **2017**, *82*, E97–E110. [CrossRef]

27. Roden, J.A.; Gedney, S.D. Convolution PML (CPML): An efficient FDTD implementation of the CFS–PML for arbitrary media. *Microw. Opt. Technol. Lett.* **2000**, *27*, 334–339. [CrossRef]

28. Hu, Y.; Egbert, G.; Ji, Y.; Fang, G. A novel CFS-PML boundary condition for transient electromagnetic simulation using a fictitious wave domain method. *Radio Sci.* **2017**, *52*, 118–131. [CrossRef]

29. Newman, G.A. Frequency-domain modeling of airborne electromagnetic responses using staggered finite differences. *Geophys. Prospect.* **1995**, *43*, 1021–1042. [CrossRef]

30. Marchant, D. Induced Polarization Effects in Inductive Source Electromagnetic Data. Ph.D. Thesis, University of British Columbia, Vancouver, BC, Canada, 2015.

Article

Percolation and Transport Properties in The Mechanically Deformed Composites Filled with Carbon Nanotubes

Artyom Plyushch [1,2,*], Dmitry Lyakhov [3], Mantas Šimėnas [1], Dzmitry Bychanok [2], Jan Macutkevič [1], Dominik Michels [3], Jūras Banys [1], Patrizia Lamberti [4] and Polina Kuzhir [2,5]

[1] Faculty of Physics, Vilnius University, Sauletekio 9, LT-10222 Vilnius, Lithuania; mantas.simenas@ff.vu.lt (M.Š.); jan.macutkevic@ff.vu.lt (J.M.); juras.banys@ff.vu.lt (J.B.)

[2] Institute for Nuclear Problems, Belarusian State University, 220006 Minsk, Belarus; bychanok@inp.bsu.by (D.B.); kuzhir@bsu.by or polina.kuzhir@uef.fi (P.K.)

[3] Computer, Electrical and Mathematical Science and Engineering Division, 4700 King Abdullah University of Science and Technology, Thuwal 23955-6900, Saudi Arabia; dmitry.lyakhov@kaust.edu.sa (D.L.); dominik.michels@kaust.edu.sa (D.M.)

[4] Department of Information and Electrical Engineering and Applied Mathematics, University of Salerno, Via Giovanni Paolo II 132, 84084 Fisciano (SA), Italy; plamberti@unisa.it

[5] Institute of Photonics, University of Eastern Finland, Yliopistokatu 7, FI-80101 Joensuu, Finland

* Correspondence: artyom.plyushch@ff.vu.lt; Tel.: +370-6229-3229

Received: 14 January 2020; Accepted: 12 February 2020; Published: 15 February 2020

Abstract: The conductivity and percolation concentration of the composite material filled with randomly distributed carbon nanotubes were simulated as a function of the mechanical deformation. Nanotubes were modelled as the hard-core ellipsoids of revolution with high aspect ratio. The evident anisotropy was observed in the percolation threshold and conductivity. The minimal mean values of the percolation of 4.6 vol. % and maximal conductivity of 0.74 S/m were found for the isotropic composite. Being slightly aligned, the composite demonstrates lower percolation concentration and conductivity along the orientation of the nanotubes compared to the perpendicular arrangement.

Keywords: Monte Carlo simulations; percolation; conductivity; carbon nanotubes composite

---

1. Introduction

The macroscopic properties of the randomly organized media has attracted research interest for many decades. Nowadays, the the critical study points are the determination of the properties of composites filled with the carbonaceous fillers, like carbon nanotubes (CNT) [1–4], or graphene nanoplatelets (GNP) [5–8]. All these objects have high aspect ratios, and there is an interest to develop anisotropic composites using the partially oriented fillers [9–11]. There are several ways to reach the anisotropy in case of a CNT-filled composite, for instance: mechanical deformation, curing the CNT-filled polymer in external fields, or template-based techniques (see [12] and Refs therein).

Currently, there are different ways to model the macroscopic properties of carbon composites. Among them are: general effective media model [13], excluded volume theory [14], Monte Carlo models [15,16], finite and boundary element methods [17,18]. For the conductive properties modelling, the Monte Carlo based techniques are one of the most interesting due to their simplicity, extremely high tunability and good agreement with the experimental results (See [19] and Refs therein).

However, Monte Carlo models for the aligned CNT composites lack in several aspects. Firstly, many studies introduce the cut-off zenith angle as an anisotropy parameter (AP) and investigate the dependencies of the macroscopic properties on the cut-off values [20–22]. In case of the real systems, one can measure the deformation degree or the externally applied field, but not the cut-off angle.

Secondly, there is some contradiction in conductivity data for the partially oriented composites. Many report on the descent of conductivity with the increase of the AP [22–24], and several demonstrate a maximum of the conductivity with the small values of the AP [20,21,25]. To figure out the conductivity dependence, the critical concentration dependence on the anisotropy parameters should be investigated. Finally, the properties of the aligned composite should be studied along different directions, while most of the studies usually report on the behaviour along one direction [21–23].

In this paper, the CNT-filled composite after mechanical deformation, which in our model is introduced as non uniformity in angular coordinate distribution, is investigated. The percolation threshold and conductivity are computed as a function of the mechanical deformation. Two main directions (along and perpendicular to the deformation axis) are studied.

2. Model and Simulation Details

The CNTs are modelled as the ellipsoids of revolution with high aspect ratio, their semi axes are $b_1 = b_2 \ll b_3$. For the calculation the similar ellipsoids with semi axes $b_1 = 2.5$ nm and $b_3 = 37.5$ nm were taken. The representative unit cell with the volume of $(nb_3)^3$, where $n > 1$, is used.

The particles are introduced one by one into the unit cell by randomly generating Cartesian coordinates X for the centre of the ellipsoid and two spherical angular coordinates θ and ϕ for the orientation of the longest axis of the ellipsoid. Each time the non-overlap condition is checked, and the new ellipsoid is stored only if it does not intersect the walls of the unit cell and does not penetrate into any of the already existing particles. After the predefined number of particle is reached, the connectivity is checked along the selected direction using the Dijkstra's protocol [26].

The algorithm for the percolation computation is organised as a Tabu search method and it stops when the percolation is achieved [27–29]. Until the percolation is reached, the number of particles increases in each cycle. The detailed modelling procedure is described in our previous paper [30]. The periodic boundary conditions are applied for the calculation of the percolation concentration and conductivity.

The connection criteria for the nanotubes is obtained as follows. The dependence of the tunnelling resistance on the distance between the nanotubes given as [8,31]:

$$\rho = \frac{h^2}{e^2 \sqrt{2mb}} \exp \left(\frac{4\pi d}{h} \sqrt{2mb} \right) \tag{1}$$

where b is the potential barrier, and d is the distance, e and m denote the elementary change and electron mass, respectively. The reciprocal value $1/\rho$ is the tunnelling conductivity. The conductivity drastically decreases with the distance increment and reaches already small values of 10^{-4} S/m for 2 nm separation. Thus, for the percolation computations the separation of 2 nm is considered as the connection criteria, while for the piezoresistivity computation direct values obtained from Equation (1) were used. In the last case, the tunnelling barrier of $b = 0.75$ eV.

To introduce anisotropy, the composite was considered as mechanically stretched. The probability density function (PDF) for the CNT angular distribution was derived using the following assumptions: (i) the mechanical deformation oriented along the z-axis, (ii) the centre of the ellipsoid keeps its position after the deformation, and (iii) the Poison's shrinkage in the perpendicular directions was neglected. The last assumption is justified, since the independence of the percolation value on the unit cell volume was demonstrated previously [30].

Thus, after the deformation, X and φ remain unchanged. The PDF for the θ angle was obtained as the modified function for the mechanically deformed composite filled with cylinders [32]:

$$\psi(\theta) = \frac{k^2}{[k^2 \cos^2 \theta + \sin^2 \theta]^{3/2}} \tag{2}$$

where k is the deformation coefficient introduced as the ratio of the final and initial lengths of the unit cell, $k = \frac{l}{l_0}$. For $k = 1$, the PDF function (2) will return uniform distribution, while for $k > 1$ some non uniformity will appear (see Figure 1). In order to apply the function to spherical coordinates, the final distribution will be given taking into account the Jacobian as $\psi(\theta)sin(\theta)$.

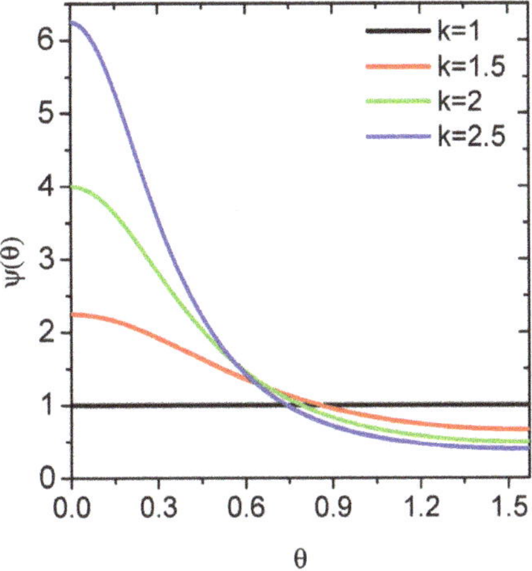

Figure 1. Probability density function (PDF) for the zenith angle for different deformations (k) for the carbon nanotubes (CNTs).

The conductivities are computed according to the following protocol. Firstly, the resistance of the nanotubes is taken as infinite, so the total conductivity of the composite is governed by the inter-tube tunnelling (Equation (1)). Next, the direction for the conductivity computation is selected and the nanotubes near the initial and final boarders are collected. The Dijkstra algorithm is used to trace the paths of the minimal resistance between the initial and final tubes. The array $R(t,l)$ (where t and l stand for the initial and final indexes of the tube, respectively) is computed. To implement the periodic boundary conditions the array $B(l,t)$ of the boundary resistances was introduced. The total resistance for the selected fixed l is follows:

$$Tot(l) = \left(\sum_t \frac{1}{R(t,l)} \right)^{-1} + \left(\sum_{t'} \frac{1}{B(l,t')} \right)^{-1} \tag{3}$$

And finally the total conductivity of the composite is computed as

$$\sigma = \left(\sum_l \frac{1}{Tot(l)} \right) \tag{4}$$

The algorithm is implemented using PGI CUDA FORTRAN standards [33]. The big enough number of the realizations (500–600) was collected for each particular case.

3. Results and Discussion

3.1. Percolation Computations

The comparison of the nanotubes angular distribution for the isotropic and heavily deformed composite is presented on Figure 2. For the visualisation purposes, the ellipsoids with small aspect ratio of 5 and the cubic unit cell with side length of 60 nm were used.

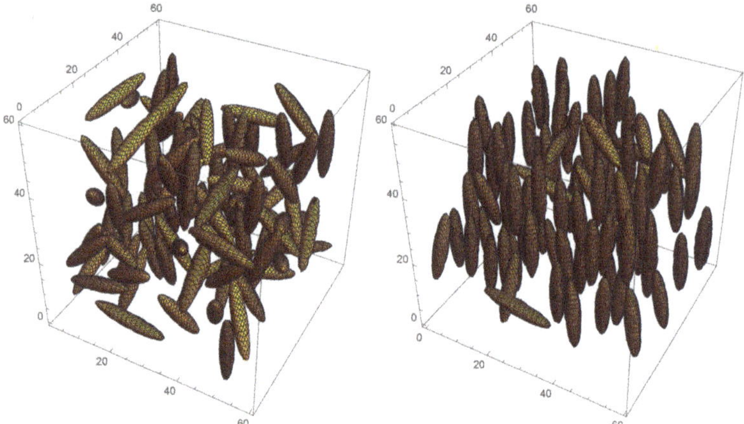

Figure 2. Visualisation of the CNT-filled composite with $k = 1$ and $k = 10$.

For the computations of the percolation threshold, the cubic unit cell with $n = 3$ was considered. Firstly, the system with uniform and fully aligned ($\theta = 0$, $k = \infty$) angular distribution of the ellipsoids was studied. The empirical cumulative probability distribution (CDF) for the percolation concentration p_c in non-deformed and aligned composites is presented in Figure 3. As it was expected, the distribution of the uniform composite coincides in different directions. Upon deformation, the percolation concentration increases in both directions, and the slope of CDF changes as well.

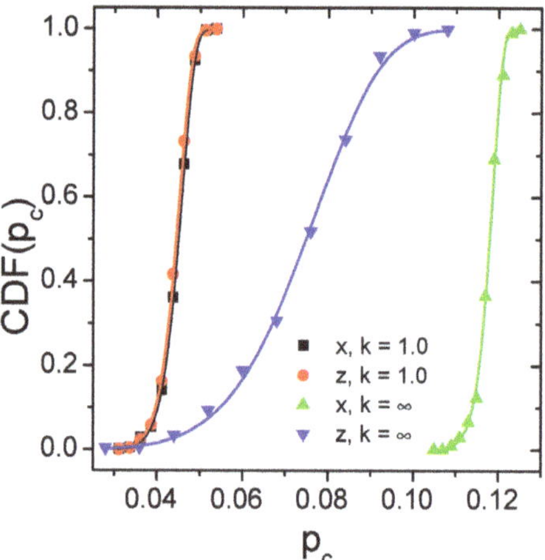

Figure 3. Empirical cumulative probability function for the percolation concentration in deformed and non-deformed composites (symbols), and Weibull cumulative probability distributions (CDFs) (5) (solid curves).

Obtained distribution follows the Weibull law [34]:

$$CDF_W(p_c, \lambda, m) = 1 - e^{-(p_c/\lambda)^m} \tag{5}$$

where λ is the scale factor, and m is the slope or Weibull modulus. The mean value of percolation concentration is $< p_c > = \lambda\Gamma(1 + 1/m) \approx \lambda$, where Γ is the Euler gamma function. The values for λ of 0.12 and 0.079, and k of 53.69 and 6.00 were obtained for fully oriented composite in the $x-$ and $z-$directions, correspondingly.

The dependencies of λ and m parameters on the deformation coefficient were also studied. The results were collected in Figure 4. For both directions, the Weibull scale factor λ increases with the deformation, so the minimal concentration is achieved for non deformed composite. This conclusion is supported with the measurement results [22–24].

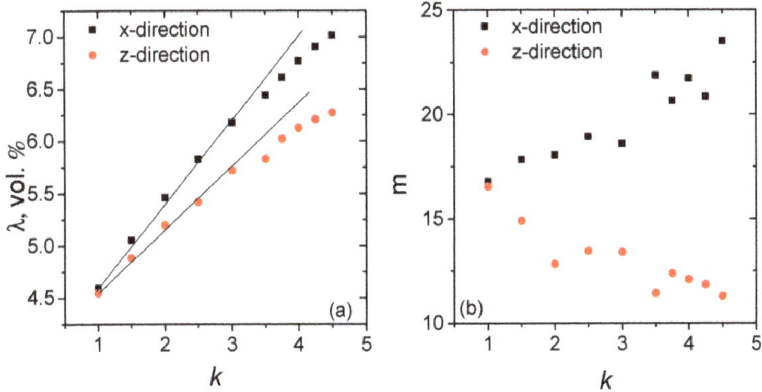

Figure 4. Scale factor (**a**) and Weibull modulus (**b**) as a functions of the deformation.

Anisotropic properties of the deformed composite are confirmed by distinct behaviour of λ and m along the $x-$ and $z-$directions. For low values of k (up to 3), the dependencies of λ *versus* k demonstrate linear increase both along x and z with the lower slope along the $z-$direction. In contrast to the scale factor λ, the Weibull modulus m demonstrates the decrease in the direction of the deformation and increase in the perpendicular one. This can be understood in terms of very different dimensionality of the inclusions network. Being isotropically distributed, the network of ellipsoids has 3D dimensionality, and, upon the deformation, the dimensionality changes toward 1D along the $z-$direction and 2D along the perpendicular one. As a result, a very small number of the particles (three, in the studied situation) may percolate with non-zero probability along z for fully aligned system. It is expected that both λ and k will asymptotically approach the above mentioned values for $k = \infty$.

3.2. Piezoresistivity Computations

Firstly, the dependence of the conductivity distribution on the unit cell size and the isotropy of the non deformed composite were investigated checked (see Figure 5). The peak values of the conductivity coincide for different cell sizes, while the distribution becomes broader with the decrease of the cell size. The conductivity distributions for the composites above the percolation threshold are close to lognormal in agreement with literature [35]. We conclude, that the usage of the periodic boundary conditions (PBC) in Equations (3) and (4) allowed us to obtain the independence on the unit cell size. For further computations, the cell with $n = 4$ was used.

The conductivity dependence on the concentration p and deformation k is presented in Figure 6. The concentration dependencies of σ (Figure 6a) follows the power law [36]. After the deformation, σ decreases both along $x-$ and $z-$directions, and this is supported by the experimental data [37–39]. Significant anisotropy (more than the order of magnitude) was observed for the deformed composite. It was previously shown, that the $z-$direction is preferable for the percolation in the deformed composites in comparison with x. But, being percolated, the composite has lower conductivity along z, than one along the perpendicular direction. In contrast to the percolation, where the appearance of only one conductive path is necessary, the conductivity strongly depends on the number of conductive paths through the unit cell and the total tunnelling distance.

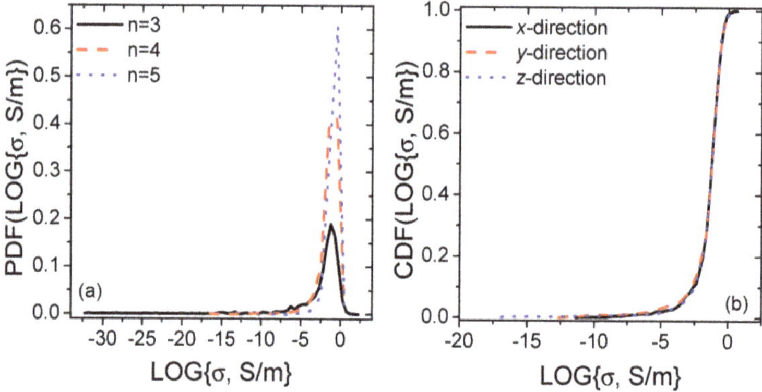

Figure 5. (**a**) Empirical PDF of the conductivity of the isotropic composite with 5 vol. % of CNTs with different cell size *n*. (**b**) Empirical CDF of the conductivity in different directions for the isotropic composite with 5 vol. % of CNTs with *n* = 4. 1500 realisations were collected.

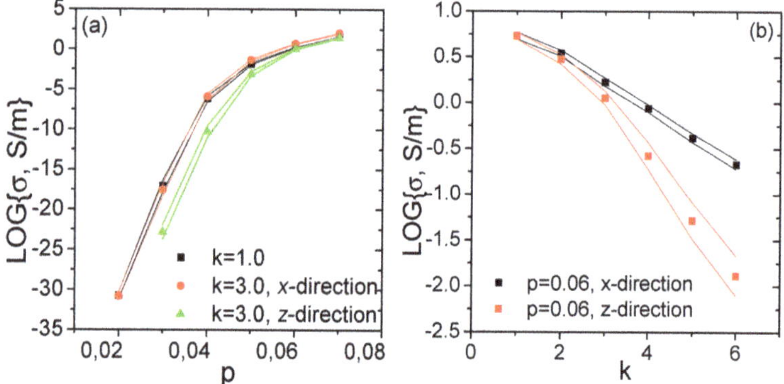

Figure 6. Conductivity dependence (**a**) on the concentration for the initial and deformed composite in different directions, and (**b**) on the deformation for the samples with 6 vol. % of the CNTs. Symbols stands for the mean values, and lines denote the 95 % confidence interval width.

Note that in the microwave and THz frequency regions the partial alignment of the nanotubes will lead to the increase of the conductivity and imaginary part of the permittivity along the deformation [32,40]. But in studied case the situation is opposite. That is related to very different mechanisms of the interaction of the electromagnetic radiation with the media at different frequencies. At microwave frequencies, the mechanism is the polarisation of the CNT [41]. In case of the direct current and quasistatic frequency range, where the tunnelling mechanism is dominant [42], the conductivity decreases due to the tunnelling distance increase.

Boundary Conditions Impact

In the introduction part it was mentioned, that there is the contradiction in the modelling results presented in literature, and several authors observe the maximum of the conductivity in the *z*-direction for slightly aligned composites. In our case the maximum is observed, if the conductivity is calculated without the PBC in Equations (3) and (4) (see Figure 7). The maximum is reached for $k \approx 1.5$. The conductivity along the x-direction decreases, similarly to presented in Figure 6b.

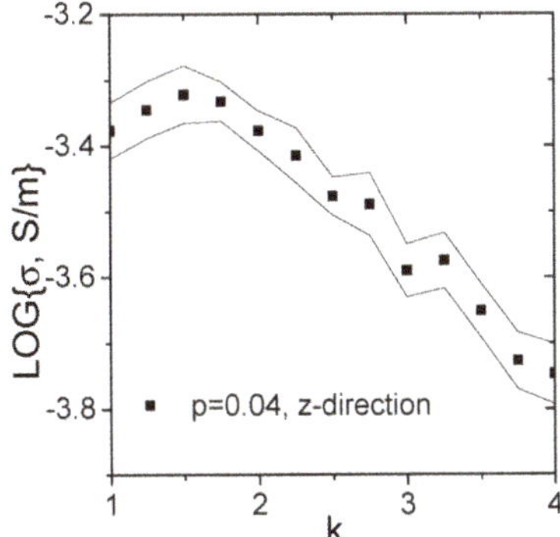

Figure 7. Conductivity dependence on the deformation for the samples with 4 vol. % of the CNTs, computed without boundary conditions. Symbols stands for the mean values, and lines denote the 95% confidence interval width.

Both models with and without PBC should provide similar results for the infinitely large unit cell. But in the case of the PBC model, the independence of the conductivity on the cell size is demonstrated. Thus, the appearance of the maximum of the conductivity is the cell size related effect of the non-PBC model. Both PBC and non-PBC models are suitable for applications. For the thin films filled with aligned nanotubes the non-PBC model is preferable, while with the composite size increase, the PBC model should be used.

4. Conclusions

The composite filled with carbon nanotubes after the mechanical deformation was simulated. The nanotubes were modelled as randomly distributed non-overlapping ellipsoids. The mechanical stretching of the composite was introduced as the non uniformity of the angular distribution of the ellipsoids. It was shown that the uniform composite provides the higher conductivity and its percolation concentration is the lower, than that of the aligned composite. The anisotropy of the macroscopic properties was investigated. The percolation concentration and conductivity are lower along the direction of the partial orientation of the nanotubes in comparison with the perpendicular one. That can be understood in terms of the conductive paths tunnelling distance and the total number of the conductive paths formed in different directions.

The impact of the periodic boundary conditions was cleared. It was demonstrated, that very different behaviour of the conductivity upon the CNT alignment presented in literature may be explained by the boundary conditions. It was proved, that the model with the periodic boundary conditions provide more relevant conductivity results since the conductivity becomes unit cell size-independent.

The presented model may be used for strain sensor development. The utilisation of the model as pre-experimental step allows to find out the optimal conditions for the composite synthesis, taking into account the nanotube aspect ratio, concentration, the matrix-related properties (in terms of the tunnelling barrier value). At the same time, it allows to predict the behaviour of the main macroscopic parameters on the deformation.

Author Contributions: A.P. code developing, writing the draft; P.K. J.M. and M.Š. writing the final paper; D.L. and D.M. GPU computations; P.K., D.B., P.L. and J.B. conceptualization. All authors have read and agreed to the published version of the manuscript.

Funding: This research was partially funded by H2020 RISE 734164 Graphene 3D, H2020 RISE Project 823728 DiSeTCom. PK is supported by Horizon 2020 IF TURANDOT project 836816 and the Academy of Finland Flagship Programme, Photonics Research and Innovation (PREIN), decision 320166.

Acknowledgments: Massive GPU computations were performed on KAUST's Ibex HPC. We thank KAUST Supercomputing Core Lab team and especially to Mohsin Shaikh and Passant Hafez for assistance with execution tasks on Volta and Pascal GPU nodes.

Conflicts of Interest: The authors declare no conflict of interest.

Abbreviations

The following abbreviations are used in this manuscript:

PBC	Periodic boundary conditions
CNT	Carbon nanotube
PDF	Probability density function
CDF	Cumulative distribution function

References

1. Foygel, M.; Morris, R.; Anez, D.; French, S.; Sobolev, V. Theoretical and computational studies of carbon nanotube composites and suspensions: Electrical and thermal conductivity. *Phys. Rev. B* **2005**, *71*, 104201. [CrossRef]
2. Fang, C.; Zhang, J.; Chen, X.; Weng, G.J. A monte carlo model with equipotential approximation and tunneling resistance for the electrical conductivity of carbon nanotube polymer composites. *Carbon* **2019**, *146*, 125–138. [CrossRef]
3. Ni, X.; Hui, C.; Su, N.; Cutler, R.; Liu, F. A 3D percolation model for multicomponent nanocarbon composites: The critical role of nematic transition. *Nanotechnology* **2019**, *30*, 185302. [CrossRef]
4. Hu, Y.; Huang, J.C. Mechanical Properties and Reinforcement Mechanisms of Carbon Nanotube Composites with Amine Functional Groups Based on Molecular Dynamics. *J. Phys. Conf. Ser.* **2019**, *1176*, 052054. [CrossRef]
5. Liu, M.; Kinloch, I.A.; Young, R.J.; Papageorgiou, D.G. Modelling mechanical percolation in graphene-reinforced elastomer nanocomposites. *Compos. Part B Eng.* **2019**, *178*, 107506. j.compositesb.2019.107506. [CrossRef]
6. Sagalianov, I.; Vovchenko, L.; Matzui, L.; Lazarenko, O. Synergistic enhancement of the percolation threshold in hybrid polymeric nanocomposites based on carbon nanotubes and graphite nanoplatelets. *Nanoscale Res. Lett.* **2017**, *12*, 140. [CrossRef]
7. Sagalianov, I.Y.; Lazarenko, O.A.; Vovchenko, L.L.; Matzui, L.Y. Monte-Carlo study of the percolation in a binary composites: Hardcore and softcore models comparison. In Proceedings of the 2017 IEEE 7th International Conference on Nanomaterials: Application & Properties (NAP), Zatoka, Ukraine, 10–15 September 2017.
8. Oskouyi, A.; Sundararaj, U.; Mertiny, P. Tunneling conductivity and piezoresistivity of composites containing randomly dispersed conductive nano-platelets. *Materials* **2014**, *7*, 2501–2521. [CrossRef] [PubMed]
9. Mai, C.K.; Liu, J.; Evans, C.M.; Segalman, R.A.; Chabinyc, M.L.; Cahill, D.G.; Bazan, G.C. Anisotropic thermal transport in thermoelectric composites of conjugated polyelectrolytes/single-walled carbon nanotubes. *Macromolecules* **2016**, *49*, 4957–4963. [CrossRef]
10. Pan, T.W.; Kuo, W.S.; Tai, N.H. Tailoring anisotropic thermal properties of reduced graphene oxide/multi-walled carbon nanotube hybrid composite films. *Compos. Sci. Technol.* **2017**, *151*, 44–51. [CrossRef]
11. Chen, J.; Wang, L.; Gui, X.; Lin, Z.; Ke, X.; Hao, F.; Li, Y.; Jiang, Y.; Wu, Y.; Shi, X.; others. Strong anisotropy in thermoelectric properties of CNT/PANI composites. *Carbon* **2017**, *114*, 1–7. [CrossRef]
12. Xie, X.L.; Mai, Y.W.; Zhou, X.P. Dispersion and alignment of carbon nanotubes in polymer matrix: A review. *Mater. Sci. Eng. R Rep.* **2005**, *49*, 89–112. [CrossRef]

13. Radzuan, N.A.M.; Sulong, A.B.; Sahari, J. A review of electrical conductivity models for conductive polymer composite. *Int. J. Hydrog. Energy* **2017**, *42*, 9262–9273. [CrossRef]

14. Sun, Y.; Bao, H.D.; Guo, Z.X.; Yu, J. Modeling of the electrical percolation of mixed carbon fillers in polymer-based composites. *Macromolecules* **2008**, *42*, 459–463. [CrossRef]

15. Román, S.; Lund, F.; Bustos, J.; Palza, H. About the relevance of waviness, agglomeration, and strain on the electrical behavior of polymer composites filled with carbon nanotubes evaluated by a Monte-Carlo simulation. *Mater. Res. Express* **2018**, *5*, 015044. [CrossRef]

16. Zhao, N.; Kim, Y.; Koo, J.H. 3D Monte Carlo simulation modeling for the electrical conductivity of carbon nanotube-incorporated polymer nanocomposite using resistance network formation. *Mater. Sci.* **2018**, *2*, 1. [CrossRef]

17. Liu, Y.; Chen, X. Continuum models of carbon nanotube-based composites using the boundary element method. *Electron. J. Bound. Elem.* **2003**, *1*. [CrossRef]

18. Liu, Y.; Nishimura, N.; Otani, Y. Large-scale modeling of carbon-nanotube composites by a fast multipole boundary element method. *Comput. Mater. Sci.* **2005**, *34*, 173–187. [CrossRef]

19. Berhan, L.; Sastry, A. Modeling percolation in high-aspect-ratio fiber systems. I. Soft-core versus hard-core models. *Phys. Rev. E* **2007**, *75*, 041120. [CrossRef]

20. Gong, S.; Zhu, Z.; Meguid, S. Anisotropic electrical conductivity of polymer composites with aligned carbon nanotubes. *Polymer* **2015**, *56*, 498–506. [CrossRef]

21. Du, F.; Fischer, J.E.; Winey, K.I. Effect of nanotube alignment on percolation conductivity in carbon nanotube/polymer composites. *Phys. Rev. B* **2005**, *72*, 121404. [CrossRef]

22. Zeng, X.; Xu, X.; Shenai, P.M.; Kovalev, E.; Baudot, C.; Mathews, N.; Zhao, Y. Characteristics of the electrical percolation in carbon nanotubes/polymer nanocomposites. *J. Phys. Chem. C* **2011**, *115*, 21685–21690. [CrossRef]

23. Ghazavizadeh, A.; Baniassadi, M.; Safdari, M.; Atai, A.; Ahzi, S.; Patlazhan, S.; Gracio, J.; Ruch, D. Evaluating the effect of mechanical loading on the electrical percolation threshold of carbon nanotube reinforced polymers: A 3D Monte-Carlo study. *J. Comput. Theor. Nanosci.* **2011**, *8*, 2087–2099. [CrossRef]

24. Li, C.; Chou, T.W. Electrical conductivities of composites with aligned carbon nanotubes. *J. Nanosci. Nanotechnol.* **2009**, *9*, 2518–2524. [CrossRef] [PubMed]

25. Bao, W.; Meguid, S.; Zhu, Z.; Meguid, M. Modeling electrical conductivities of nanocomposites with aligned carbon nanotubes. *Nanotechnology* **2011**, *22*, 485704. [CrossRef]

26. Dijkstra, E.W. A note on two problems in connexion with graphs. *Numer. Math.* **1959**, *1*, 269–271. [CrossRef]

27. Glover, F. Tabu search—part II. *ORSA J. Comput.* **1990**, *2*, 4–32. [CrossRef]

28. Glover, F. Tabu search—part I. *ORSA J. Comput.* **1989**, *1*, 190–206. [CrossRef]

29. Glover, F.; Laguna, M. Tabu Search. In *Handbook of Combinatorial Optimization*; Springer: Berlin, Germany, 2013; pp. 3261–3362.

30. Plyushch, A.; Lamberti, P.; Spinelli, G.; Macutkevič, J.; Kuzhir, P. Numerical Simulation of the Percolation Threshold in Non-Overlapping Ellipsoid Composites: Toward Bottom-Up Approach for Carbon Based Electromagnetic Components Realization. *Appl. Sci.* **2018**, *8*, 882. [CrossRef]

31. Simmons, J.G. Generalized formula for the electric tunnel effect between similar electrodes separated by a thin insulating film. *J. Appl. Phys.* **1963**, *34*, 1793–1803. [CrossRef]

32. Bychanok, D.; Kanygin, M.; Okotrub, A.V.; Shuba, M.; Paddubskaya, A.; Pliushch, A.; Kuzhir, P.; Maksimenko, S. Anisotropy of the electromagnetic properties of polymer composites based on multiwall carbon nanotubes in the gigahertz frequency range. *Jetp Lett.* **2011**, *93*, 607. [CrossRef]

33. CUDA FORTRAN Available online: https://developer.nvidia.com/cuda-fortran (accessed on 14 February 2020).

34. Weibull, W.; others. A statistical distribution function of wide applicability. *J. Appl. Mech.* **1951**, *18*, 293–297.

35. Dagan, G. Higher-order correction of effective permeability of heterogeneous isotropic formations of lognormal conductivity distribution. *Transp. Porous Media* **1993**, *12*, 279–290. [CrossRef]

36. Stauffer, D.; Aharony, A. *Introduction to Percolation Theory*; CRC Press: Boca Raton, FL, USA, 1994.

37. Zhang, R.; Baxendale, M.; Peijs, T. Universal resistivity–strain dependence of carbon nanotube/polymer composites. *Phys. Rev. B* **2007**, *76*, 195433. [CrossRef]

38. Alig, I.; Skipa, T.; Engel, M.; Lellinger, D.; Pegel, S.; Pötschke, P. Electrical conductivity recovery in carbon nanotube–polymer composites after transient shear. *Phys. Status Solidi B* **2007**, *244*, 4223–4226. [CrossRef]

39. Hu, N.; Karube, Y.; Arai, M.; Watanabe, T.; Yan, C.; Li, Y.; Liu, Y.; Fukunaga, H. Investigation on sensitivity of a polymer/carbon nanotube composite strain sensor. *Carbon* **2010**, *48*, 680–687. [CrossRef]
40. Bychanok, D.; Shuba, M.; Kuzhir, P.; Maksimenko, S.; Kubarev, V.; Kanygin, M.; Sedelnikova, O.; Bulusheva, L.; Okotrub, A. Anisotropic electromagnetic properties of polymer composites containing oriented multiwall carbon nanotubes in respect to terahertz polarizer applications. *J. Appl. Phys.* **2013**, *114*, 114304. [CrossRef]
41. Shuba, M.; Slepyan, G.Y.; Maksimenko, S.; Thomsen, C.; Lakhtakia, A. Theory of multiwall carbon nanotubes as waveguides and antennas in the infrared and the visible regimes. *Phys. Rev. B* **2009**, *79*, 155403. [CrossRef]
42. Li, C.; Thostenson, E.T.; Chou, T.W. Dominant role of tunneling resistance in the electrical conductivity of carbon nanotube–based composites. *Appl. Phys. Lett.* **2007**, *91*, 223114. [CrossRef]

 *applied sciences*

Article

Super-Gain Optical Parametric Amplification in Dielectric Micro-Resonators via BFGS Algorithm-Based Non-Linear Programming

Özüm Emre Aşırım * and Mustafa Kuzuoğlu

Department of Electrical and Electronics Engineering, Middle East Technical University, 06800 Ankara, Turkey; kuzuoglu@metu.edu.tr
* Correspondence: e176154@metu.edu.tr; Tel.: +90-5323329262

Received: 6 February 2020; Accepted: 28 February 2020; Published: 4 March 2020

Abstract: The goal of this paper is to show that super-gain optical parametric amplification can be achieved even in a small micro-resonator using high-intensity ultrashort pump waves, provided that the frequencies of the ultrashort pulses are tuned to maximize the intracavity magnitude of the wave to be amplified, which we call the stimulus wave. In order to accomplish this, we have performed a dispersion analysis via computational modeling of the electric polarization density in terms of the non-linear electron cloud motion and we have concurrently solved the electric polarization density and the wave equation for the electric field. Based on a series of non-linear programming-integrated finite difference time-domain simulations, we have identified the optimal pump wave frequencies that simultaneously maximize the stored electric energy density and the polarization density inside a micro-resonator by using the Broyden–Fletcher–Goldfarb–Shanno (BFGS) optimization algorithm. When the intracavity energy and the polarization density (which acts as an energy coupling coefficient) are simultaneously high, an input wave can be strongly amplified by efficiently drawing energy from a highly energized cavity. Therefore, we propose that micrometer-scale achievement of super-gain optical parametric amplification is possible in a micro-resonator via high-intensity ultrashort "pump wave" pulses, by determining the optimal frequencies that concurrently maximize the stored electric energy density and the polarization density in a dielectric interaction medium.

Keywords: optical parametric amplification; non-linear wave mixing; micro-resonator; optimization

1. Introduction

Electromagnetic wave amplification by non-linear wave mixing has been studied for decades with most of the research focusing on the field of non-linear optics. It has been well-established that a low power input wave can be amplified via mixing with a high-intensity pump wave, in a medium that demonstrates a high degree of non-linearity. The fundamental theory of intra-cavity non-linear optical amplification is about the energy transfer from the pump wave-energized resonator to the input wave as a consequence of non-linear optical coupling [1–3]. The more elaborate theory of non-linear wave amplification involves a dispersion analysis that takes into account the frequency dependence of the non-linearity of the interaction medium. However, based on our in-depth investigations, a direct gain factor maximization approach via a non-linear constrained optimization algorithm is lacking. The background research about the theory of non-linear wave mixing has been mostly experimental rather than computational, especially for the purpose of optical amplification. The major reason of this tendency is that this topic is usually studied in the micrometer- or nanometer-wavelength range, but the required interaction medium length to observe a strong non-linear effect is in the millimeter or centimeter range [4]. This requires a tremendous computational cost for acquiring meaningful results, particularly for wave amplification purposes.

Another incentive for the experimental investigation of the topic is the desire to discover new materials that exhibit unusually high non-linearity under excitation. More recently, researchers focus on determining the resonance frequencies of the non-linear electrical response of commonly used materials in photonics such as graphene and gallium nitride. This approach has been mostly successful and resulted in an improvement of efficiencies in many applications in photonics. However, for the purpose of optical amplification, achieving a resonant non-linear optical response usually fails as the dielectric absorption around any non-linear resonance peak is quite strong to prevent significant amplification. Moreover, the well-established concept of optical parametric amplification yields a negligible gain factor in a few micrometers-long resonator. The required interaction medium length for high-gain optical parametric amplification is usually in the order of centimeters. Therefore, optical amplification by non-linear wave mixing is not so feasible for utilization in micro-resonators, optical transistors, micro-modulators, and many other device models in integrated photonics such as those mentioned in [5,6]. However, if one can perform super-gain optical parametric amplification in the micrometer scale, much more powerful macroscale devices can be produced for high-power applications, by engineering the individual micro-scale devices to operate in an array form to maximize optical interference, such as high-power welding machines and super-intensity lasers that are employed in particle accelerators and fusion reactors. A major scientific advancement can be in the field of optical antennas via supercontinuum generation for achieving ultra-wideband operation.

In this article, we have performed a numerical analysis that provides evidence for the high-gain amplification of a low-power stimulus wave, via intense pump waves of ultrashort duration, inside a several micrometers long micro-resonator, by maximizing the electric energy density of the pump wave in the resonator.

Since the order of amplification that can be achieved via non-linear wave mixing depends on the amount of stored electric energy density and the pump wave-initiated polarization density, which acts as a non-linear coupling coefficient, we will first present the energy storage dynamics for a resonator and we will recall the definitions of the electric polarization density and the gain factor of an amplification process. Then we will present the basic background for wave propagation in non-linear dispersive media and we will formulate our energy maximization (optimization) problem based on the partial differential equations given to express wave propagation in non-linear dispersive media. Finally, we will present two numerical experiments and analyze their results.

2. Polarization Density and the Cavity Quality (Q) Factor

The Q factor indicates the amount of the stored energy inside a resonator for a given round trip loss. A high Q value usually signifies that the resonator is low loss, i.e., has a low loss factor, which means that high energy can be trapped efficiently in the resonator. The Q factor depends on the resonator length, the reflectivities of the resonator walls, the frequency of the propagating wave, the total absorption coefficient of the interaction medium between the resonator walls, and any sort of diffraction or scattering loss that may take place inside a resonator. The Q factor of a resonator is defined as:

$$CAVITY\ QUALITY\ (Q)FACTOR = 2\pi \frac{Stored\ energy}{Energy\ dissipated\ per\ round\ trip} = fT_{rt}\frac{2\pi}{\zeta} = \frac{4fL\pi}{\zeta c}. \tag{1}$$

T_{rt} : *Round trip time, f : Wave frequency, ζ : Fractional power loss per round trip c : Speed of light, L : Cavity length*

The electric energy density in a resonator depends on the magnitude of the electric field and the polarizability of the interaction medium. For a given resonator configuration, the stored electric energy density in a homogenous isotropic interaction medium is given as [7]:

$$W_e = Stored\ energy\ density = \frac{1}{2}ED = \frac{1}{2}E(\varepsilon_\infty E + P) = \frac{1}{2}\varepsilon_\infty E^2 + \frac{1}{2}EP. \tag{2}$$

D : Electric flux density, P : Polarization density, E : Electric field intensity, ε_∞ : Background (infinite spectral band)permittivity

The electrical polarization density (P) in an interaction medium is a microscopic parameter that indicates both the volume density of electrons and the displacement of an electron with respect to the position of the nucleus [3]. For a given electron density of a medium, the more an electron is allowed to displace from its initial position (and the nucleus) the higher the polarization density becomes. It is a measure of the electrical excitability of a material by an incident electromagnetic wave [1]. Polarization density is an important parameter for energy storage in a resonator as more energy can be stored in a cavity with a highly polarizable interaction medium. This is because, as each electron displaces further from the nucleus, their potential energy becomes higher, and an abundance of electrons in the medium will lead to a higher potential energy being stored in the medium [4]. For a highly non-linear medium, the polarization density itself depends on the magnitude of the electric field [1,4]. For these reasons, we will frequently use the term polarization density in our analyses.

3. Wave Propagation in Non-Linear Dispersive Media

As previously stated, non-linear interaction takes place when a wave that propagates in a medium have a very high intensity. This degree of non-linearity inducing high intensities is usually possible with ultra-short duration pulses and can be practically generated from a mode locked laser or a Q-switched laser, which have durations on the scale of picoseconds or femtoseconds. Since such high-intensity pulses have ultra-short durations, we must perform a dispersion analysis.

Impulse response of the polarization density of many dielectric media last much longer in duration than the pulse durations of such high intensity pulses [8,9].

The inclusion of dispersion in the analysis means that we need to solve for the polarization density of the interaction medium separately. In fact, it is much more accurate to solve the wave equation in parallel with the non-linear electron cloud motion equation for a non-linear dispersive medium, since parameters such as the resonance frequency, damping rate, atom density, and atomic diameter have typical values for most solid media, which makes the simulation results more realistic. To determine the time variation of the electric field in a non-linear dispersive medium, we need to solve the following equations (see Figure 1) [1]:

$$\nabla^2 E - \mu_0\varepsilon_\infty\frac{\partial^2 E}{\partial t^2} = \mu_0\sigma\frac{\partial E}{\partial t} + \mu_0\frac{\partial^2 P}{\partial t^2}. \tag{3}$$

$$\frac{\partial^2 P}{\partial t^2} + \gamma\frac{\partial P}{\partial t} + \omega_0{}^2 P - \frac{\omega_0{}^2}{Ned}P^2 - \frac{\omega_0{}^2}{N^2 e^2 d^2}P^3 = \frac{Ne^2}{m}E. \tag{4}$$

P : Polarization density , e : Electron charge, m : Electron mass, E : Electric field (V/m)

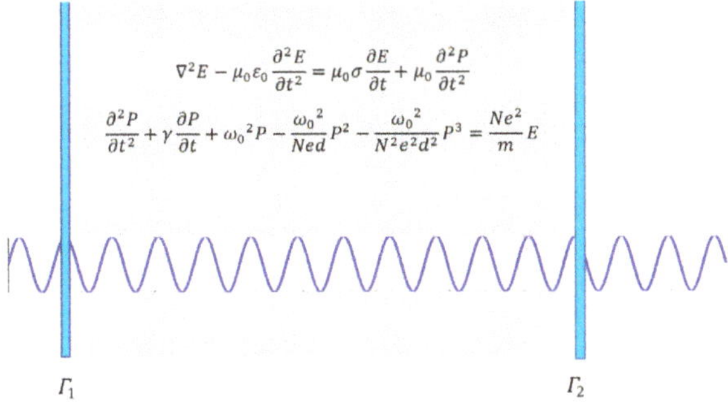

Figure 1. An interaction medium placed inside a cavity.

In Equation (4) we have accounted up to the third order of the polarization density as higher order terms are negligibly small. For a dielectric medium, the electric conductivity is almost zero ($\sigma \approx 0$). Some typical values for solid dielectric media are [1]; *Resonance frequency* : $f_0 = 1.1 \times 10^{15} Hz$, *Damping rate* : $\gamma = 1 \times 10^9$ *Hz*, *Electron density* : $N = 3.5 \times 10^{28}/m^3$, *Atomic diameter* : $d = 0.3$ *nanometer*

Electromagnetic wave amplification by non-linear wave mixing involves two waves, namely the low intensity stimulus (input) wave to be amplified, and the high-intensity pump wave. Let us first consider the high-intensity pump wave E_2 without the presence of the low intensity stimulus wave E_1. In this case, the pair of equations that model the propagation of E_2 is given as:

$$\nabla^2(E_2) - \mu_0 \varepsilon_\infty \frac{\partial^2(E_2)}{\partial t^2} = \mu_0 \sigma \frac{\partial(E_2)}{\partial t} + \mu_0 \frac{\partial^2 P_2}{\partial t^2}. \tag{5a}$$

$$\frac{\partial^2 P_2}{\partial t^2} + \gamma \frac{\partial P_2}{\partial t} + \omega_0^2 (P_2) - \frac{\omega_0^2}{Ned}(P_2)^2 - \frac{\omega_0^2}{N^2 e^2 d^2}(P_2)^3 = \frac{Ne^2}{m}(E_2) \tag{5b}$$

P_2 : *Polarization density due to the electric field* E_2, P_1 : *Polarization density due to the electric field* E_1

Now assume that E_1 and E_2 are simultaneously propagating in the same medium, in that case the pair of equations that describe the total electric field propagation is written as:

$$\nabla^2(E_1 + E_2) - \mu_0 \varepsilon_\infty \frac{\partial^2(E_1 + E_2)}{\partial t^2} = \mu_0 \sigma \frac{\partial(E_1 + E_2)}{\partial t} + \mu_0 \frac{\partial^2(P_1 + P_2)}{\partial t^2}. \tag{6a}$$

$$\frac{\partial^2(P_1+P_2)}{\partial t^2} + \gamma \frac{\partial(P_1+P_2)}{\partial t} + \omega_0^2(P_1 + P_2) - \frac{\omega_0^2}{Ned}(P_1 + P_2)^2 - \frac{\omega_0^2}{N^2 e^2 d^2}(P_1 + P_2)^3 = \frac{Ne^2}{m}(E_1 + E_2). \tag{6b}$$

Our aim is to solve for the low amplitude wave E_1 in the presence of the high-amplitude wave E_2, i.e., we want to solve for the stimulus wave E_1 when there is an energy coupling from E_2. To model this problem, we subtract Equation (5a,b) from Equation (6a,b) respectively [1], which yields:

$$\nabla^2(E_1) - \mu_0 \varepsilon_\infty \frac{\partial^2(E_1)}{\partial t^2} = \mu_0 \sigma \frac{\partial(E_1)}{\partial t} + \mu_0 \frac{\partial^2(P_1)}{\partial t^2}. \tag{7a}$$

$$\frac{\partial^2(P_1)}{\partial t^2} + \gamma \frac{\partial(P_1)}{\partial t} + \omega_0^2(P_1) - \frac{\omega_0^2}{Ned}\{P_1^2 + 2P_1 P_2\} - \frac{\omega_0^2}{N^2 e^2 d^2}\{P_1^3 + 3P_1^2 P_2 + 3P_1 P_2^2\} = \frac{Ne^2}{m}(E_1) \tag{7b}$$

Equations (5a,b) and (7a,b) show that E_2 and E_1 are coupled to each other. The pump wave E_2 acts as a source field distribution for electromagnetic wave propagation, with the stimulus wave E_1 acting as the carrier distribution in reference to the model presented in [10]. Based on Equation (7a,b), we will examine if it is feasible to amplify the low-intensity electric field E_1, via energy coupling from the microresonator that is energized by the high-intensity electric field E_2 (see Figure 2).

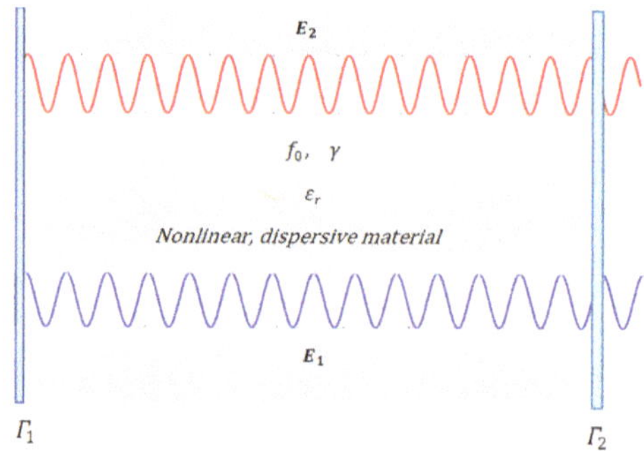

Figure 2. Concurrent propagation of the stimulus and the pump waves in a micro-resonator.

From Equation (7a), we can see that the stimulus wave E_1 has a source term P_1, which is the polarization density of the stimulus wave. The source term P_1 is also dependent on the source term (polarization density) of the high-intensity pump wave P_2. This means that in order to solve for the stimulus wave E_1, we need to solve for the pump wave E_2 as the equations for the stimulus wave and the pump wave are coupled to each other via the coupling term P_2. Therefore, the stimulus wave can be solved by simultaneously solving all these four equations (Equations (5a,b) and (7a,b)).

Our goal is to maximize the stimulus wave magnitude at a given stimulus wave frequency; $\max(|E_1(f_{st} = f')|)$. In order to achieve this, we will make a dispersion analysis that is based on the high-intensity pump wave frequency. By sweeping the excitation frequency of the pump wave (f_p) in a large spectral interval that extends from the far-infrared region to the near-ultraviolet region, we can investigate the pump wave frequency dependent variation of the maximum stimulus wave magnitude for $f_{st} = f'$, and we can select the optimal pump wave frequency value that maximizes the magnitude of the stimulus wave for $f_{st} = f'$. Mathematically our optimization problem can be stated as:

$$\text{Given that } f_{min} < f_p < f_{max} ; \quad \boldsymbol{max} \left| E_1(f_{st} = f') \right| \text{ based on the following equations}$$

$$\nabla^2\left(E_2(f_p)\right) - \mu_0\varepsilon_\infty \frac{\partial^2\left(E_2(f_p)\right)}{\partial t^2} = \mu_0\sigma\frac{\partial\left(E_2(f_p)\right)}{\partial t} + \mu_0\frac{\partial^2 P_2}{\partial t^2} \tag{8a}$$

$$\frac{\partial^2 P_2}{\partial t^2} + \gamma\frac{\partial P_2}{\partial t} + \omega_0^2(P_2) - \frac{\omega_0^2}{Ned}(P_2)^2 - \frac{\omega_0^2}{N^2e^2d^2}(P_2)^3 = \frac{Ne^2}{m}\left(E_2(f_p)\right) \tag{8b}$$

$$\nabla^2\left(E_1(f_p)\right) - \mu_0\varepsilon_\infty \frac{\partial^2\left(E_1(f_p)\right)}{\partial t^2} = \mu_0\sigma\frac{\partial\left(E_1(f_p)\right)}{\partial t} + \mu_0\frac{\partial^2 (P_1)}{\partial t^2} \tag{8c}$$

$$\frac{\partial^2 (P_1)}{\partial t^2} + \gamma\frac{\partial (P_1)}{\partial t} + \omega_0^2(P_1) - \frac{\omega_0^2}{Ned}\left\{P_1^2 + 2P_1P_2\right\} - \frac{\omega_0^2}{N^2e^2d^2}\left\{P_1^3 + 3P_1^2P_2 + 3P_1P_2^2\right\}$$
$$= \frac{Ne^2}{m}\left(E_1(f_p)\right) \tag{8d}$$

where

$$E_2(x = x_{input}, t) = A_p \cos\left(2\pi f_p t + \psi_p\right)\left(u(t) - u\left(t - \Delta T_p\right)\right) V/m \quad (u(t) : Unit\ step\ function)$$

$$E_1(x = x_{input}, t) = A_{st} \cos(2\pi f_{st} t + \psi_{st})(u(t) - u(t - \Delta T_{st}))\frac{V}{m} \ , \ A_p \gg A_{st}, \ \Delta T_p \ll \Delta T_{st}$$

If we are using N ultrashort high-intensity pulse excitations to amplify the low-intensity stimulus wave,

$$E_2(x = x_{input}, t) = \sum_{i=1}^{N} A_i \cos(2\pi f_i t + \psi_i)(u(t) - u(t - \Delta T_i)) V/m \quad (u(t) : Unit\ step\ function)$$

$$E_1(x = x_{input}, t) = A_{st} \cos(2\pi(f_{st})t + \psi_{st})(u(t) - u(t - \Delta T_{st})) \ V/m$$

$$\text{where } A_i \gg A_{st}, \ \Delta T_i \ll \Delta T_{st}, \ i = 1, 2, \ldots, N$$

Then the dispersion analysis-based optimization problem is stated as follows:

$$\boldsymbol{f_p} = \{ f_1 , f_2 , \ldots, f_N \}$$

$$\boldsymbol{f_{min}} = \left\{ f_{min,1} , f_{min,2} , \ldots, f_{min,N} \right\}, \ \boldsymbol{f_{max}} = \left\{ f_{max,1} , f_{max,2} , \ldots, f_{max,N} \right\}$$

$$\text{Given that } \boldsymbol{f_{min}} < \boldsymbol{f_p}$$
$$< \boldsymbol{f_{max}} ; \quad \boldsymbol{max} \left| E_1(f_{st} \right.$$
$$\left. = f') \right| \text{ based on the following equations}$$

$$\nabla^2\big(E_2(f_p)\big) - \mu_0\varepsilon_\infty \frac{\partial^2\big(E_2(f_p)\big)}{\partial t^2} = \mu_0\sigma\frac{\partial\big(E_2(f_p)\big)}{\partial t} + \mu_0\frac{\partial^2 P_2}{\partial t^2} \tag{9a}$$

$$\frac{\partial^2 P_2}{\partial t^2} + \gamma\frac{\partial P_2}{\partial t} + \omega_0^2(P_2) - \frac{\omega_0^2}{Ned}(P_2)^2 - \frac{\omega_0^2}{N^2e^2d^2}(P_2)^3 = \frac{Ne^2}{m}\big(E_2(f_p)\big) \tag{9b}$$

$$\nabla^2\big(E_1(f_p)\big) - \mu_0\varepsilon_\infty \frac{\partial^2\big(E_1(f_p)\big)}{\partial t^2} = \mu_0\sigma\frac{\partial\big(E_1(f_p)\big)}{\partial t} + \mu_0\frac{\partial^2(P_1)}{\partial t^2} \tag{9c}$$

$$\frac{\partial^2(P_1)}{\partial t^2} + \gamma\frac{\partial(P_1)}{\partial t} + \omega_0^2(P_1) - \frac{\omega_0^2}{Ned}\{P_1^2 + 2P_1P_2\} - \frac{\omega_0^2}{N^2e^2d^2}\{P_1^3 + 3P_1^2P_2 + 3P_1P_2^2\}$$
$$= \frac{Ne^2}{m}\big(E_1(f_p)\big) \tag{9d}$$

Note that in this case the pump wave comprises N ultrashort pulses instead of a single ultrashort pulse. Therefore, we have a multiparameter optimization problem.

The physics behind the efficient amplification of the stimulus wave involves the simultaneous maximization of the stored electric energy density and the polarization density originated by the pump wave in the resonator (see Figure 3). This can be explained in two steps, first we need to maximize the stored energy in the cavity in order to transfer a high amount of energy to the stimulus wave. Second, we need to maximize the polarization density of the pump wave, which acts as the energy coupling coefficient based on Equation (7b).

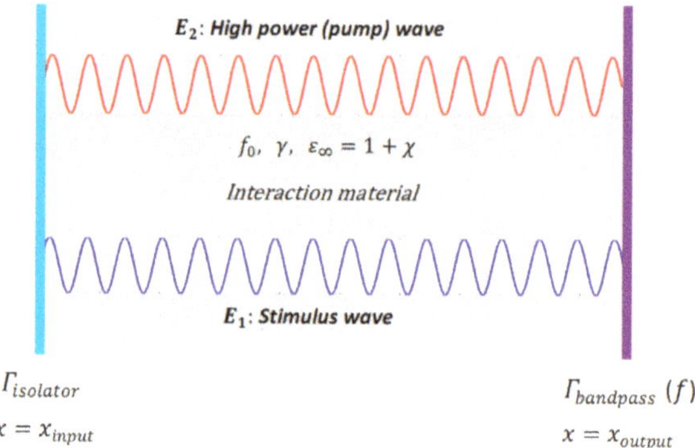

Figure 3. Configuration of the cavity.

Even if we maximize the stored electric energy density in a resonator, if the non-linear coupling coefficient P_2 is not high, then we cannot efficiently transfer the accumulated energy into the stimulus wave and high-gain amplification of the stimulus wave does not occur.

In order to perform the optimization of the stimulus wave magnitude at a given stimulus wave frequency $\big(max\, |E_1(f_{st} = f')|\big)$, we need an efficient optimization algorithm. In the next section, we will use the computationally efficient quasi-Newton Broyden–Fletcher–Goldfarb–Shanno (BFGS) algorithm to perform the maximization of the stimulus wave magnitude at a desired frequency.

4. Optimization of Optical Parametric Amplification Gain Factor in a Micro-Resonator

Assume that we are using N ultrashort high power pulses to amplify the stimulus wave. These ultrashort pulses have similar pulse energies so that each of them affects the amplification performance in the same degree. The ultrashort high-power pulses are summed up to form the pump wave. At a given spatial input point, the pump wave and the stimulus wave are given as:

$$E_p\left(x = x_{input}, t\right)$$
$$= \sum_{i=1}^{N} A_i \cos(2\pi v_i t + \psi_i)(u(t) - u(t - \Delta T_i))\ V$$
$$/m \quad (u(t) : Unit\ step\ function)$$

$$E_{st}\left(x = x_{input}, t\right) = A_{st} \cos(2\pi(v_{st})t + \psi_{st})(u(t) - u(t - \Delta T_{st}))\ V/m$$

$$where\ A_i \gg A_{st},\ \Delta T_i \ll \Delta T_{st},\ i = 1, 2, \ldots, N$$

We want to tune the frequencies $(v_1, v_2, \ldots, v_N)$ of these ultrashort pulses so that the gain factor is maximized. In order to do that, we use the BFGS algorithm, so that the Hessian matrix of each iteration is recursively updated. The BFGS algorithm is one of the quasi-Newton methods that are used to compute the Hessian matrix. Recursive computation reduces the computational cost of the optimization by eliminating the need to compute the second derivative at each iteration. We will use the BFGS algorithm because of its accuracy and simplicity. The most general form of the quasi-Newton method is given as [11]:

$$x^{(k)} = x^{(k-1)} - \alpha_k\left(H^{(k-1)}\right)^{-1}\left(\nabla f\left(x^{(k-1)}\right)\right), \quad k = 1, 2, 3, \ldots \tag{10}$$

$f\left(x^{(k-1)}\right) : Cost\ function,\ x^{(k)} : Optimization\ parameters,\ H^{(k-1)} : Hessian\ matrix,\ \alpha_k : Step\ size$

Quasi-Newton methods, like steepest descent, require only the gradient of the objective function to be supplied at each iteration. The Hessian is updated by analyzing successive gradient vectors. The whole BFGS algorithm is as described below [11].

Given a starting point x_0, convergence tolerance $\varepsilon > 0$, inverse Hessian approximation H_0;
$$k \leftarrow 0;$$
while $\nabla f_k > \varepsilon$;
   Compute search direction
$p_k = -H_k \nabla f_k$;
Set $x_{k+1} = x_k + \alpha_k\,p_k$, *where* α_k *is computed from a line search procedure to satisfy the Wolfe conditions.*
Define $s_k = x_{k+1} - x_k$ *and* $y_k = \nabla f_{k+1} - \nabla f_k$;
Compute H_{k+1} *using*;
$H_{k+1} = \left(I - \rho_k s_k y_k^T\right) H_k \left(I - \rho_k y_k s_k^T\right) + \rho_k s_k s_k^T \quad (BFGS)$
Where $\rho_k = \frac{1}{y_k^T s_k}$
$$k \leftarrow k + 1;$$
end   *(while)*

The step size α_k can be computed from a line search procedure to satisfy the Wolfe conditions:

$$f(x_k + \alpha_k\,p_k) \leq f(x_k) + c_1\,\alpha_k \nabla f_k^T p_k \tag{11}$$

$$\left|\nabla f(x_k + \alpha_k\,p_k)^T p_k\right| \leq c_2\left|\nabla f_k^T p_k\right| 0 < c_1 < c_2 < 1 \tag{12}$$

Alternatively, the step size α_k can be computed from the so-called backtracking approach [11]:
  Choose $\alpha > 0$, $\rho \in (0, 1)$, $c \in (0, 1)$
repeat until $f(x_k + \alpha\,p_k) \leq f(x_k) + c\alpha\nabla f_k^T p_k$
$\alpha \leftarrow \rho\alpha$
end

Since our problem is the amplification of a stimulus (input) wave via non-linear wave mixing with a high-power pump wave, for this problem, the optimization parameters are the frequencies of the N ultrashort pulses $v_1, v_2, \ldots, v_N$ that constitute the total pump wave. Assume that E_1 is the low power stimulus wave to be amplified, and E_2 is the high-power pump wave, which is the combination of N ultrashort pulses. The general formulation for the maximization of the stimulus wave magnitude (gain factor) is summarized as follows:

Optimization parameters: $\mathbf{v} = [v_1, v_2, \ldots, v_N]$, **Cost function to be maximized:** $f = |E_1(\mathbf{v})|$
Constraints: $g_1(\mathbf{v}) \le c_1, g_2(\mathbf{v}) \le c_2, \ldots, g_N(\mathbf{v}) \le c_N$
Equations:

$$\nabla^2(E_2(\mathbf{v})) - \mu_0 \varepsilon_\infty \frac{\partial^2(E_2(\mathbf{v}))}{\partial t^2} = \mu_0 \sigma \frac{\partial(E_2(\mathbf{v}))}{\partial t} + \mu_0 \frac{d^2 P_2}{\partial t^2} \tag{13}$$

$$\frac{d^2 P_2}{dt^2} + \gamma \frac{dP_2}{dt} + \omega_0{}^2 P_2 - \frac{\omega_0{}^2 P_2{}^2}{Ned} - \frac{\omega_0{}^2 P_2{}^3}{N^2 e^2 d^2} = \frac{Ne^2(E_2(\mathbf{v}))}{m} \tag{14}$$

$$\nabla^2(E_1(\mathbf{v})) - \mu_0 \varepsilon_\infty \frac{\partial^2(E_1(\mathbf{v}))}{\partial t^2} = \mu_0 \sigma \frac{\partial(E_1(\mathbf{v}))}{\partial t} + \mu_0 \frac{d^2(P_1)}{\partial t^2} \tag{15}$$

$$\frac{d^2(P_1)}{dt^2} + \gamma \frac{d(P_1)}{dt} + \omega_0{}^2(P_1) - \frac{\omega_0{}^2(P_1{}^2 + 2P_1 P_2)}{Ned} - \frac{\omega_0{}^2(P_1{}^3 + 3P_1{}^2 P_2 + 3P_1 P_2{}^2)}{N^2 e^2 d^2} = \frac{Ne^2(E_1(\mathbf{v}))}{m} \tag{16}$$

This problem is a constrained optimization problem, we can convert this problem into an unconstrained optimization problem by modifying the cost function via the addition of penalty functions. In the case of a maximization problem, these penalty functions yield a decrease in the cost function when the constraints are violated. In our case, the penalty for violating the constraints is adjusted to yield a quadratic decrease:

Augmented cost function: (penalty function method)

$$f = |E_1(\mathbf{v})| - L \left\{ \sum_{i=1}^{N} \delta_i(g_i(\mathbf{v}) - c_i) \right\}^q, \delta_i = \left\{ \begin{array}{ll} 0 & if\ g_i(\mathbf{v}) \le c_i \\ > 0 & if\ g_i(\mathbf{v}) > c_i \end{array} \right\} \tag{17}$$

q: positive valued penalty exponent, L: Positive valued penalty constant, δ_i: penalty coefficients
Optimization process:

$$v_{k+1} = v_k + \alpha_k p_k$$

$$p_k = -H_k \nabla f_k$$

$$s_k = v_{k+1} - v_k$$

$$y_k = \nabla f_{k+1} - \nabla f_k$$

$$H_{k+1} = \left(I - \rho_k s_k y_k{}^T\right) H_k \left(I - \rho_k y_k s_k{}^T\right) + \rho_k s_k s_k{}^T \ (BFGS)$$

$$\nabla f = \begin{bmatrix} \frac{f(v_1+\epsilon, v_2, \ldots, v_N) - f(v_1, v_2, \ldots, v_N)}{\epsilon} \\ \frac{f(v_1, v_2+\epsilon, \ldots, v_N) - f(v_1, v_2, \ldots, v_N)}{\epsilon} \\ \cdot \\ \cdot \\ \cdot \\ \frac{f(v_1, v_2, \ldots, v_N+\epsilon) - f(v_1, v_2, \ldots, v_N)}{\epsilon} \end{bmatrix}, \ \rho_k = \frac{1}{y_k{}^T s_k}$$

The convergence rate of the BFGS algorithm is super-linear, but our formulated problem is a non-linear optimization problem (non-linear programming). Therefore, the convergence is not reached immediately. Furthermore, the recursive computation of the Hessian matrix slows down the convergence rate. For these reasons, the computation of the optimum frequency values takes a great amount of time.

5. Finite-Difference Time-Domain Solution of the Gain Factor Optimization Problem in Optical Parametric Amplification

Equations (5a,b) and (7a,b) can be discretized using the finite-difference time-domain (FDTD) method as stated in Equations (18a,b) and (18c,d), for every iteration k of our optimization problem.

Our initial goal is to discretize Equation (5a,b) in order to get $E_{2,k}(i, j+1)$ i.e., the value of $E_{2,k}$ at a certain spatial coordinate at the next time step. Since $E_{2,k}$ and $P_{2,k}$ are coupled to each other, initially we attempt to solve for $P_{2,k}(i, j+1)$ and then substitute its value into the wave equation for $E_{2,k}(i, j+1)$. These two equations are solved recursively for every time step and for all spatial points in a given one-dimensional solution domain. In order to increase the accuracy of our obtained solution, we should select Δt and Δx as small as we can [12,13]. Then we may proceed on the discretization of Equation (7a,b) and substitute the previously obtained value of $P_{2,k}(i, j)$ from Equation (5a,b), for the solution of $E_{1,k}(i, j+1)$ in Equation (7a,b). Finally, we update the values of the optimization parameters based on the BFGS algorithm, and we repeat the entire procedure for every iteration of the optimization process until the desired gain factor is achieved (see Figure 4).

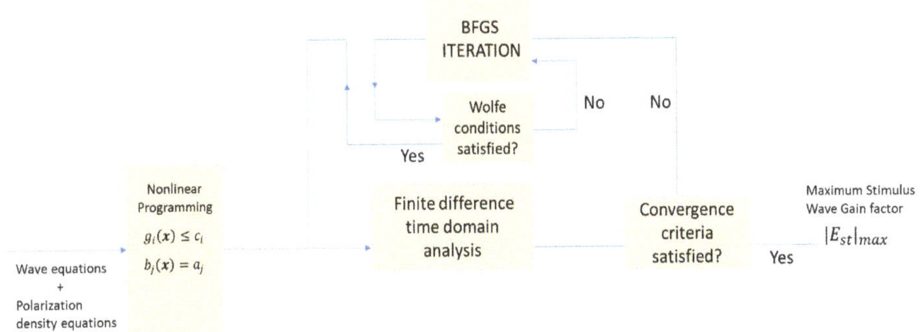

Figure 4. Flowchart diagram of Broyden–Fletcher–Goldfarb–Shanno (BFGS)-based non-linear programming integrated in finite-difference time-domain (FDTD) analysis.

FDTD Equations:

$$\frac{E_{2,k}(i+1,j)-2E_{2,k}(i,j)+E_{2,k}(i-1,j)}{\Delta x^2} - \mu_0\varepsilon_\infty(i,j)\frac{E_{2,k}(i,j+1)-2E_{2,k}(i,j)+E_{2,k}(i,j-1)}{\Delta t^2} = $$
$$\mu_0\sigma(i,j)\frac{E_{2,k}(i,j)-E_{2,k}(i,j-1)}{\Delta t} + \mu_0\frac{P_{2,k}(i,j+1)-2P_{2,k}(i,j)+P_{2,k}(i,j-1)}{\Delta t^2} \tag{18a}$$

$$\frac{P_{2,k}(i,j+1)-2P_{2,k}(i,j)+P_{2,k}(i,j-1)}{\Delta t^2} + \gamma\frac{P_{2,k}(i,j)-P_{2,k}(i,j-1)}{\Delta t} + 4\pi^2 f_0^2\big(P_{2,k}(i,j)\big)-$$
$$\frac{4\pi^2 f_0^2}{Ned}\big(P_{2,k}(i,j)\big)^2 - \frac{4\pi^2 f_0^2}{N^2 e^2 d^2}\big(P_{2,k}(i,j)\big)^3 = \frac{Ne^2}{m}\big(E_{2,k}(i,j)\big). \tag{18b}$$

$$\frac{E_{1,k}(i+1,j)-2E_{1,k}(i,j)+E_{1,k}(i-1,j)}{\Delta x^2} - \mu_0\varepsilon_\infty(i,j)\frac{E_{1,k}(i,j+1)-2E_{1,k}(i,j)+E_{1,k}(i,j-1)}{\Delta t^2} = $$
$$\mu_0\sigma(i,j)\frac{E_{1,k}(i,j)-E_{1,k}(i,j-1)}{\Delta t} + \mu_0\frac{P_{1,k}(i,j+1)-2P_{1,k}(i,j)+P_{1,k}(i,j-1)}{\Delta t^2} \tag{18c}$$

$$\frac{P_{1,k}(i,j+1)-2P_{1,k}(i,j)+P_{1,k}(i,j-1)}{\Delta t^2} + \gamma\frac{P_{1,k}(i,j)-P_{1,k}(i,j-1)}{\Delta t} + 4\pi^2 f_0^2\big(P_{1,k}(i,j)\big)$$
$$-\frac{4\pi^2 f_0^2}{Ned}\Big\{\big(P_{1,k}(i,j)\big)^2 + 2P_{1,k}(i,j)P_{2,k}(i,j)\Big\} - \frac{4\pi^2 f_0^2}{N^2 e^2 d^2}\Big\{\big(P_{1,k}(i,j)\big)^3 \tag{18d}$$
$$+3\big(P_{1,k}(i,j)\big)^2 P_{2,k}(i,j) + 3P_{1,k}(i,j)\big(P_{2,k}(i,j)\big)^2\Big\} = \frac{Ne^2}{m}\big(E_{1,k}(i,j)\big).$$

x: spatial coordinate, t: time, k: iteration number, $E_k(x,t) = E_k(i\Delta x, j\Delta t) \rightarrow E_k(i, j)$ $E_{2,k}$: *High intensity pump wave at iteration* k, $E_{1,k}$: *Stimulus wave at iteration* k

Constraints

$$b_1(v) = a_1 g_1(v) \le c_1$$

$$b_2(v) = a_2\, g_2(v) \le c_2$$

$$b_N(v) = a_N g_N(v) \le c_N$$

Cost function: $\big|E_1(v)\big|$

Augmented cost function: (penalty function method)

$$f = \left|E_1(\nu)\right| + L_1\left\{\sum_{i=1}^{N} \delta_i(g_i(\nu) - c_i)^2\right\} + L_2\left\{\sum_{j=1}^{M} \zeta_j\big(b_j(\nu) - a_j\big)^2\right\}$$

$$\delta_i = \left\{ \begin{array}{ll} 0 & if \;\; g_i(\nu) \le c_i \\ > 0 & if \;\; g_i(\nu) > c_i \end{array} \right\}$$

Iterations:

$$\nu_{k+1} = \nu_k + \alpha_k p_k$$

$$p_k = -H_k \nabla f_k$$

$$s_k = x_{k+1} - x_k$$

$$y_k = \nabla f_{k+1} - \nabla f_k$$

$$H_{k+1} = \left(I - \rho_k s_k y_k{}^T\right)H_k\left(I - \rho_k y_k s_k{}^T\right) + \rho_k s_k s_k{}^T \;\; \text{(BFGS)}$$

$$\rho_k = \frac{1}{y_k{}^T s_k}$$

Wolfe conditions for α_k: $f(x_k + \alpha_k \, p_k) \le f(x_k) + c_1 \, \alpha_k \nabla f_k{}^T p_k$

$$\left|\nabla f(x_k + \alpha_k \, p_k)^T p_k\right| \le c_2\left|\nabla f_k{}^T p_k\right| \;\; 0 < c_1 < c_2 < 1$$

6. Numerical Experiments

6.1. Double-Frequency Tuning for Gain Factor Optimization

Assume that a 250 THz ($\lambda_{free\;space}$ = 1.2 µm) infra-red stimulus wave E_{st} and a high power pump wave E_{hp} that is composed of two high-intensity ultrashort pulses (frequencies are to be determined) are propagating inside a low-loss (high Q) cavity that has two reflecting walls (see Figure 5). The wall on the left side can be thought as an optical isolator, which fully transmits from its left side and almost fully reflects from its right side. The wall on the right side represents an optical band-pass filter with a frequency-dependent reflection coefficient $\Gamma(f)$. Both waves are generated at x = 0 µm, at the time instant t = 0 s. The waves and the parameters of the gain medium are as given below:

$$E_{hp}(x = 0\mu m, t) = \sum_{i=1}^{2} A_i \cos(2\pi f_i t + \psi_i)(u(t) - u(t - \Delta T_i)) \quad (u(t): Unit\;step\;function)$$

$$Where \; A_1 = 2 \times 10^8, \;\; A_2 = 1.5 \times 10^8, \;\; \psi_1 = 0, \;\; \psi_2 = 0, \;\; \Delta T_1 = 0.5ps, \;\; \Delta T_2 = 1ps$$

$$E_{st}(x = 0\mu m, t) = 1 \times \sin\big(2\pi\big(2.5 \times 10^{14}\big)t\big) \; V/m, \quad for \;\; 0 \le t \le 10ps$$

$$Dielectric\;constant\;of\;the\;gain\;medium = \; \varepsilon_\infty = 1 + \chi = 12 \quad (\mu_r = 1)$$

$$Resonance\;frequency\;of\;the\;gain\;medium: \;\; f_0 = 500THz$$

$$Damping\;coefficient\;of\;the\;gain\;medium: \;\; \gamma = 1 \times 10^9 Hz$$

$$Time\;interval\;and\;duration\;of\;simulation: \;\; 0 \le t \le 10ps$$

$$Spatial\;range\;of\;the\;gain\;medium: 0\mu m < x < 10\mu m$$

$$Right\;cavity\;wall\;location: \;\; x = 10\mu m; Left\;cavity\;wall\;location: \;\; x = 0\mu m$$

$$Electron\;density\;of\;the\;gain\;medium: \;\; N = 3.5 \times 10^{28}/m^3; \;\; Atomic\;diameter: d = 0.3\;nm$$

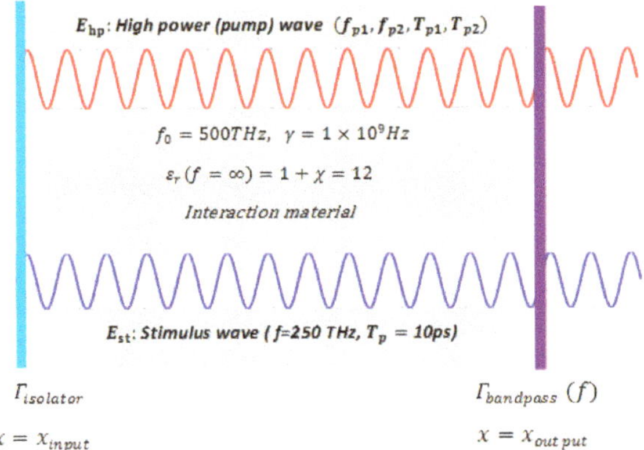

Figure 5. Configuration of the cavity and the parameters for Section 6.1.

Our problem: find the optimal pump wave pulse frequencies f_{p1}, f_{p2} that maximize the magnitude of the monochromatic stimulus wave in the cavity ($|E_{st}(f_{st} = 250\ THz)|$), for $10\ THz < \{f_{p1}, f_{p2}\} < 1000THz$ (THz to UV), and for $0\mu m < x < 10\mu m, 0 \le t \le 10ps$, such that

$$\nabla^2 E_{hp}\big(f_{p1},\ f_{p2}\big) - \mu_0 \varepsilon_\infty \frac{\partial^2 E_{hp}\big(f_{p1},\ f_{p2}\big)}{\partial t^2} = \mu_0 \sigma \frac{\partial E_{hp}\big(f_{p1},\ f_{p2}\big)}{\partial t} + \mu_0 \frac{\partial^2 P_{hp}}{\partial t^2}$$

$$\frac{\partial^2 P_{hp}}{\partial t^2} + \gamma \frac{\partial P_{hp}}{\partial t} + \omega_0^2 \big(P_{hp}\big) - \frac{\omega_0^2}{Ned}\big(P_{hp}\big)^2 - \frac{\omega_0^2}{N^2 e^2 d^2}\big(P_{hp}\big)^3 = \frac{Ne^2}{m} E_{hp}\big(f_{p1},\ f_{p2}\big).$$

$$\nabla^2 E_{st}\big(f_{p1},\ f_{p2}\big) - \mu_0 \varepsilon_\infty \frac{\partial^2 E_{st}\big(f_{p1},\ f_{p2}\big)}{\partial t^2} = \mu_0 \sigma \frac{\partial E_{st}\big(f_{p1},\ f_{p2}\big)}{\partial t} + \mu_0 \frac{\partial^2 P_{st}}{\partial t^2}$$

$$\frac{\partial^2 (P_{st})}{\partial t^2} + \gamma \frac{\partial (P_{st})}{\partial t} + \omega_0^2 (P_{st}) - \frac{\omega_0^2}{Ned}\big\{P_{st}^2 + 2P_{st}P_{hp}\big\} - \frac{\omega_0^2}{N^2 e^2 d^2}\big\{P_{st}^3 + 3P_{st}^2 P_{hp} + 3P_{st}P_{hp}^2\big\}$$
$$= \frac{Ne^2}{m} E_{st}\big(f_{p1},\ f_{p2}\big)$$

Our aim is to maximize the magnitude of the stimulus wave at its original frequency $f_{st} = 250THz$ (monochromatic form). This is a precaution against any degree of spectral broadening that the stimulus wave may go through while being amplified. Therefore, our cost function is chosen as (at any spatial point $x = x'$ inside the cavity):

$$Q = \; \big|E_{st}(f_{st} = 250THz)\big|$$
$$= \left| \int_{2.5\times10^{14}-\Delta f}^{2.5\times10^{14}+\Delta f} \left\{ \int_0^{\Delta T} \{E_{st}(x = x', t)e^{-i(2\pi\Omega)t}\}dt \right\} e^{i(2\pi\Omega)t} d\Omega \right|$$

where $\Delta T = 10ps, 0 \le t \le 10ps, \big(2.5 \times 10^{14} - \Delta f\big) < \Omega < \big(2.5 \times 10^{14} + \Delta f\big), \Delta f = 1THz$

Initial conditions:

$$P_{hp}(x,0) = P_{hp}'(x,0) = E_{hp}(x,0) = E_{hp}'(x,0) = P_{st}(x,0) = P_{st}'(x,0) = E_{st}(x,0) = E_{st}'(x,0) = 0$$

Boundary and excitation conditions:

$$E_{hp}(x = 0\ \mu m, t) = \sum_{i=1}^{2} A_i \cos(2\pi f_i t + \psi_i)(u(t) - u(t - \Delta T_i))$$

where $A_1 = 2 \times 10^{14}$, $A_2 = 1.5 \times 10^{14}$, $\psi_1 = 0$, $\psi_2 = 0$, $\Delta T_1 = 0.5 ns$, $\Delta T_2 = 1 ns$

$$E_{st}(x = 0\mu m, t) = 1 \times \frac{\sin\left(2\pi\left(2.5 \times 10^{14}\right)t\right)\ V}{m}, \quad for\ \ 0 \le t \le 10 ps$$

$$E_{hp}(x = 15\mu m, t) = E_{st}(x = 15\mu m, t) = 0 \quad for\ 0 < t < 10 ps$$

Absorbing boundary condition (perfectly matched layer):

$$\sigma(x) = \{(x - (L - \Delta))\sigma_0, \quad (L - \Delta) \le x < L\ \}, \ for\ L = 15\mu m, \quad \Delta = 2.5\mu m, \quad \sigma_0 = 4.5 \times 10^8 S/m$$

Optical isolator condition: full reflection at $x = 0\mu m$

$$\Gamma(x = 0\mu m, t) = 1 \quad (Reflection\ coefficient\ is\ equal\ to\ 1)$$

Optical bandpass filter condition: frequency dependent reflection at $x = 10\mu m$

$$\left|\Gamma(f')\right| = 1 - e^{-\left(\frac{(f' - 250THz)}{\sqrt{2}THz}\right)^2}$$

Cost function to be maximized:

$$Q\left(f_{p1}, f_{p2}\right) = \left|E_{st}(\ f_{st} = 250THz)\right| - \delta_1\left(f_{p1} - f_{max}\right)^2 - \delta_2\left(f_{min} - f_{p1}\right)^2 - \delta_3\left(f_{p2} - f_{max}\right)^2 - \delta_4\left(f_{min} - f_{p2}\right)^2$$

where

$$\delta_1 = \begin{cases} 0 & if\ \ f_{p1} \le f_{max} \\ \frac{|E_{st}(\ f_{st} = 250THz)|}{10^{27}} & if\ \ f_{p1} > f_{max} \end{cases}, \quad \delta_2 = \begin{cases} 0 & if\ \ f_{p1} \ge f_{min} \\ \frac{|E_{st}(\ f_{st} = 250THz)|}{10^{27}} & if\ \ f_{p1} < f_{min} \end{cases}$$

$$\delta_3 = \begin{cases} 0 & if\ \ f_{p2} \le f_{max} \\ \frac{|E_{st}(\ f_{st} = 250THz)|}{10^{27}} & if\ \ f_{p2} > f_{max} \end{cases}, \quad \delta_4 = \begin{cases} 0 & if\ \ f_{p2} \ge f_{min} \\ \frac{|E_{st}(\ f_{st} = 250THz)|}{10^{27}} & if\ \ f_{p2} < f_{min} \end{cases}$$

Optimization algorithm (BFGS):

$$Set\ H_1 = \begin{bmatrix} 1 & 0 \\ 0 & 1 \end{bmatrix}, \ f_{p1,0} = 100THz, \ f_{p1,1} = 102THz,$$

$$f_{p2,0} = 100THz, \qquad f_{p2,1} = 103THz, \ \alpha_1 = 0.5$$

$$\nabla Q_k = \begin{bmatrix} \frac{Q\left(f_{p1,k}, f_{p2,k}\right) - Q\left(f_{p1,k-1}, f_{p2,k}\right)}{f_{p1,k} - f_{p1,k-1}} \\ \frac{Q\left(f_{p1,k}, f_{p2,k}\right) - Q\left(f_{p1,k}, f_{p2,k-1}\right)}{f_{p2,k} - f_{p2,k-1}} \end{bmatrix}$$

$$p_k = -H_k \nabla Q_k$$

$$f_{p,\ k+1} = f_{p,k} + \alpha_k p_k, \ f_{p,k} = \begin{bmatrix} f_{p1,k} \\ f_{p2,k} \end{bmatrix}$$

$$s_k = f_{p,k+1} - f_{p,k}$$

$$\nabla Q_{k+1} = \begin{bmatrix} \frac{Q\left(f_{p1,k+1}, f_{p2,k}\right) - Q\left(f_{p1,k}, f_{p2,k}\right)}{f_{p1,k+1} - f_{p1,k}} \\ \frac{Q\left(f_{p1,k}, f_{p2,k+1}\right) - Q\left(f_{p1,k}, f_{p2,k}\right)}{f_{p2,k+1} - f_{p2,k}} \end{bmatrix}$$

$$y_k = \nabla Q_{k+1} - \nabla Q_k$$

$$\rho_k = \frac{1}{y_k^T s_k}$$

$$H_{k+1} = \left(I - \rho_k s_k y_k^T\right) H_k \left(I - \rho_k y_k s_k^T\right) + \rho_k s_k s_k^T \quad \text{(BFGS)}$$

$$I : Identity\ matrix$$

In order to satisfy the Wolfe conditions, the step size at each iteration is chosen as:

$$\alpha_k = C^{\left(\log \left|\frac{Q(f_{p,k})}{Q(f_{p,k}) - Q(f_{p,k-1})}\right|\right) / \left(\left|\frac{Q(f_{p,k})}{Q(f_{p,k}) - Q(f_{p,k-1})}\right|\right)} \tag{19}$$

where C is just a constant ($1 < C < 1.5$) and α_k is the step size at iteration k. This formula (Equation (19)) was determined by trial and error and was found to satisfy Wolfe's conditions automatically at each iteration. This saves us from the huge computational cost of running another iteration loop to determine the step size at each iteration of the optimization process. In this simulation $C = 1.445$. Based on the above formulations, the maximum stimulus wave amplitude that has been reached in the cavity (for $0 < t < 10$ ps) is determined as $Gain_{max} = \left|E_{st}(f_{st} = 250THz)\right|_{max} = 4.67 \times 10^8\,V/m$, which corresponds to the frequencies $f_{p1} = 387.2\ THz$, $f_{p2} = 319.4\ THz$ (see Table 1), or to the free space wavelengths $\lambda_{p1} = 0.939\ \mu m$, $\lambda_{p2} = 0.775\ \mu m$ as the ultrashort pulses can be practically generated outside the resonator. In this simulation, the gain factor of the amplification is defined as:

$$Gain_{max} = \left|E_{st}(f_{st} = 250THz)\right|_{max} = \frac{Amplitude\ of\ the\ stimulus\ wave\ at\ t=t_{max}\ for\ x=x'}{Amplitude\ of\ the\ stimulus\ wave\ at\ t=0\ for\ x=x'}.$$

where x' is chosen as an intra-cavity point ($x' = 5.73 \mu m$) and $t_{max} = 10\ picoseconds$.

$$W_{e,p} = Stored\ electric\ energy\ density\ via\ pump\ wave = \frac{1}{2}\varepsilon_\infty E_{pump}^2 + \frac{1}{2}E_{pump}P_{pump}.\left(\frac{Joules}{m^3}\right)$$

$$P_{pump} : Polarization\ density\ created\ by\ the\ pump\ wave\left(\frac{Coulomb}{m^2}\right),\quad E_{pump} : Pump\ wave\ electric\ field\ intensity$$

$$\varepsilon_\infty : Background\ (infinite\ spectral\ band)\,permittivity$$

Table 1. BFGS algorithm-based optimization process.

f_{p1}	f_{p2}	$Gain_{max}$	$W_{e,p}$ $(\frac{J}{m^3})$	$P_{pump}(\frac{C}{m^2})$	k (Iteration #)
100 THz	100 THz	0.84	1.29×10^7	0.08	1
107.2 THz	115.7 THz	6.23	1.80×10^7	0.09	4
154.9 THz	156.6 THz	4.44	3.36×10^7	0.08	7
218.6 THz	199.5 THz	313.52	4.48×10^7	0.10	10
198.3 THz	214.5 THz	37.16	1.67×10^7	0.10	13
229.5 THz	243.0 THz	240.58	5.97×10^7	0.09	16
263.1 THz	227.2 THz	646.72	5.54×10^7	0.10	19
322.7 THz	278.9 THz	1.57×10^3	6.12×10^7	0.12	22
396.0 THz	299.8 THz	4.28×10^4	9.39×10^8	0.15	25
391.6 THz	293.4 THz	9.16×10^4	1.76×10^8	0.16	28
380.7 THz	311.7 THz	3.85×10^5	1.26×10^8	0.20	30
383.4 THz	317.2 THz	8.11×10^4	6.40×10^7	0.20	32
386.0 THz	318.4 THz	6.32×10^6	2.89×10^8	0.23	34
386.8 THz	318.8 THz	9.79×10^7	2.63×10^8	0.27	36
387.2 THz	319.3 THz	3.96×10^8	2.51×10^8	0.29	38
387.2 THz	319.4 THz	4.67×10^8	2.95×10^8	0.29	39

As we can see from Table 1, the optimal ultrashort pulse frequencies correspond to very high stored electric energy density and high polarization density. The stored electric energy density and the polarization density must be simultaneously high for a significant stimulus wave amplification. The stored electric energy density indicates the achievable order of stimulus wave amplification [14,15], and the polarization density acts as a coupling coefficient, which is a measure of how much stored electric energy can be coupled to the stimulus wave.

The time variation of the spectrally broadened (polychromatic) stimulus wave between t = 6.6 picoseconds and t = 10 picoseconds is shown in Figure 6. From the figure, we can see that the polychromatic stimulus wave reaches an amplitude of approximately 8×10^8 V/m.

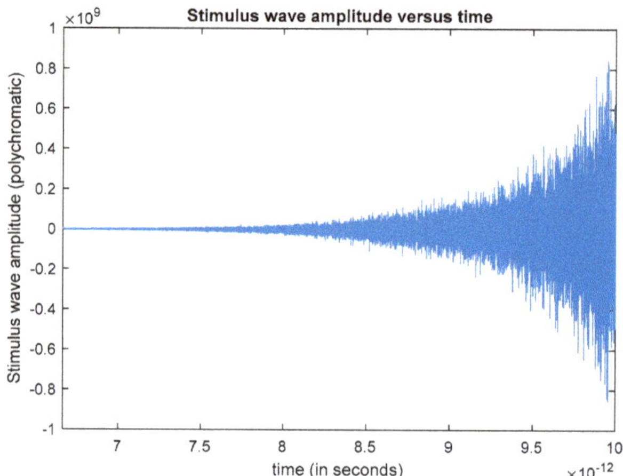

Figure 6. Stimulus wave amplification (in polychromatic form) inside the cavity at x = 5.73 μm.

6.2. Triple-Frequency Tuning for Gain Factor Optimization

Assume that a 300 THz ($\lambda_{free\ space}$ =1 μm) infra-red stimulus wave E_{st} and a high-power pump wave E_{hp} that is composed of three ultrashort pulses (frequencies are to be determined), are generated to propagate in a micro-resonator with two reflecting walls. The wall on the left side can be thought as an optical isolator, which fully transmits from its left side and almost fully reflects from its right side. The wall on the right side represents an optical band-pass filter with a frequency-dependent reflection coefficient $\Gamma(f)$ (see Figure 7). Both waves are generated at x = 0 μm and at the time instant t = 0 ps. The waves and the parameters of the gain medium are as given below:

$$E_{hp}(x = 0\mu m, t) = \sum_{i=1}^{3} A_i \cos(2\pi f_i t + \psi_i)(u(t) - u(t - \Delta T_i))(u(t) : Unit\ step\ function)$$

where $A_1 = 1.5 \times 10^8$, $A_2 = 2 \times 10^8$, $A_3 = 2.5 \times 10^8$, $\Delta T_1 = 4ps$, $\Delta T_2 = 2ps$, $\Delta T_2 = 1ps$

$$E_{st}(x = 0\mu m, t) = 1 \times \sin\left(2\pi\left(3 \times 10^{14}\right)t\right)\ V/m, \quad for\ \ 0 \leq t \leq 40ps$$

$$Dielectric\ constant\ of\ the\ gain\ medium = \varepsilon_\infty = 1 + \chi = 10 \quad (\mu_r = 1)$$

$$Resonance\ frequency\ of\ the\ gain\ medium:\ \ f_0 = 600THz$$

$$Damping\ rate\ of\ the\ gain\ medium:\ \ \gamma = 1 \times 10^{10}Hz$$

$$Duration\ of\ simulation:\ \ 0 \leq t \leq 40ps;\ Range\ of\ the\ gain\ medium: 0\mu m < x < 10\mu m$$

$$Right\ cavity\ wall\ location:\ \ x = 10\mu m;\ Left\ cavity\ wall\ location:\ \ x = 0\mu m$$

$$Electron\ density\ of\ the\ medium:\ \ N = 3.5 \times 10^{28}/m^3;\ \ Atomic\ diameter\ : d = 0.3\ nm$$

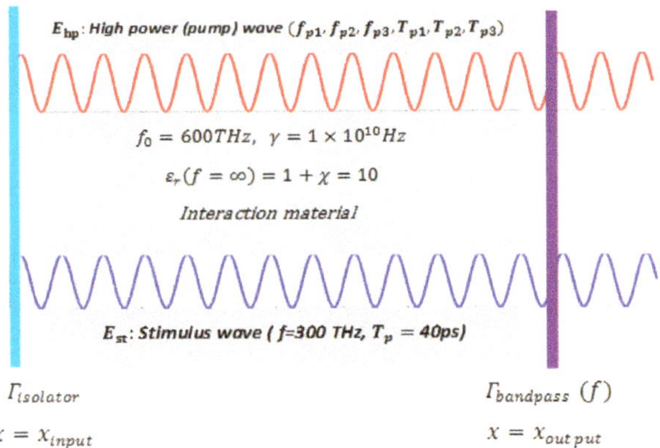

Figure 7. Configuration of the cavity and the parameters for Section 6.2.

Our problem: Find the optimal pump wave pulse frequencies f_{p1}, f_{p2}, f_{p3} that maximize the magnitude of the monochromatic stimulus wave in the cavity $(|E_{st}(f_{st} = 300THz)|)$, for 50 THz $< \{f_{p1}, f_{p2}, f_{p3}\} < 500\ THz, 0\mu m < x < 10\mu m, 0 \le t \le 40\ ps$, such that

$$\nabla^2 E_{hp}\big(f_{p1},\ f_{p2},\ f_{p3}\big) - \mu_0 \varepsilon_\infty \frac{\partial^2 E_{hp}\big(f_{p1},\ f_{p2},\ f_{p3}\big)}{\partial t^2} = \mu_0 \sigma \frac{\partial E_{hp}\big(f_{p1},\ f_{p2},\ f_{p3}\big)}{\partial t} + \mu_0 \frac{\partial^2 P_{hp}}{\partial t^2}$$

$$\frac{\partial^2 P_{hp}}{\partial t^2} + \gamma \frac{\partial P_{hp}}{\partial t} + \omega_0^2 \big(P_{hp}\big) - \frac{\omega_0^2}{Ned}\big(P_{hp}\big)^2 - \frac{\omega_0^2}{N^2 e^2 d^2}\big(P_{hp}\big)^3 = \frac{Ne^2}{m} E_{hp}\big(f_{p1},\ f_{p2},\ f_{p3}\big).$$

$$\nabla^2 E_{st}\big(f_{p1},\ f_{p2},\ f_{p3}\big) - \mu_0 \varepsilon_\infty \frac{\partial^2 E_{st}\big(f_{p1},\ f_{p2},\ f_{p3}\big)}{\partial t^2} = \mu_0 \sigma \frac{\partial E_{st}\big(f_{p1},\ f_{p2},\ f_{p3}\big)}{\partial t} + \mu_0 \frac{\partial^2 P_{st}}{\partial t^2}$$

$$\frac{\partial^2 (P_{st})}{\partial t^2} + \gamma \frac{\partial (P_{st})}{\partial t} + \omega_0^2 (P_{st}) - \frac{\omega_0^2}{Ned}\big\{P_{st}^2 + 2P_{st}P_{hp}\big\} - \frac{\omega_0^2}{N^2 e^2 d^2}\big\{P_{st}^3 + 3P_{st}^2 P_{hp} + 3P_{st}P_{hp}^2\big\}$$
$$= \frac{Ne^2}{m} E_{st}\big(f_{p1},\ f_{p2},\ f_{p3}\big)$$

It is important to emphasize that our aim is to maximize the magnitude of the stimulus wave at its original frequency $f_{st} = 300\ THz$ (monochromatic form). This is a precaution against any degree of spectral broadening that the stimulus wave may go through while being amplified. A plain attempt to maximize the magnitude of the stimulus wave $(|E_{st}|)$ independent of the original excitation frequency (f_{st}), may result in an amplified stimulus wave with different frequency components. In fact, these different frequency components might be even much more dominant than the original excitation frequency of the stimulus wave. Therefore, our cost function is chosen as (at any spatial point $x = x'$)

$$Q = \big|E_{st}(f_{st} = 300THz)\big|$$
$$= \left| \int_{3\times10^{14}-\Delta f}^{3\times10^{14}+\Delta f} \left\{ \int_0^{\Delta T} \{E_{st}(x = x', t)e^{-i(2\pi\Omega)t}\}dt \right\} e^{i(2\pi\Omega)t}d\Omega \right|$$

where $\Delta T = 40ps,\ 0 \le t \le 40ps,\ \big(3 \times 10^{14} - \Delta f\big) < \Omega < \big(3 \times 10^{14} + \Delta f\big),\ \Delta f = 0.5THz$

Initial conditions:

$$P_{hp}(x,0) = P_{hp}'(x,0) = E_{hp}(x,0) = E_{hp}'(x,0) = P_{st}(x,0) = P_{st}'(x,0) = E_{st}(x,0) = E_{st}'(x,0) = 0$$

Boundary and excitation conditions:

$$E_{hp}(x = 0\mu m, t) = \sum_{i=1}^{3} A_i \cos(2\pi f_i t + \psi_i)(u(t) - u(t - \Delta T_i))$$

where $A_1 = 1.5 \times 10^8$, $A_2 = 2 \times 10^8$, $A_3 = 2.5 \times 10^8$, $\Delta T_1 = 4ps$, $\Delta T_2 = 2ps$, $\Delta T_2 = 1ps$

$$E_{st}(x = 0\mu m, t) = 1 \times \frac{\sin\left(2\pi(3\times10^{14})t\right)}{m} \frac{V}{}, \ for \ 0 \le t$$
$$\le 40ps, \ E_{hp}(x = 15\mu m, t) = E_{st}(x = 15\mu m, t) = 0 \quad for \ 0 < t$$
$$< 40ps$$

Absorbing boundary condition (perfectly matched layer):

$$\sigma(x) = \{ (x - (L - \Delta))\sigma_0, \quad (L - \Delta) \le x < L \ \}, \ for \ L = 15\mu m, \ \Delta = 2.5\mu m, \ \sigma_0 = 4.5 \times 10^8 S/m$$

Optical isolator condition: full reflection at $x = 0\mu m$; $\Gamma(x = 0\mu m, t) = 1 (Reflection \ coefficient = 1)$

Optical bandpass filter condition: frequency-dependent reflection at $x = 10\mu m$;

$$\left|\Gamma(f')\right| = 1 - e^{-\left(\frac{(f'-300THz)}{\sqrt{2}THz}\right)^2}, \quad for \ x = 10\mu m, \ 0 < t < 40ps$$

Augmented cost function to be maximized:

$$Q\left(f_{p1}, f_{p2}, f_{p3}\right) = \left|E_{st}(f_{st} = 300THz)\right|$$
$$- \sum_{i=1}^{3} \left\{ \delta_{i,1}\left(f_{p,i} - f_{max}\right)^2 + \delta_{i,2}\left(f_{min} - f_{p,i}\right)^2 \right\}$$

$$\delta_{i,1} = \left\{ \begin{array}{ll} 0 & if \ f_{p,i} \le f_{max} \\ \frac{\left|E_{st}(f_{st}=300THz)\right|}{10^{27}} & if \ f_{p,i} > f_{max} \end{array} \right\}, \ \delta_{i,2} = \left\{ \begin{array}{ll} 0 & if \ f_{p,i} \ge f_{min} \\ \frac{\left|E_{st}(f_{st}=300THz)\right|}{10^{27}} & if \ f_{p,i} < f_{min} \end{array} \right\}$$

Optimization algorithm (BFGS):

$$Set \ \mathbf{H_1} = \begin{bmatrix} 1 & 0 & 0 \\ 0 & 1 & 0 \\ 0 & 0 & 1 \end{bmatrix}, \ f_{p1,0} = 200THz, \ f_{p1,1} = 205THz$$

$$f_{p2,0} = 150THz, \ f_{p2,1} = 153THz, \ f_{p3,0} = 100THz, \ f_{p3,1} = 102THz, \ \alpha_1 = 0.5$$

$$\nabla Q_k = \begin{bmatrix} \frac{Q\left(f_{p1,k}, f_{p2,k}, f_{p3,k}\right) - Q\left(f_{p1,k-1}, f_{p2,k}, f_{p3,k}\right)}{f_{p1,k} - f_{p1,k-1}} \\ \frac{Q\left(f_{p1,k}, f_{p2,k}, f_{p3,k}\right) - Q\left(f_{p1,k}, f_{p2,k-1}, f_{p3,k}\right)}{f_{p2,k} - f_{p2,k-1}} \\ \frac{Q\left(f_{p1,k}, f_{p2,k}, f_{p3,k}\right) - Q\left(f_{p1,k}, f_{p2,k}, f_{p3,k-1}\right)}{f_{p3,k} - f_{p3,k-1}} \end{bmatrix}$$

$$p_k = -H_k \nabla Q_k, \quad f_{p, \ k+1} = f_{p,k} + \alpha_k p_k, \ f_{p,k} = \begin{bmatrix} f_{p1,k} \\ f_{p2,k} \\ f_{p3,k} \end{bmatrix}, \quad s_k = f_{p,k+1} - f_{p,k}$$

$$\nabla Q_{k+1} = \begin{bmatrix} \frac{Q\left(f_{p1,k+1}, f_{p2,k}, f_{p3,k}\right) - Q\left(f_{p1,k}, f_{p2,k}, f_{p3,k}\right)}{f_{p1,k+1} - f_{p1,k}} \\ \frac{Q\left(f_{p1,k}, f_{p2,k+1}, f_{p3,k}\right) - Q\left(f_{p1,k}, f_{p2,k}, f_{p3,k}\right)}{f_{p2,k+1} - f_{p2,k}} \\ \frac{Q\left(f_{p1,k}, f_{p2}, f_{p3,k+1}\right) - Q\left(f_{p1,k}, f_{p2,k}, f_{p3,k}\right)}{f_{p3,k+1} - f_{p3,k}} \end{bmatrix}$$

$$y_k = \nabla Q_{k+1} - \nabla Q_k \quad \rho_k = \frac{1}{y_k^T s_k}$$

$$BFGS\ update\ \ \mathbf{H}_{k+1} = \left(\mathbf{I} - \rho_k s_k y_k^T\right)\mathbf{H}_k\left(\mathbf{I} - \rho_k y_k s_k^T\right) + \rho_k s_k s_k^T$$

$$\mathbf{I} : Identity\ matrix$$

In order to satisfy the Wolfe conditions, the step size at each iteration is chosen as:

$$\alpha_k = C^{\left(\log\left|\frac{Q(f_{p,k})}{Q(f_{p,k}) - Q(f_{p,k-1})}\right|\right) / \left(\left|\frac{Q(f_{p,k})}{Q(f_{p,k}) - Q(f_{p,k-1})}\right|\right)}$$

where C is just a constant $(1 < C < 1.5)$ and α_k is the step size at iteration k. In this simulation C = 1.45. Based on the above formulations, the maximum stimulus wave amplitude that has been reached in the cavity (for 0 < t < 40 ps) is determined as $Gain_{max} = \left|E_{st}(f_{st} = 300THz)\right|_{max} = 6.34 \times 10^8 V/m$, which corresponds to the frequencies $f_{p1} = 290.8THz, f_{p2} = 410.6THz, f_{p3} = 209.7THz$ (see Table 2), or to the free-space wavelengths $\lambda_{p1} = 1.032\mu m, \lambda_{p2} = 0.731\mu m, \lambda_{p3} = 1.431\mu m$ as the ultrashort pulses can be practically generated outside the resonator. In this simulation, the gain factor of the amplification is defined as:

$$Gain_{max} = \left|E_{st}(f_{st} = 300THz)\right|_{max}$$
$$= \frac{Amplitude\ of\ the\ stimulus\ wave\ at\ t=t_{max}\ for\ x=x'}{Amplitude\ of\ the\ stimulus\ wave\ at\ t=0\ for\ x=x'}$$

where x' is chosen as an intra-cavity point $(x' = 5\mu m)$ and $t_{max} = 40 picoseconds$.

$$W_{e,p} = Stored\ electric\ energy\ density\ via\ pump\ wave = \frac{1}{2}\varepsilon_\infty E_{pump}^2 + \frac{1}{2}E_{pump}P_{pump}.\ \left(\frac{Joules}{m^3}\right)$$

$$P_{pump} : Polarization\ density\ created\ by\ the\ pump\ wave\left(\frac{Coulomb}{m^2}\right)$$

$$E_{pump} : Pump\ wave\ electric\ field\ intensity,\ \ \varepsilon_\infty : Background\ permittivity$$

Table 2. BFGS algorithm-based optimization process.

$Gain_{max}$	f_{p1}	f_{p2}	f_{p3}	$W_{e,p}\ (\frac{J}{m^3})$	$P_{pump}(\frac{C}{m^2})$	k (Iteration #)
2.84	200 THz	150 THz	100 THz	1.41×10^7	0.07	1
5.47	205 THz	153Hz	102 THz	1.32×10^7	0.09	2
4.43	216.2 THz	165.4 THz	115.1 THz	2.13×10^7	0.06	5
9.28	239.5 THz	189.8 THz	133.5 THz	3.76×10^7	0.08	8
57.78	297.2 THz	228.3 THz	168.8 THz	6.61×10^7	0.09	11
129.51	296.0 THz	344.1 THz	159.2 THz	6.18×10^7	0.10	14
396.63	284.1 THz	392.0 THz	202.5 THz	7.98×10^7	0.10	17
3049.7	287.6 THz	399.3 THz	204.6 THz	8.53×10^7	0.12	20
4.2×10^4	289.6 THz	403.4 THz	206.1 THz	1.07×10^8	0.17	23
7.6×10^5	290.7 THz	408.7 THz	208.9 THz	1.72×10^8	0.19	26
1.4×10^7	291.9 THz	410.2 THz	209.4 THz	1.49×10^8	0.23	31
2.9×10^8	291.1 THz	410.8 THz	209.9 THz	2.36×10^8	0.22	34
6.3×10^8	290.8 THz	410.6 THz	209.7 THz	2.94×10^8	0.23	36

As we can see from Table 2, the optimal ultrashort pulse frequencies correspond to a very high stored electric energy density and a high polarization density. If we have a look at Table 2, we can see that the stored electric energy density and the polarization density must be simultaneously high for a significant stimulus wave amplification. The stored electric energy density indicates the achievable order of stimulus wave amplification. The polarization density serves as an energy-coupling coefficient and is a measure of the stored electric energy that can be coupled to the stimulus wave.

7. Verification of Our Computational Model

Our numerical model is verified by using the experimentally validated formulas of the topics *sum frequency generation via non-linear wave mixing* and *optical amplification via non-linear wave mixing* as described in the examples below.

Example 1. *Sum frequency generation via non-linear wave mixing (frequency up-conversion)*

This example is about the generation of a higher frequency component (ω_3), by mixing of two monochromatic waves with frequencies ω_1 and ω_2 via non-linear coupling, such that $\omega_3 = \omega_2 + \omega_1$.

The pump wave E_2 is originated at x = 2.5 μm, with an amplitude of A_2 V/m and a frequency of 180 THz.

$$E_2(x = 2.5\mu m, t) = A_2 \times \sin\left(2\pi\left(1.8 \times 10^{14}\right)t + \varphi_2\right) \ V/m$$

The input wave E_1 is originated at x = 2.5 μm, with an amplitude of A_1 V/m and a frequency of 120 THz (see Figure 8).

$$E_1(x = 2.5\mu m, t) = A_1 \times \sin\left(2\pi\left(1.2 \times 10^{14}\right)t + \varphi_1\right) \ V/m \ (assume \ \varphi_1 = 0, \ \varphi_2 = 0 \ )$$

Spatial and temporal simulation parameters $0 \le x \le 10\mu m$, $0 \le t \le 60ps$

Resonance frequency of the medium : $f_0 = 1.1 \times 10^{15} Hz$

Damping rate of the medium : $\gamma = 1 \times 10^{12} Hz$

Dielectric coefficient of the medium $(\varepsilon_\infty) = 1 + \chi = 10$ $(\mu_r = 1)$

Left PML (left absorption layer) is from $x = 0$ to $x = 2.25\mu m$

Right PML (right absorption layer) is from $x = 7.75\mu m$ to $x = 10\mu m$

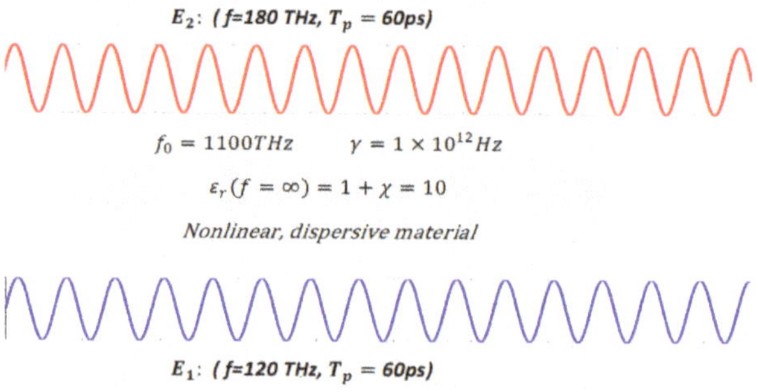

Figure 8. Configuration for the frequency up-conversion in example 1.

The experimentally verified theoretical formula for frequency up-conversion efficiency is given as [1,4]

$$\eta_{theoretical} = \frac{\omega_3}{\omega_2}\left(\sin \sqrt{2d^2n^3\omega_3{}^2(cn\varepsilon_0 A_2{}^2)L^2} \ \right)^2 = \frac{\omega_3}{\omega_2}\left(\sin \sqrt{2d^2n^4\omega_3{}^2 c\varepsilon_0 A_2{}^2 L^2} \ \right)^2$$

$$\omega_2 = Frequency \ of \ the \ pump \ wave \ , \quad \omega_1 = Frequency \ of \ the \ input \ wave$$
$$d = Material \ non-linearity \ coefficient, \ n = \ Refractive \ index \tag{20}$$
$$A_2 = Pump \ wave \ amplitude \ A_1 = \ Input \ wave \ amplitude, \ L = \ Length \ of \ the \ non-linear \ media$$
$$\omega_3 = \omega_1 + \omega_2 = Frequency \ of \ the \ upconverted \ wave$$

Our computational model, which is based on the non-linear electron motion equation, is implemented via FDTD discretization. Coupled with the wave equation, the total wave $E = E_1 + E_2$ is computed from:

$$\frac{E(i+1,j)-2E(i,j)+E(i-1,j)}{\Delta x^2} - \mu_0 \varepsilon_\infty(i,j)\frac{E(i,j+1)-2E(i,j)+E(i,j-1)}{\Delta t^2}$$
$$= \mu_0 \sigma(i,j)\frac{E(i,j)-E(i,j-1)}{\Delta t} + \mu_0 \frac{P(i,j+1)-2P(i,j)+P(i,j-1)}{\Delta t^2}. \tag{21a}$$

$$\frac{P(i,j+1)-2P(i,j)+P(i,j-1)}{\Delta t^2} + \gamma\frac{P(i,j)-P(i,j-1)}{\Delta t} + \omega_0^2(P(i,j)) - \frac{\omega_0^2}{Ned}(P(i,j))^2$$
$$- \frac{\omega_0^2}{N^2 e^2 d^2}(P(i,j))^3 = \frac{Ne^2}{m}(E(i,j)). \tag{21b}$$

For a time interval of $0 \le t \le t_{max}$, the computational formula for frequency up-conversion efficiency is:

$$\eta_{computational} = \frac{\text{Intensity of the } \omega_3 \text{ frequency component of the total wave at t} = t_{max}}{\text{Intensity of the } \omega_2 \text{ frequency component of the total wave at t} = 0} \tag{22}$$

In this example, we have used the following values for each efficiency formula:

$$\omega_2 = (2\pi \times 180)THz, \omega_1 = (2\pi \times 120)THz$$

L = length of the interaction media = 3.33 μm (from x = 3.33 μm to 6.66 μm)
$\omega_3 = frequency\ of\ the\ upconverted\ wave = 2\pi \times 300THz$, n= $Refractive\ index = \sqrt{10}$
d = material non-linearity coefficient = 6.3×10^{-22} (The theoretical and the computational results agree for this value of d for a sample pump wave amplitude of $A_2 = 10^9 V/m$ (see Figure 9). Our aim is to see if the results also agree for all the other pump wave amplitudes for this value of d)

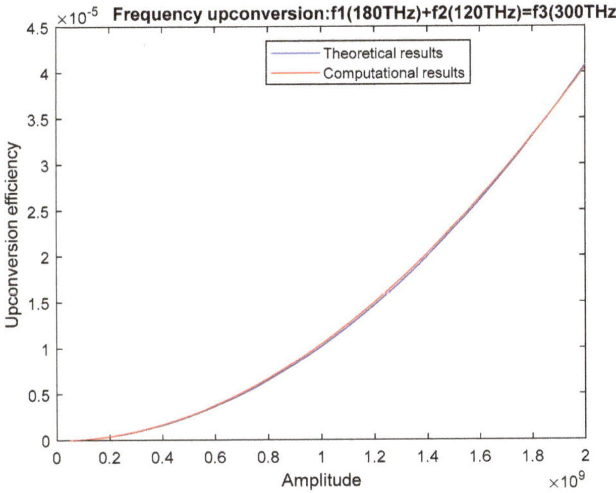

Figure 9. Comparison of the frequency up-conversion efficiencies for $f_3 = 300$ THz and d = 6.3×10^{-22}, versus the pump wave amplitude.

A_2 = pump wave amplitude (varied from $5 \times 10^7 V/m$ to $2 \times 10^9 V/m$)
A_1 = input wave amplitude = $A_2/100$ ($A_1 \ll A_2$)

Example 2. *Optical amplification via non-linear wave mixing (parametric amplification).*

The high-intensity pump wave E_2 is generated at x = 2.5 μm. It has an amplitude of A_2 V/m and a frequency of 220 THz.

$$E_2(x = 2.5\mu m, t) = A_2 \times \sin\left(2\pi\left(2.2 \times 10^{14}\right)t + \varphi_2\right) \ V/m$$

The input wave E_1 is generated at x = 2.5 μm. It has an amplitude of A_1 V/m and a frequency of 140 THz (see Figure 10)).

$$E_1(x = 2.5\mu m, t) = A_1 \times \sin\left(2\pi\left(1.4 \times 10^{14}\right)t + \varphi_1\right) \ V/m \ (assume\ that\ \varphi_1 = 0,\ \varphi_2 = 0)$$

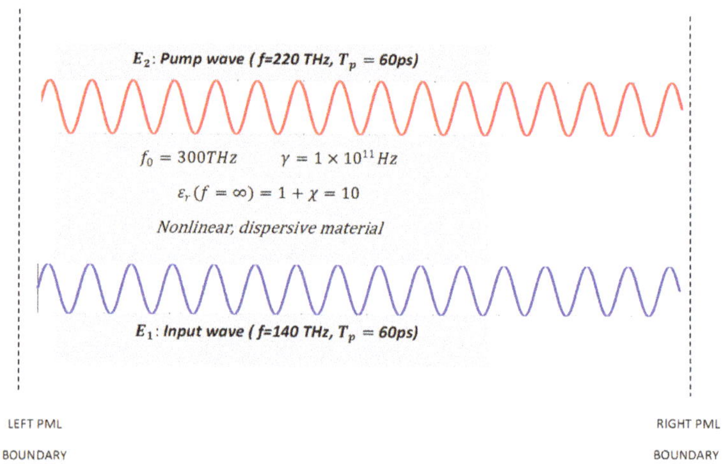

Figure 10. Configuration for optical parametric amplification in example 2.

Range of independent simulation variables: $0 \leq x \leq 13\mu m, \ 0 \leq t \leq 60ps$
Resonance frequency of the interaction medium : $f_0 = 3 \times 10^{14} Hz$
Damping coefficient of the interaction medium : $\gamma = 1 \times 10^{11} Hz$
Dielectric constant of the interaction medium $(\varepsilon_\infty) = 1 + \chi = 10 \quad (\mu_r = 1)$
Spatial range of the interaction medium : $4\mu m < x < 9\mu m$
Left perfectly matched layer (left absorption layer) is from $x = 0$ *to* $x = 2.25\mu m$
Right perfectly matched layer (right absorption layer) is from $x = 10.75\mu m$ *to* $x = 13\mu m$

Our goal is to amplify the input wave via energy coupling from the high-power pump wave. The computational results are obtained by solving (21a,b) and by computing the ratio of the spectral intensity of the input wave for $\omega = \omega_1$ at $t = t_{max}$ to the input wave intensity for $\omega = \omega_1$ at $t = 0$:

$$\eta_{computational} = \frac{\text{Intensity of the } \omega_1 \text{ frequency component of the total wave at } t = t_{max}}{\text{Intensity of the } \omega_1 \text{ frequency component of the total wave at } t = 0} \tag{23}$$

The experimentally verified theoretical formula for parametric amplification, which is derived from the solution of the non-linear wave equation that is based on material non-linearity coefficient, is given as [1,4]:

$$\eta_{theoretical} = \cos h^2\left(Ld \ \sqrt{\omega_1(\omega_2 - \omega_1)\eta^3} \ \sqrt{0.5cn\varepsilon_0 E_{pump}^2}\right) \tag{24}$$

L : *Length of the interaction medium* $= 5\mu m, \ d$: *Nonlinearity coefficient* $= 1.2 \times 10^{-21} C/V^2$

$$\eta = \textit{Intrinsic impedance} = 119.2\Omega \ (ohm)$$

The resulting amplification of the input wave (gain factor) is plotted with respect to the pump wave amplitude based on both the theoretical and the computational formulations (see Figure 11). Notice that the gain factor increases quadratically with the pump wave amplitude. The resulting gain is quite small, as parametric amplification practically requires an interaction medium length on the order of centimeters [16–20], whereas in this example we have an interaction medium length of 5 µm.

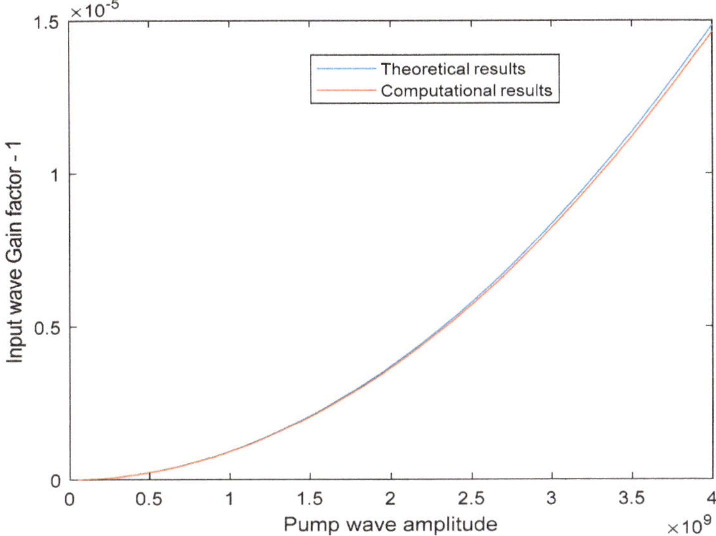

Figure 11. Comparison of the parametric amplification rates with respect to the pump wave amplitude.

8. Discussion

It is important to note that although the presented algorithm is very accurate in obtaining the highest gain factor and determining the optimal frequencies for the ultrashort pump wave pulses, the parameters of the gain medium may limit the maximum achievable gain factor. For example, in our presented numerical experiments, the background permittivity values were set as $\varepsilon_\infty = 10$ and $\varepsilon_\infty = 12$ respectively, these values correspond to the background permittivity of many solid dielectric media in the near infra-red and the visible frequency range. A very large gain factor is also achievable for smaller values of the background permittivity at the expense of a slight increase in ultrashort pump wave intensity to compensate for the decreased stored energy in the micro-resonator. Similarly, for a gain material with a higher background permittivity, the required pump wave intensity can be lowered in order to achieve the same gain factor. Another important parameter that heavily influences the gain factor is the polarization damping rate or simply the damping rate γ. A damping rate that is greater than $\gamma = 10^{11}$ slows down the rate of amplification via non-linear wave mixing and results in a lower gain factor [21].

The frequencies of the ultrashort pump wave pulses can be practically tuned from the far IR (infrared) range to the near UV (ultraviolet) range. Commercial IR laser sources such as the Neodymium-YAG laser, or the Helium-Neon laser can be used in the near IR range and also in the visible range via frequency doubling. A visible laser light can be used in combination with a near IR laser light to generate a UV laser light via the process of sum frequency generation. A far IR laser light and even a THz laser light can be generated using two near IR laser lights of slightly different frequencies, through difference frequency generation process.

Optical amplification via non-linear wave mixing offers the advantage of wide-band amplification of a monochromatic stimulus wave, practically in a frequency interval ranging from the far IR to the near UV part of the spectrum [1,4,19]. Hence, the numerical algorithm presented in Figure 4, can be used to amplify a monochromatic stimulus wave of any frequency that lies between the far IR to the near UV part of the spectrum.

Our main aim in this article is to show that super-gain optical parametric amplification can be achieved in a simple fabry-perot type micro-resonator. For more complicated resonators including microring and microdisc resonators, the one-dimensional model presented in this article must be extended to a two-dimensional model. Our algorithm, which is presented in Figure 4, can be extended to a two-dimensional or a three-dimensional micro-resonator analysis using the technique that is presented in the article referenced in [22]. This article explains how, using FDTD, Maxwell's 2-D and 3-D curl equations in vector form can be effectively stepped in time simultaneously with a system of auxiliary differential equations, in order to create an integrated model that is capable of accurately simulating a medium involving instantaneous non-linearity, dispersive non-linearity, and multiple linear dispersions. Concerning implementation, the Greene-Taflove algorithm given in [22] can easily be incorporated into our algorithm during the discretization of the equations via finite difference time domain method. Hence the algorithm presented in this article can be employed in the analysis of arbitrary-shaped 2-D and 3-D micro-resonators.

9. Conclusions

For high-gain stimulus wave amplification via non-linear coupling in a micro-resonator, a high stored energy and a high polarization density is required. In order to maximize the intracavity energy and the polarization density, for a given resonance frequency (f_0), the frequencies of the high-intensity ultrashort pulses that comprise the pump wave must be tuned accordingly. This can be accomplished via a computationally cost-efficient optimization algorithm such as the quasi-Newton BFGS algorithm. The optimization algorithm must be embedded in the numerical discretization method by choosing the stimulus wave magnitude as the cost function and by modifying it according to the problem constraints.

If the optimal frequencies of the high-intensity ultrashort pulses are set, it is possible to amplify a low-intensity stimulus wave with a very large gain coefficient even inside a micrometer-scale resonator.

Author Contributions: Conceptualization, Ö.E.A.; Methodology, Ö.E.A.; Software, Ö.E.A.; Validation, Ö.E.A, M.K.; Formal analysis, Ö.E.A.; Investigation, Ö.E.A.; Resources, Ö.E.A.; Data curation, Ö.E.A, M.K.; Writing—Original draft preparation, Ö.E.A.; Writing—Review and editing, Ö.E.A, M.K.; Visualization, Ö.E.A.; Supervision, M.K. All authors have read and agreed to the published version of the manuscript.

Funding: This research received no external funding.

Conflicts of Interest: The authors declare no conflict of interest.

References

1. Boyd, R.W. *Nonlinear Optics*; Academic Press: Cambridge, MA, USA, 2008; pp. 105–107.
2. Fox, M. *Optical Properties of Solids*; Oxford University Press: Oxford, UK, 2002; pp. 237–239.
3. Balanis, C.A. *Advanced Engineering Electromagnetics*; John Wiley & Sons: Hoboken, NJ, USA, 1989; pp. 66–67.
4. Saleh, B.E.; Teich, M.C. *Fundamentals of Photonics*; Wiley-Interscience: Hoboken, NJ, USA, 2007; pp. 885–917.
5. Xu, K. Integrated silicon directly modulated light source using p-well in standard CMOS technology. *IEEE Sens. J.* **2016**, *16*, 6184–6191. [CrossRef]
6. Donley, E.A.; Heavner, T.P.; Levi, F.; Tataw, M.O.; Jefferts, S.R. Double-pass acousto-optic modulator system. *Rev. Sci. Instrum.* **2005**, *76*, 036112. [CrossRef]
7. Silfvast, W.T. *Laser Fundamentals*; Cambridge University Press: Cambridge, UK, 2004; pp. 24–35.
8. Satsuma, J.; Yajima, N.B. Initial value problems of one-dimensional self-modulation of nonlinear waves in dispersive media. *Prog. Theor. Phys. Suppl.* **1974**, *55*, 284–306. [CrossRef]
9. Taflove, A.; Hagness, S.C. *Computational Electrodynamics: The Finite-Difference Time-Domain Method*; Artech House: Pimlico, UK, 2005; pp. 353–361.

10. Xu, K. Silicon MOS optoelectronic micro-nano structure based on reverse-biased PN junction. *Phys. Status Solidi-A Appl. Mater. Sci. (PSS-A)* **2019**, *216*, 1800868. [CrossRef]

11. Nocedal, J.; Wright, S. *Numerical Optimization*; Springer: Berlin/Heidelberg, Germany, 2006; pp. 135–149.

12. Maksymov, I.S.; Sukhorukov, A.A.; Lavrinenko, A.V.; Kivshar, Y.S. Comparative Study of FDTD-Adopted Numerical Algorithms for Kerr Nonlinearities. *IEEE Antennas Wirel. Propag. Lett.* **2011**, *10*, 143–146. [CrossRef]

13. Ozgun, O.; Kuzuoglu, M. Near-field performance analysis of locally-conformal perfectly matched absorbers via Monte Carlo simulations. *J. Comput. Phys.* **2007**, *227*, 1225–1245. [CrossRef]

14. Ciriolo, A.G.; Negro, M.; Devetta, M.; Cinquanta, E.; Faccialà, D.; Pusala, A.; De Silvestri, S.; Stagira, S.; Vozzi, C. Optical parametric amplification techniques for the generation of high-energy few-optical-cycles IR pulses for strong field applications. *Appl. Sci.* **2017**, *7*, 265. [CrossRef]

15. Migal, E.A.; Potemkin, F.V.; Gordienko, V.M. Highly efficient optical parametric amplifier tunable from near-to mid-IR for driving extreme nonlinear optics in solids. *Opt. Lett.* **2017**, *42*, 5218–5221. [CrossRef] [PubMed]

16. Wu, C.; Fan, J.; Chen, G.; Jia, S. Symmetry-breaking-induced dynamics in a nonlinear microresonator. *Opt. Express* **2019**, *27*, 28133–28142. [CrossRef] [PubMed]

17. Lee, S.B.; Bark, H.S.; Jeon, T.I. Enhancement of THz resonance using a multilayer slab waveguide for a guided-mode resonance filter. *Opt. Express* **2019**, *27*, 29357–29366. [CrossRef] [PubMed]

18. Aşırım, Ö.E.; Kuzuoğlu, M. Optimization of optical parametric amplification efficiency in a microresonator under ultrashort pump wave excitation. *Int. J. Electromagn. Appl.* **2019**, *9*, 14–34.

19. El Dirani, H.; Youssef, L.; Petit-Etienne, C.; Kerdiles, S.; Grosse, P.; Monat, C.; Pargon, E.; Sciancalepore, C. Ultralow-loss tightly confining Si3N4 waveguides and high-Q microresonators. *Opt. Express* **2019**, *27*, 30726–30740. [CrossRef]

20. Izadi, M.A.; Nouroozi, R. Adjustable propagation length enhancement of the surface plasmon polariton wave via phase sensitive optical parametric amplification. *Sci. Rep.* **2018**, *8*, 15495. [CrossRef]

21. Aşırım, Ö.E.; Kuzuoğlu, M. Numerical study of resonant optical parametric amplification via gain factor optimization in dispersive microresonators. *Photonics* **2020**, *7*, 5. [CrossRef]

22. Greene, J.H.; Taflove, A. General vector auxiliary differential equation finite-difference time-domain method for nonlinear optics. *Opt. Express* **2006**, *14*, 8305–8310. [CrossRef]

 *applied sciences*

Article

A Simple Analytical Solution for the Designing of the Birdcage RF Coil Used in NMR Imaging Applications

Young Cheol Kim [1], Hyun Deok Kim [2,*], Byoung-Ju Yun [2] and Sheikh Faisal Ahmad [1,*]

[1] Institute of Advanced Convergence Technology, Kyungpook National University, 80 Daehak-ro, Buk-gu, Daegu 41566, Korea; yckim@knu.ac.kr
[2] School of Electronics Engineering, College of IT Engineering, Kyungpook National University, 80 Daehak-ro, Buk-gu, Daegu 41566, Korea; bjisyun@ee.knu.ac.kr
* Correspondence: hdkim@knu.ac.kr (H.D.K.); faisalthestar@knu.ac.kr (S.F.A.); Tel.: +82-53-940-8678 (S.F.A.)

Received: 13 January 2020; Accepted: 25 March 2020; Published: 26 March 2020

Abstract: A novel analytical solution for the designing of the birdcage RF coil has been demonstrated in this paper. A new concept of dominant resonance path has been introduced in this paper which is used to identify the specific closed current loop in the birdcage RF coil which is responsible for the dominant resonance frequency mode. This concept is used to determine the precise numerical values of the lumped capacitance deployed in the legs and/or end-rings of the birdcage RF coil for its proper operation at the desired resonance frequency. The analytical solution presented in this paper has been established by performing the two-port network based equivalent circuit modeling of the birdcage RF coil. The proposed analytical solution uses T-matrix theory and develops a relationship between the input impedance of the birdcage coil and the impedances of its leg and end-ring segments. The proposed analytical solution provides the information about the resonance frequency spectrum of the birdcage RF coil and solves the issue of its interfacing with external circuits without affecting its resonance characteristics. Based upon the proposed analysis and designing strategy presented in this paper, the low pass, high pass and band pass configurations of the birdcage RF coil were successfully implemented with FPCB (Flexible Printed Circuit board) technique for small volume NMR imaging applications at 1.5 T and 3.0 T MRI system. The results obtained for the implemented birdcage coils using the proposed analysis and designing technique are in closed agreement with already established methods.

Keywords: MRI system; birdcage coil; birdcage configurations; coil capacitance; analytical solution; equivalent circuit modelling; T-matrix theory; 3D-EM simulation; small volume RF coil

---

1. Introduction

NMR imaging is highly in demand for the medical diagnostics and various other nonclinical and research applications due to its ability of providing very high-quality anatomical images non-invasively [1–3]. A highly sensitive and sophisticated radio frequency (RF) coil (or resonator) is responsible for obtaining these high-quality images only in a highly uniform magnetic field environment of an MRI system [4,5]. One such resonator is known as birdcage RF coil which is a popular choice for the nuclear magnetic resonance (NMR) based volume imaging for medical diagnostics [6]. Having the features like excellent magnetic field homogeneity, high signal-to-noise ratio, design flexibility and the ability to be designed for multi-resonance operation, the birdcage coil is a widely used RF resonator for whole volume NMR imaging applications at high as well as ultra-high field MRI systems [7–13].

The birdcage coil is a closed ladder network which is composed of identical cascaded segments of inductive and capacitive elements [14]. There exist multiple closed current loops in a birdcage coil which are responsible for the simultaneous existence of a definite number of resonance frequency signals. In general, a birdcage coil with N number of cascaded segments can be used to support maximum $N/2 + 2$

resonance frequency modes [15]. Several analytical and numerical solutions have been presented which explain the working of birdcage RF coil quite efficiently and provide the information about its resonance frequency spectrum [15–20]. Most of the existing solutions either involve some tedious techniques with complex mathematical formulation or limited to the determination of the resonance characteristics of the birdcage coil. However, in practical applications, it is required to connect the external circuits like impedance matching circuit (with the input/output ports) and detuning circuits (in the end-ring and/or leg segments) with the birdcage RF coil. This process can seriously affect its resonance frequency spectrum [9]. The existing methods however do not address this fundamental issue of external circuit interfacing with the birdcage coil.

In this paper, we have introduced a new concept of the dominant resonance path for the birdcage RF coil which is a simple and intuitive designing technique. The method works by identifying the closed current loop in the birdcage RF coil who is responsible of causing the dominant resonance frequency which is used for NMR imaging applications [21]. The proposed strategy provides rather convenient way to determine the lumped capacitances for the leg and end-ring segments of the birdcage RF coil for its proper operation at desired dominant resonance frequency. We have also devised a simple analytical solution for the birdcage RF coil by establishing it equivalent circuit model which is based upon the transmission matrix theory [21,22]. The proposed analytical solution provides the mathematical expression for the input impedance Z_{in} of the birdcage RF coil as seen from the port in terms of the impedances of its leg and end-ring segments. Along with explaining the resonance frequency characteristics of the birdcage RF coil, the proposed analytical solution also provides the impedance at any desired position in the coil for the desired resonance frequency. This methodology solves the problems of external circuit interfacing with birdcage coil without disturbing its resonance characteristics.

The paper is divided into two main sections. The first section is about the *analytical solution for the birdcage RF coil* which explains the concept of dominant resonance path in the birdcage RF coil. It is then followed by a novel analytical solution which is derived using simple equivalent circuit modeling-and T-matrix theory. Second section is the *design and analysis of the birdcage RF coil* which contains the results (analytical, simulated and measured) of the successful implementation of the low pass, high pass and band pass configurations of the birdcage RF coil for small volume NMR imaging applications at 1.5 T and 3 T MRI systems by using the proposed analytical solution.

2. Analytical Solution for Birdcage RF Coil

2.1. Configuration of the Birdcage RF Coil

The line diagram of a conventional *N*-segments birdcage RF coil structure and the block diagram for its equivalent circuit model are shown in the Figure 1.

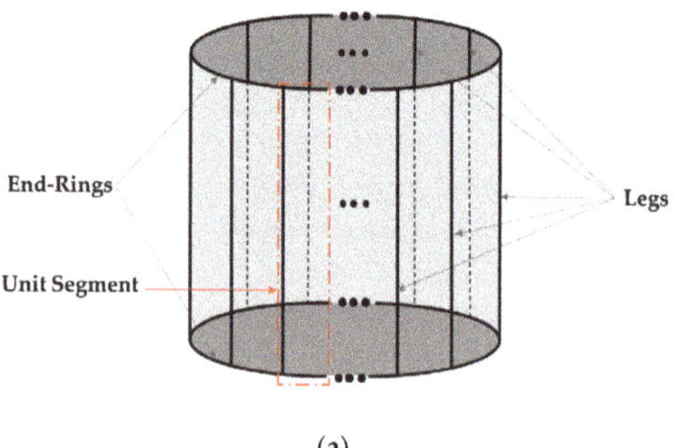

(a)

Figure 1. *Cont.*

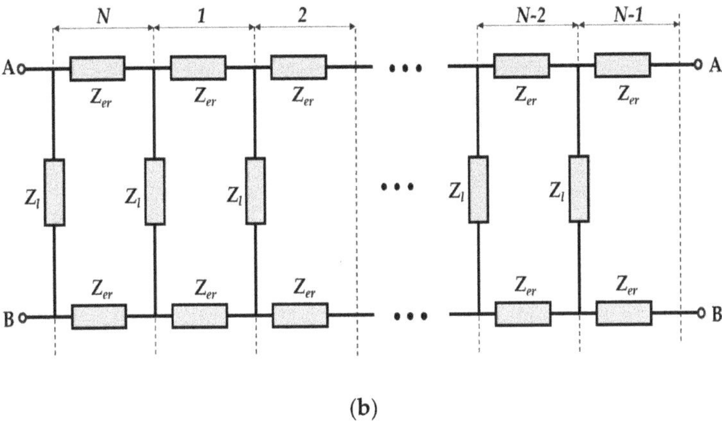

(b)

Figure 1. *N*-segments birdcage RF coil: (**a**) Line diagram; (**b**) Block diagram of the equivalent circuit model.

A conventional birdcage RF is a three-dimensional cylindrical structure which is composed of N (even) number of segments which are connected to each other and are precisely arranged in the axial and azimuthal plans as shown in Figure 1a. The axial element of a segment is known as leg (or rung) which is connected to the conductor segments known as end-ring on both sides in the azimuthal planes. Each segment of a birdcage RF coil contains identical legs with impedance Z_l and identical end-rings with impedance Z_{er} as shown in Figure 1b. The impedances Z_l and Z_{er} are composed of inductive and capacitive elements with negligible resistance (practically less than 1 ohm). The inductive elements of a segment can be the cylindrical wires or the rectangular conductors (of finite thickness or thin foils). These are arranged in the axial and azimuthal planes. The capacitive elements in a segment are the lumped capacitors (with non-magnetic characteristics) which can be inserted in its leg or/and end-rings. Based on the capacitor position in a segment, three configurations of the birdcage RF coil can be realized. These are known as low pass (LP), high pass (HP), and band pass (BP) [14]. A single segment of each configuration is shown in Figure 2.

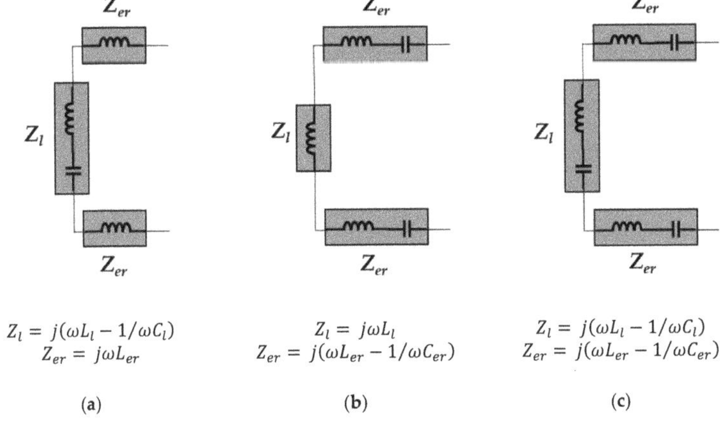

$$Z_l = j(\omega L_l - 1/\omega C_l)$$
$$Z_{er} = j\omega L_{er}$$

$$Z_l = j\omega L_l$$
$$Z_{er} = j(\omega L_{er} - 1/\omega C_{er})$$

$$Z_l = j(\omega L_l - 1/\omega C_l)$$
$$Z_{er} = j(\omega L_{er} - 1/\omega C_{er})$$

(**a**)   (**b**)   (**c**)

Figure 2. Equivalent circuit diagrams of a single segment of the birdcage RF coil configurations: (**a**) Low pass (LP); (**b**) High pass (HP); (**c**) Band pass (BP).

Due to circular ladder structure with multiple cascaded identical segments, there does exist multiple resonance frequencies (some time referred as resonance modes) simultaneously in a conventional birdcage RF coil. The total number of these resonance frequencies mainly depend upon the number of legs (N), the inductances of leg (L_1) and end-ring segment (L_2), and the lumped capacitance in the leg (C_1) and in the end-ring segment (C_2). The first generalized analytic solution to determine the m possible resonance frequency modes (f_m) of an N leg birdcage RF coil was developed by Hayes C.E. in 1985 with the aid of its equivalent circuit analysis is given in Equation (1) [14].

$$f_m = \frac{1}{2\pi} \sqrt{\frac{\frac{2}{C_1} \sin^2 \frac{\pi m}{N} + \frac{1}{C_2}}{L_1 \sin^2 \frac{\pi m}{N} + L_2}} \quad (m = 0,\, 1,\, 2,\, \dots, N/2) \tag{1}$$

The above equation is although for a band pass configuration which however can be converted for low pass and high pass configurations by removing the terms containing C_1 and C_2 respectively. The numerical values of all possible resonance frequency modes of a birdcage coil obtained using Equation (1) are approximate as the terms L_1 and L_2 in Equation (1) represents the self-inductances of leg and end-ring segments. A more accurate but rather complex solution was provided by Jiaming J. in 1991 who also included the mutual inductance effect in the calculations of L_1 and L_2 [23].

2.2. Dominant Resonance Path

Basic principal behind the establishment of the birdcage coil was the development of a circular ladder structure which should be composed of the phase delay lines who can establish a unique phase pattern of the axial current on each coil leg [14]. The inductance (L) and capacitance (C) parameters of leg and end-ring segments are adjusted in such a manner that an ideal sinusoidal distribution of currents in terms of their intensities is realized on the legs of birdcage coil. An ideal sinusoidal intensity profile of currents along with their respective direction in different legs for the dominant resonance mode of an 8-leg birdcage RF coil is shown in Figure 3.

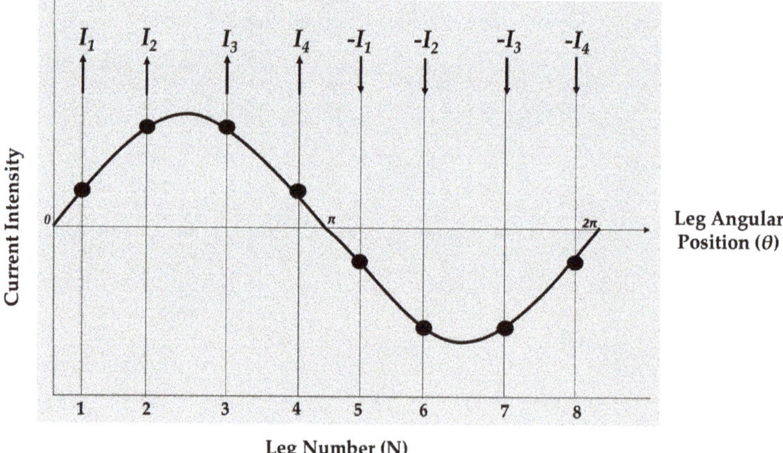

Figure 3. Sinusoidal legs currents distribution on an 8-legs birdcage coil in dominant resonance mode.

This sinusoidal legs currents distribution is responsible of producing the homogeneous non-zero magnetic field everywhere inside the birdcage RF coil [24]. The resonance caused by this current distribution is known as the dominant resonance mode. The direction of currents in different legs of the birdcage RF coil as shown in Figure 3 provides the idea of the closed current loop which is responsible for the dominant resonance mode. The total path length of such closed current loop in the birdcage RF coil can be given by the following Equation (2).

$$P = 2(\pi r + l) \tag{2}$$

where r is radius of the coil and l is the length of its leg. The above path length equation provides the information about the segments of the birdcage RF coil which are involved in the creation of the closed current loop whose resonance frequency is used for NMR imaging operation. A dominant resonance closed current loop consists of two coil legs which are joined by $N/2$ consecutive end-ring segments on each end-ring as shown in Figure 4a.

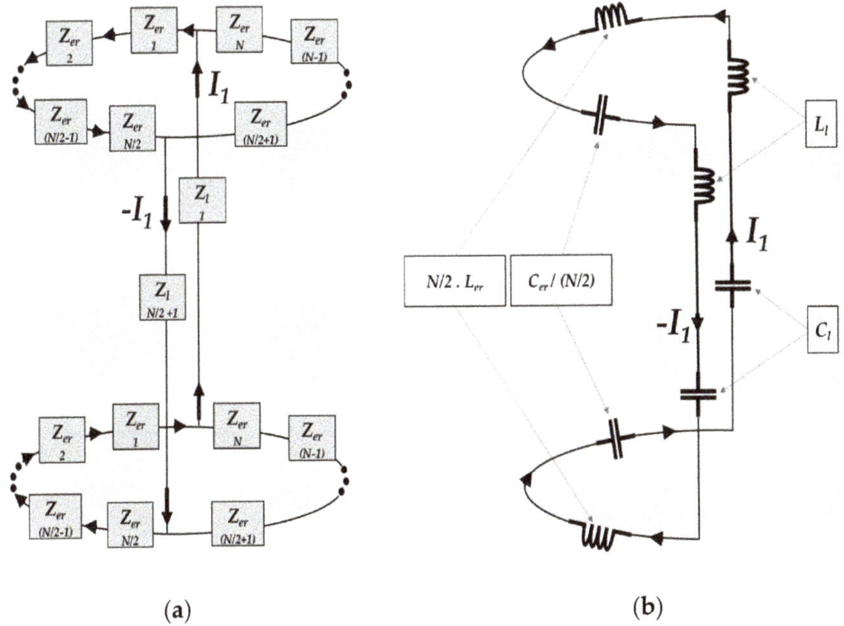

(a)　　　　(b)

Figure 4. A closed current loop in the birdcage RF coil representing: (**a**) The dominant resonance path; (**b**) Total lumped inductance and capacitance of end-rings and legs in the dominant resonance path.

There exists total $N/2$ closed current loops of path length P in birdcage RF coil. Any leg of the birdcage RF coil which is involved in the establishment of a closed current loop dose not involved in the other simultaneously existing $N/2-1$ closed current loops under dominant resonance condition. The desired dominant resonance mode required in the birdcage coil is the resonance of this closed current loop which can be determined as

$$f = \frac{1}{2\pi \sqrt{L_T \cdot C_T}} \tag{3}$$

where L_T and C_T are the total inductance and capacitance of the dominant resonance current loop. The total inductance L_T is the sum of the inductances of two rungs and N end-ring segments which are connected in series as shown in Figure 4b. It can be given as follows.

$$L_T = 2L_l + N \cdot L_{er} \tag{4}$$

Here the L_l and L_{er} are the total inductances of the legs and end-ring segments of a birdcage coil respectively. The numerical value of the self-inductance of a conductor with cylindrical or rectangular geometry can be determined by using the already established following relationships [23].

For a cylindrical conductor of length s and radius r,

$$L = \frac{\mu_0\, s}{2\pi} \cdot \left[\ln \frac{2s}{r} - 1 \right] \tag{5}$$

For a rectangular strip conductor of length s and width w,

$$L = \frac{\mu_0\, s}{2\pi} \cdot \left[\ln \frac{2s}{w} + \frac{1}{2} \right] \tag{6}$$

The total inductance L_T of the dominant resonance loop for all three configurations of the birdcage RF coil which is determined by using Equation (4) remains unchanged. However, a single generalized equation cannot be established to determine the total capacitance C_T of the dominant resonance loop for all three configurations of the birdcage coil. As for the low pass coil, capacitors are present only in the legs, for high pass only in the end rings while for the band pass in both locations, so the relationship to determine the total capacitance in each case can be given as follows,

$$\text{Band Pass }\; C_T = \frac{C_L \cdot C_{ER}}{NC_L + 2C_{ER}} \tag{7}$$

$$\text{Low Pass }\;\; C_T = \frac{C_l}{2} \tag{8}$$

$$\text{High Pass }\;\; C_T = \frac{C_{er}}{N} \tag{9}$$

The lumped capacitors required in the legs and end-rings of the low pass and high pass birdcage RF coil can easily be determined using Equations (8) and (9) respectively for the given coil dimensions and required resonance frequency. However, for the band pass birdcage coil, the end-ring capacitor value is usually needed to be assumed while the leg capacitor value is calculated using Equation (7). Moreover, there exist different combinations of leg and end-ring capacitors for the band pass birdcage coil and the one which causes more homogeneous magnetic field distribution is selected.

2.3. Equivalent Circuit Analysis

Most of the analytical solutions for the birdcage RF coil which are commonly developed by using the basic circuit analysis, transmission line theory, numerical electromagnetics or any other mathematical technique are limited to the determination of the resonance frequency modes only. However, the analytical solutions proposed in this paper provide a comprehensive mathematical formulation for the input port impedance in the leg or end-ring segment of the birdcage RF coil (regardless of its configuration) as a function of frequency. This is used to compute the numerical value of the impedance at any desired position for any desired frequency. This solves the problem of external circuit interfacing with the birdcage coil without effecting its resonance characteristics. The proposed method is based upon the determination of a single equivalent transmission matrix (T_e-matrix) for the total equivalent circuit of the birdcage RF coil.

A conventional birdcage coil is an RF resonator which is composed of identical cascaded segments of inductive and capacitive elements. The conductor segments are the sources of inductance in the birdcage coil. The numerical value of the equivalent inductance (self and mutual) of the conductor segments with respect to its geometry (cylindrical wires or rectangular strips) which is determined by using mathematical equations is used as the lumped inductance [24]. While the lumped capacitors are the capacitive elements of the birdcage coil circuit. The block diagram of a unit segment of the birdcage RF coil consisting of the leg impedance Z_l and the end-ring impedance Z_{er} is shown in Figure 5a. The two ports equivalent circuit of the unit segment which is required to compute the T-matrix of the single segment of birdcage RF coil is also shown in Figure 5b.

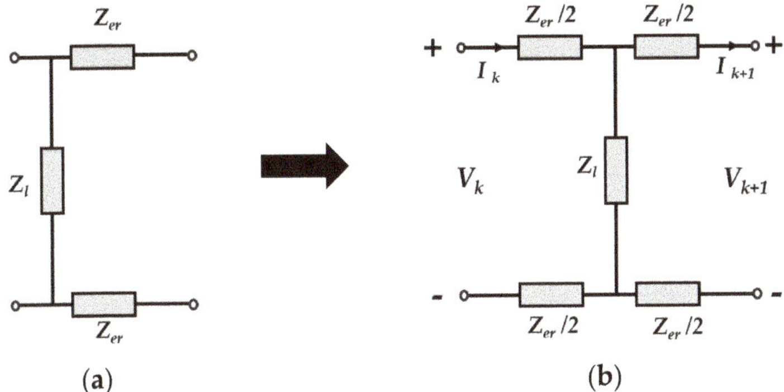

Figure 5. The block diagrams of: (**a**) The unit segment of a conventional birdcage RF coil; (**b**) Two port equivalent circuit of the unit segment.

The impedance of leg segments Z_l and the impedance of the end-ring segments Z_{er} can be given as follows,

$$Z_l = R_l + j\left(2\pi f \cdot L_l - \frac{1}{2\pi f \cdot C_l}\right) \tag{10}$$

$$Z_{er} = R_{er} + j\left(2\pi f \cdot L_{er} - \frac{1}{2\pi f \cdot C_{er}}\right) \tag{11}$$

The input and output voltage and currents of the two-port network of Figure 5b can be related to each other via following matrix equation.

$$\begin{bmatrix} V_k \\ I_k \end{bmatrix} = T \begin{bmatrix} V_{k+1} \\ I_k + 1 \end{bmatrix} \tag{12}$$

where T represents the transmission (ABCD) matrix of a unit segment of the birdcage RF coil that can be obtained by using the following matrix Equation (13).

$$T = \begin{bmatrix} A & B \\ C & D \end{bmatrix} = \begin{bmatrix} 1 + \frac{Z_{er}}{Z_l} & 2Z_{er} + \frac{Z_{er}^2}{Z_l} \\ \frac{1}{Z_l} & 1 + \frac{Z_{er}}{Z_l} \end{bmatrix} \tag{13}$$

In a conventional birdcage RF coil, N number of such identical segments as shown if Figure 5 are connected in cascade with each other. As the symmetry conditions prevail in this circuit, a single transmission matrix T_e can be defined for the equivalent circuit of birdcage RF coil which can represent its overall transmission characteristics. However, the equivalent transmission matrix T_e is a product of the N-1 identical transmission matrices T because the Nth segment (which can be chosen arbitrarily) of the birdcage RF coil is used to establish an interface with the receiver port of the MRI apparatus. Thus, its transmission matrix is not included in the computation of T_e. The single equivalent transmission matrix T_e can be expressed by Equation (14).

$$T_e = T^{(N-1)} = \begin{bmatrix} A_e & B_e \\ C_e & D_e \end{bmatrix} \tag{14}$$

Since an arbitrary transmission matrix can be converted to a two-port network, the block diagram of the equivalent circuit of a birdcage RF coil containing two-port equivalent circuit model of the $N-1$ segments connected in cascade with the two-port equivalent circuit model of the Nth segment is shown in Figure 6.

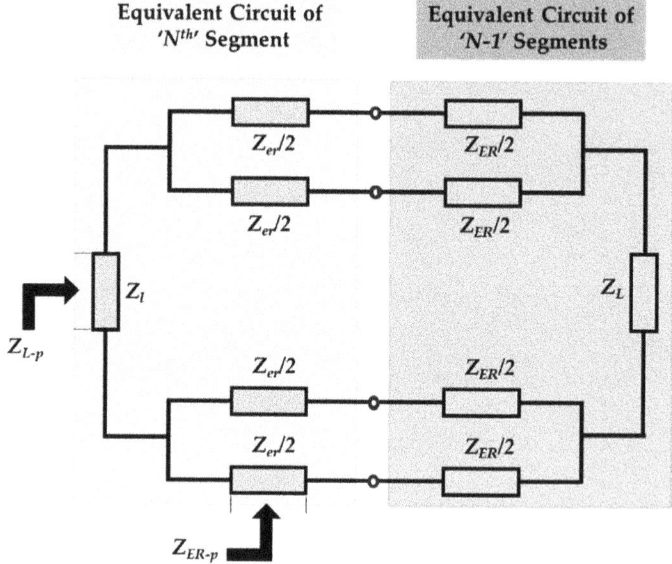

Figure 6. Equivalent circuit model of N segments birdcage RF coil by converting each segment into a two-port network.

The equivalent circuit impedances Z_{ER} and Z_L can be determined by using the elements of the equivalent transmission matrix T_e with the help of following Equations (15) and (16) respectively.

$$Z_L = \frac{1}{C_e} \tag{15}$$

$$Z_{ER} = \frac{A_e - 1}{C_e} \tag{16}$$

In a birdcage RF coil port can be established by interfacing the external circuit across any leg or end-ring segment but it exhibits the similar frequency characteristics. However, the port impedance as viewed from the external circuit would be different at both positions which results in different expressions for the general analytical solution. By considering the port connected across the end-ring segment $Z_{er}/2$, the general analytical solution can be developed by calculating the equivalent impedance of the circuit model shown in Figure 6 as follows.

$$Z_a = \frac{(Z_{er} + Z_{ER})}{2} \tag{17}$$

$$Z_b = \frac{(Z_{er} + Z_{ER})}{4} \tag{18}$$

$$Z_c = \frac{(Z_{er} + Z_{ER})}{4} + (Z_l + Z_L) \tag{19}$$

$$Z_{eq} = \frac{(Z_1 \cdot Z_3)}{(Z_1 + Z_3)} = \frac{(Z_{er} + Z_{ER}) \cdot [4(Z_l + Z_L) + (Z_{er} + Z_{ER})]}{8(Z_l + Z_L) + 6(Z_{er} + Z_{ER})} \tag{20}$$

The final expression of the general analytical solution for the case when the port is created in the end-ring segment is given by the Equation (21).

$$Z_{ER\_p} = \frac{1}{2} \cdot \frac{Z_{er} \cdot \left(Z_{ER} + 2Z_{eq}\right)}{Z_{er} + Z_{ER} + 2Z_{eq}} \tag{21}$$

In a similar way the general analytical solution for the case when the port is created in the leg segment can be derived and the final expression is given by the Equation (22).

$$Z_{L\_p} = \frac{Z_l \cdot (2Z_L + Z_{er} + Z_{ER})}{2(Z_l + Z_L) + (Z_{er} + Z_{ER})} \tag{22}$$

The proposed method allows the input impedance to be obtained directly from the leg or end-ring of the birdcage RF coil. Precisely speaking, unlike the existing solutions for analyzing the relationship between the components values and the resonant frequency, the input impedance can be easily obtained at any point of the birdcage coil. Therefore, the method is not only useful to determine the resonance frequencies of the birdcage RF coil, but also to provide the port impedance which is an essential parameter to be known for interfacing any external circuitry like impedance matching circuit, detuning circuit etc. to the birdcage RF coil.

3. Design and Analysis of the Birdcage RF Coil

3.1. Design Based on Analytical Solution

Birdcage RF coils are designed in various sizes with respect to their length and diameter in according with the specified applications. The main tasks in the designing of a birdcage RF coil is the choice of coil configurations, number of legs, dimensions of legs and end-ring segments, lumped capacitance, and port impedance matching at desired resonance frequency. Legs and end-ring segments conductors are the sources of inductance in the birdcage coil. Their count and dimensions are the controlling factors of the magnetic field homogeneity as well as SAR inside the coil. While on the other hand, the lumped capacitors are responsible for producing the resonance at required radio frequency without effecting the magnetic field distribution. A balance between the conductor size and lumped capacitance value is quite necessary for an appropriate design of the birdcage RF coil. In order to validate the proposed analytical technique, we designed the low pass, high pass and band pass configurations of the linearly polarized birdcage RF coil for ^{1}H NMR imaging in small volume applications at 1.5 T and 3 T MRI systems as shown in Figure 7.

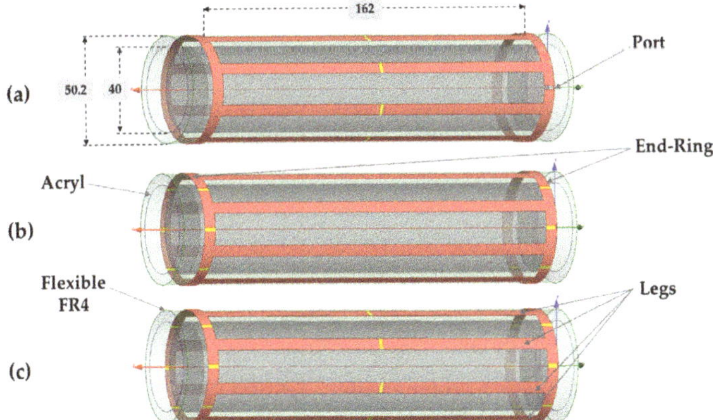

Figure 7. Configurations of the birdcage RF coil for the reception of ^{1}H NMR signal at 1.5 T and 3 T MRI systems: (**a**) Low pass; (**b**) high pass; (**c**) band pass. (All dimensions are in mm).

The designed birdcage RF coils are linearly polarized having 8 legs and 2 end-rings which are made of 35 µm thick and 5 mm wide rectangular copper strips etched on a flexible FR4 substrate ($\varepsilon_r = 4.4$) of 0.2 mm thickness. A 2 mm gap is created in the respective conductor segments of the legs and/or end-rings for SMT lumped capacitors. Total length of a single leg of the birdcage RF coil including capacitor gap is 162 mm. In order to keep the permanent 3D symmetry, the birdcage conductor pattern etched flexible FR4 was further attached on a solid acryl ($\varepsilon_r = 3.2$) cylinder of 5 mm thickness and 40 mm internal diameter. The length and diameter of the coil along with the dimension of the legs and end-rings segments are assumed to be constant for all three configurations of the 1.5 T and 3 T birdcage RF coils.

The self-inductance (referred as calculated inductance L_{cal} in this paper) of the leg and end-rings segments which is constant for all designed coils is determined using Equation (6). However, in the actual scenario there also exist very strong mutual inductance between the parallel conductor segments of the birdcage RF coil which affects the self-inductances of the coil. The self-inductance of the leg and end-ring segment which includes the effect of mutual-inductance was computed using the methods described by the birdcage builder [25]. This self-inductance is referred as effective inductance L_{eff} in this paper. For the given dimensions of the birdcage RF coil the inductance L_{cal} and L_{eff} for the leg and end-ring segment respectively are given in the Table 1 below.

Table 1. Birdcage coil segments dimensions and inductances.

Parameters	Leg	End-Ring
Length (l)	162 mm	19.8 mm
Width (w)	5 mm	5 mm
Calculated Inductance (L_{cal})	151.4 nH	10.2 nH
Effective Inductance (L_{eff})	147.5 nH	12.9 nH

These calculated and effective inductance values were further used to compute the lumped capacitances required in the legs and/or end-rings for all three configurations of the birdcage RF coil at the desired frequencies of 63.8 MHz and 127.7 MHz (for 1.5 T and 3 T MRI systems respectively) by using the Equations (7)–(9). In order to prove the validity of the proposed method, we also computed the lumped capacitances for all three configurations by using the conventional method described by Equation (1), 2D circuit simulation and 3D electromagnetic simulation. The actual values of the lumped capacitance used in the implementation of the prototypes of low pass, high pass and band pass configurations of 1.5 T and 3 T birdcage RF coils were also measured. A comparison among lumped capacitor values determined through different techniques for 1.5 T and 3 T birdcage coils are given below in the Tables 2 and 3 respectively.

Table 2. Lumped capacitor values in pF for 1.5 T birdcage coils.

Method		Low Pass		High Pass		Band Pass ($C_{er} = 200$)	
		C_l	%Error	C_{er}	Error	C_l	Error
Proposed	C_{cal}	32.4	−0.034	129.7	−0.107	92.2	−0.052
	C_{eff}	31.3	−0.023	125.3	−0.063	84	0.03
Conventional	C_{cal}	33.4	−0.044	114	0.05	77.8	0.092
	C_{eff}	32.4	−0.034	111	0.08	72.7	0.143
2D Simulation	C_{cal}	33.5	−0.045	111	0.075	73	0.161
	C_{eff}	32.5	−0.035	108	0.11	72.4	0.146
3D Simulation	C_{3D}	29.4	−0.004	120	−0.01	88	−0.01
Measured	C_{mes}	29		119		87	

Table 3. Lumped capacitor values in pF for 3 T birdcage coils.

Method		Low Pass		High Pass		Band Pass (C_{er} = 50)	
		C_l	%*Error*	C_{er}	*Error*	C_l	*Error*
Proposed	C_{cal}	8.1	−0.01	32.4	−0.034	22.9	−0.023
	C_{eff}	7.8	−0.007	31.3	−0.023	20.9	−0.003
Conventional	C_{cal}	8.35	−0.0125	28.5	0.005	19.4	0.012
	C_{eff}	8.11	−0.0101	27.7	0.013	18.2	0.024
2D Simulation	C_{cal}	8.35	−0.0125	27.7	0.013	18.8	0.018
	C_{eff}	8.12	−0.0102	26	0.03	17.8	0.028
3D Simulation	C_{3D}	7	0.001	29.5	−0.005	20.5	0.001
Measured	C_{mes}	7.1		29		20.6	

The comparison provided in above Tables 2 and 3 concludes that the numerical technique does only provide an approximate value of the lumped capacitor which is quite close to the actual value that is deployed in the prototype implementation. The measured capacitance value for the desired dominant resonance frequency is obtained by slightly tuning the numerically obtained capacitance. It is evident from the information provided in Tables 2 and 3 that in comparison to the conventional methods, the effective lumped capacitance obtained through the proposed method exhibits least numerical error and requires lesser tuning to meet the desired resonance condition. Furthermore, the proposed analytic method described in this paper is quite simple and it does not require the hit and try method for capacitance computation which is commonly adopted in conventioanl 3D EM simulations. The results obtained through such analytical technique can directly be used in the coil implementation engineering which also includes the impedance matching operation at the desired resonance frequency.

3.2. Resonance Frequency

The proposed analytical solution given in Section 1 does also provide the information about the resonance frequency modes for all three configurations (low pass, high pass and band pass) of the birdcage RF coil by plotting its input impedance Z_{ER_p} against frequency. We assumed the case where the receiver port is connected across the half end-ring segment and then determined the resonance frequencies spectrum of all three coil configurations using Equation (21). The effective inductance values of the leg and end-ring segments along with their corresponding effective capacitance values (with respect to the coil configuration) were used to compute the impedances Z_l and Z_{er}. A comparison between the input impedance obtained though the analytical solution proposed in this paper and the input impedance plot obtained through 2D circuit simulation with effective inductance and their corresponding capacitance values in ANSYS [26] for each configuration of 1.5 T and 3 T birdcage RF coils are given in the Figure 8 below.

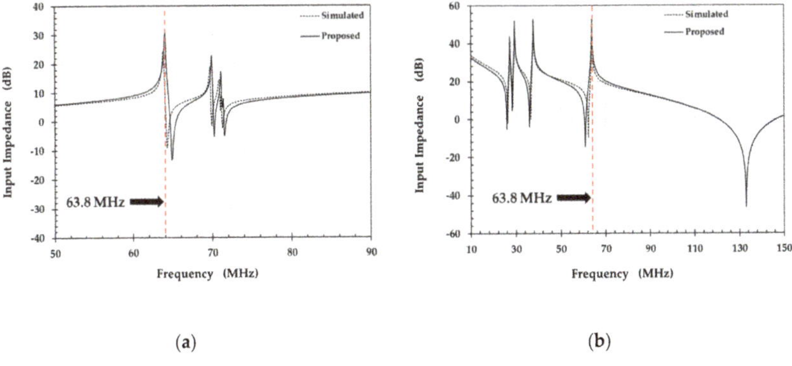

Figure 8. *Cont.*

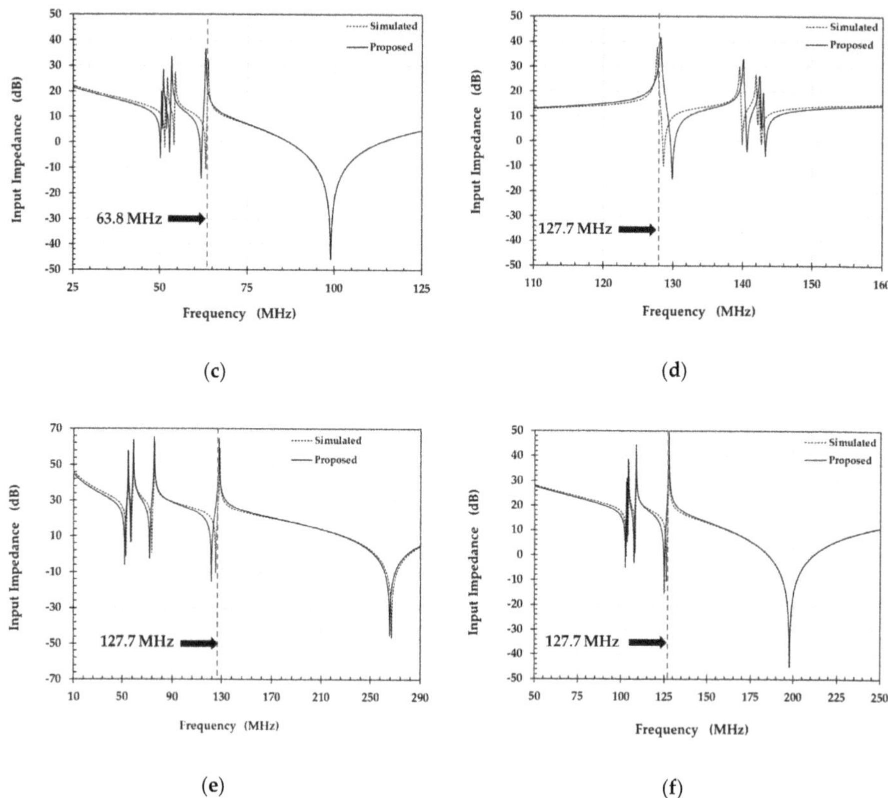

Figure 8. Input impedance obtained using proposed analytical solution and conventional 2D simulation for the different configurations of the birdcage RF coil: (**a**) 1.5T low pass, (**b**) 1.5 T high pass, (**c**) 1.5 T band pass, (**d**) 3 T low pass, (**e**) 3 T high pass, (**f**) 3 T band pass.

3.3. Impedance Matching

It can be observed in all cases of Figure 8 that the pattern of variations in the input impedance (at the dominant resonance mode as well as the higher order resonance modes) provided by the proposed analytical solution does perfectly coincide with the one which was obtained using conventional 2D circuit simulation technique. The numerical values of the input impedance Z_{in} at the desired resonance frequency is further used to design the impedance matching circuit which serves as an interface between the RF coil and receiver circuit. A conventional "L-matching" network with lumped (series) inductance and (parallel) capacitance can be used to accomplish this task. The input impedance Z_{in} serves as the load impedance which is used to obtain the required values of the lumped components of the impedance matching network connected between the desired segment (end-ring in this case) of the birdcage RF coil and the 50 ohm port of the receiver. The numerical values of the input impedance Z_{ER_p} as obtained via proposed analytical solution along with the corresponding values of the lumped components of the matching circuit for all configurations of the designed 1.5 T and 3 T birdcage RF coils are provided in the Table 4.

Table 4. Input impedance Z_{ER_p} of the designed birdcage RF coils at dominant resonance frequency and the corresponding lumped components of the matching circuit for 1.5 T and 3T MRI systems.

MRI System	Coil Configuration	Z_{ER_p} (Ohm)	C_m (pF)	L_m (nH)
	Low Pass	$34.42 - j10.86$	33.7	85.6
1.5 T	High Pass	$439.88 - j37.44$	15.3	350
	Band Pass	$68.65 - j29.33$	10.8	98.2
	Low Pass	$131.74 - j54.57$	8.3	90
3.0 T	High Pass	$13.23 + j137.45$	13.5	333
	Band Pass	$82.20 + j130.45$	16.3	136.3

The reflection coefficient S11 which is a core parameter to explains the impedance matching of an RF circuit can easily be determined with the help of input impedance by using the following Equation (23).

$$S11 = \frac{\left(Z_{ER_p} - 50\right)}{\left(Z_{ER_p} + 50\right)} \tag{23}$$

A comparison between the reflection coefficients obtained by using the proposed analytical solution in the absence as well as presence of the impedance matching circuit for all configurations of the implemented 1.5 T and 3 T birdcage RF coils is provided in Figure 9.

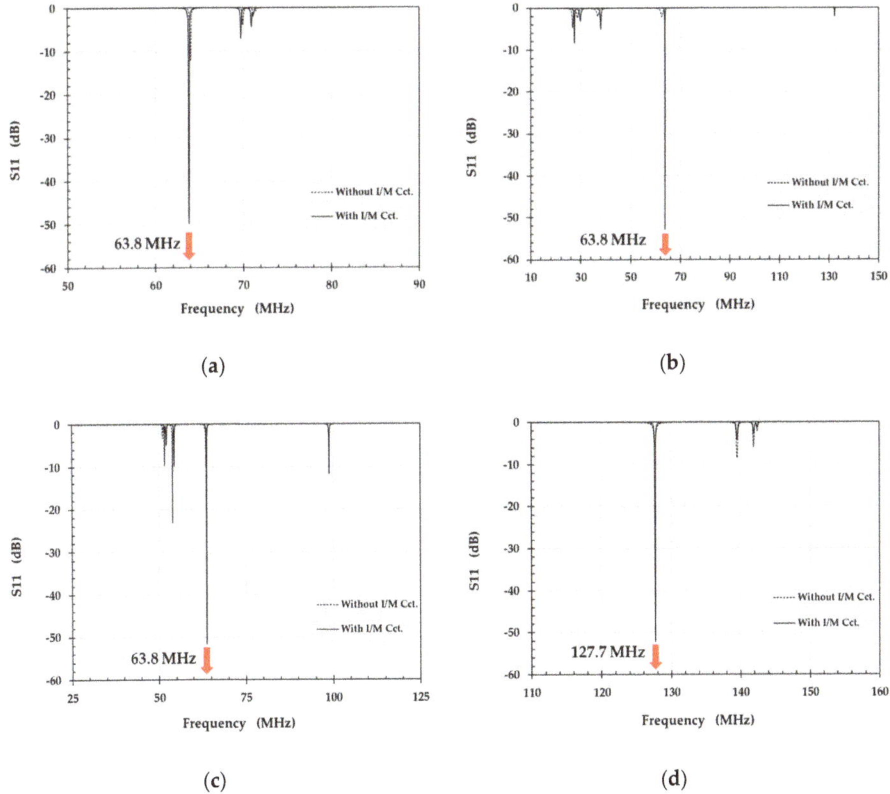

Figure 9. *Cont.*

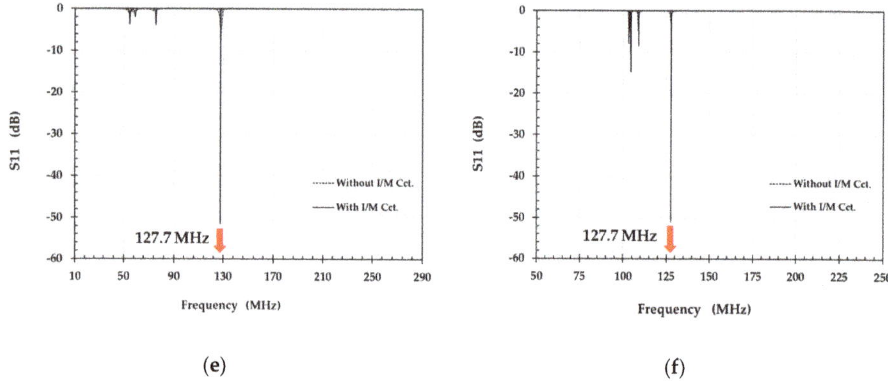

(e)   (f)

Figure 9. Reflection coefficient S11 for the different configurations of the birdcage RF coil without and with impedance matching circuit: (**a**) 1.5 T low pass, (**b**) 1.5 T high pass, (**c**) 1.5 T band pass, (**d**) 3 T low pass, (**e**) 3 T high pass, (**f**) 3 T band pass.

3.4. Magnetic Field Homogeneity

Homogeneity of the magnetic field at the desired dominant resonance frequency (63.8 MHz for 1.5 T and 127.7 MHz for 3 T) everywhere inside the designed birdcage coils using the proposed analytical solution can be observed via 3D electromagnetic simulation. The simulation models of the birdcage RF coils were created in according with the dimensional information provided in Section 3.1 using 3D electromagnetic simulation software ANSYS HFSS [26]. The magnetic field distribution inside the coil under no load conditions (assuming air inside the coil) as determined in the transverse plans at different positions for each coil configuration is shown in Figure 10 below.

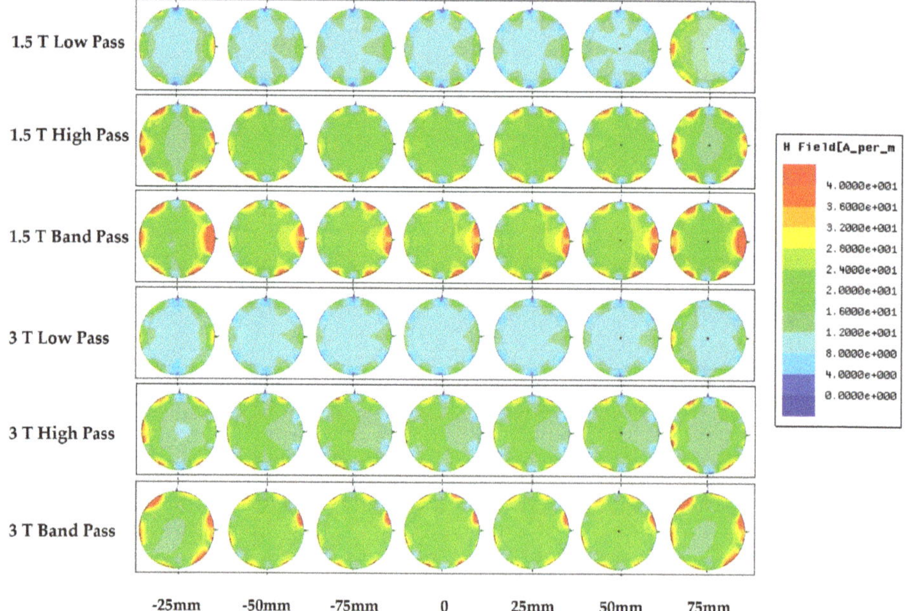

Figure 10. Magnetic field distribution at dominant resonance frequency in transvers planes at various positions for all configurations of 1.5 T and 3 T birdcage RF coils.

It can be observed in Figure 10 above that each coil configuration causes quite homogeneous magnetic field distribution at the dominant resonance frequency while the bandpass configuration produces rather strong magnetic field than others for both 1.5 T and 3 T cases.

4. Conclusions

A simple technique for the designing and analysis of the birdcage RF coil based upon dominant resonance path and a novel analytical solution was demonstrated in this paper. We introduced a new concept of the dominant resonance path in the birdcage RF coil that identifies the specific closed current loop which is responsible of causing the dominant resonance frequency that is desired for NMR imaging. The concept is used to determine the numerical values of the lumped capacitances for the leg and/or end-ring segment of the birdcage RF coil using simple mathematical formulations. We also provided the analytical solutions for the birdcage RF coil in terms of its input impedance Z_{in} by converting each of its section into a two-port network. The transmission parameters are used to determine the final analytical solution with the help of an equivalent T_e matrix for the N-1 identical cascaded segments of the birdcage RF coil. Two separate analytical solution were developed by considering the feed port or external circuit location in the end-ring segment and leg segment. Both analytical solutions efficiently explain the characteristics of the birdcage RF coil, however the commonly used end-ring feed based analytical solution was mainly discussed in this paper. We implemented the birdcage RF coil in low pass, high pass and band pass configurations with FPCB etched conductor pattern technique for small volume NMR imaging applications at 1.5 T and 3.0 T MRI system. The numerical values of the lumped capacitance for the different configurations of the birdcage RF coil provided by the dominant resonance path method were found more accurate in comparison to those which were obtained through the conventional methods. The analysis of the implemented birdcage RF coils performed with the proposed analytical solution were also in complete agreement with the conventional methods.

Author Contributions: S.F.A. and H.D.K. conceived the idea, developed the analytical solution and carried out simulations. Y.C.K. designed the coils completed the experiments. B.-J.Y. developed the programming code for the implementation of the proposed analytical solution. H.D.K. analyzed the data and S.F.A. wrote the paper. All authors have read and agreed to the published version of the manuscript.

Funding: This research was financially supported by the Ministry of Trade, Industry, and Energy (MOTIE), Korea, under the "Regional industry based organization support program" (reference number R0004072) and the BK21 Plus project funded by the Ministry of Education (21A20131600011).

Conflicts of Interest: The authors declare no conflict of interest.

References

1. Mansfield, P. Imaging by nuclear magnetic resonance. *J. Phys. E Sci. Instrum.* **1988**, *21*, 18. [CrossRef]
2. Mallard, J.; Hutchison, J.M.; Edelstein, W.; Ling, R.; Foster, M. Imaging by nuclear magnetic resonance and its bio-medical implications. *J. Biomed. Eng.* **1979**, *1*, 153–160. [CrossRef]
3. Blow, N. How to get ahead in imaging. *Nature* **2009**, *458*, 925–928. [CrossRef] [PubMed]
4. Haase, A.; Odoj, F.; Kienlin, M.V.; Warnking, J.; Fidler, F.; Weisser, A.; Nittka, M.; Rommel, E.; Lanz, T.; Kalusche, B.; et al. NMR probeheads for in vivo applications. *Concepts Magn. Reson.* **2000**, *12*, 361–388. [CrossRef]
5. Gruettrt, G.; Weisdorf, S.A.; Rajanayagan, V.; Terpsta, M.; Merkle, H.; Truwit, C.L.; Garwood, M.; Nyberg, S.L.; Uğurbil, K. Resolution improvements in in vivo 1H NMR spectra with increased magnetic field strength. *J. Magn. Reson.* **1998**, *135*, 260–264. [CrossRef]
6. Hayes, C.E. The development of the birdcage resonator: A historical perspective. *NMR Biomed.* **2009**, *22*, 908–918. [CrossRef]
7. Ahmad, S.F.; Kim, H.D. FPCB-based birdcage-type receiving coil sensor for small animal 1H 1.5T magnetic resonance imaging system. *J. Sens. Sci. Technol.* **2017**, *26*, 245–250.
8. Yeung, D.; Hutchison, J.M.S.; Lurie, D.J. An efficient birdcage resonator at 2.5 MHz using a novel multilayer self-capacitance construction technique. *MAGMA* **1995**, *3*, 163–168. [CrossRef]

9.	Ahmad, S.F.; Kim, Y.C.; Choi, I.C.; Kim, H.D. Birdcage type NMR receiver coil sensor with integrated detuning circuit for 3T MRI system. In Proceedings of the IEEE SENSORS 2015, Busan, Korea, 1–4 November 2015; IEEE: Piscataway, NJ, USA, 2015; p. 2038.

10.	Solis, S.E.; Cuellar, G.; Wang, R.R.; Tomasi, D.; Rodriguez, A.O. Transceiver 4-leg birdcage for high field MRI: Knee imaging. *Rev. Mex. Fis.* **2008**, *54*, 215–221.

11.	Pimmel, P.; Briguet, A. A Hybrid Bird Cage Resonator for Sodium Observation at 4.7 T. *Magn. Reason. Med.* **1992**, *24*, 158–162. [CrossRef]

12.	Wang, C.; Li, Y.; Wu, B.; Xu, D.; Nelson, S.J.; Vigneron, D.B.; Zhang, X. A practical multinuclear transceiver volume coil for in-vivo MRI/MRS at 7T. *Magn. Reason. Imaging* **2012**, *30*, 78–84. [CrossRef]

13.	Bernard, J.D.; Shizhe, L.; Christopher, M.C.; Gerald, D.W.; Michael, B.S. A birdcage coil tuned by RF shielding for applications at 9.4T. *J. Magn. Reason.* **1998**, *131*, 32–38.

14.	Hayes, C.E.; Edelstein, W.A.; Schenk, J.F.; Muller, O.M.; Eash, M. An efficient, highly homogeneous radiofrequency coil for whole-body NMR imaging at 1.5 T. *J. Magn. Reason.* **1985**, *63*, 622–628. [CrossRef]

15.	Leifer, M.C. Resonant modes of birdcage coil. *J. Magn. Reason.* **1997**, *124*, 51–60. [CrossRef]

16.	Tropp, J. The Theory of the bird-cage resonator. *J. Magn. Reason.* **1989**, *82*, 51–62. [CrossRef]

17.	Tropp, J. The Theory of an arbitrarily perturbed bird-cage resonator and a simple method for restoring it to full symmetry. *J. Magn. Reason.* **1991**, *95*, 235–243. [CrossRef]

18.	Pascone, R.J.; Garcia, B.J.; Ftizgerald, T.M.; Vullo, T.; Zipagan, R.; Cahill, P.T. Generalized electrical analysis of low-pass and high-pass birdcage resonators. *Magn. Reason. Imaging* **1991**, *9*, 395–408. [CrossRef]

19.	Giovannetti, G.; Landini, L.; Santarelli, F.M.; Positano, V. A fast and accurate simulator for the design of birdcage coils in MRI. *Magn. Reson. Mater. Phys. Biol. Med.* **2002**, *15*, 36–44. [CrossRef] [PubMed]

20.	Novikov, A. Advanced theory of driven birdcage resonator with losses for biomedical magnetic resonance imaging and spectroscopy. *Magn. Reason. Imaging* **2011**, *29*, 260–271. [CrossRef] [PubMed]

21.	Ahmad, S.F.; Kim, Y.C.; Choi, I.C.; Choi, S.W.; Kim, T.G.; Ahn, C.M.; Kim, H.D. Fast and efficient birdcage coil design process for high field MRI system by combining equivalent circuit model and 3D electromagnetic simulation. In Proceedings of the 2017 Asia Modelling Symposium (AMS), Kota Kinabalu, Malaysia, 4–6 December 2017; IEEE: Piscataway, NJ, USA, 2017; pp. 188–194.

22.	Kim, H.D. Analysis of the bird-cage receiver coil for MRI system employing a equivalent circuit model based on transmission matrix. *J. Korea Multimed. Soc.* **2017**, *20*, 1024–1029.

23.	Jianming, J. Analysis and design of RF coils. In *Electromagnetic Analysis and Design in Magnetic Resonance Imaging*, 1st ed.; CRC Press: Boca Raton, FL, USA, 1999; pp. 57–164.

24.	Mispelter, J.; Lupu, M. Homogeneous resonators for magnetic resonance: A review. *C. R. Chim.* **2008**, *11*, 340–355. [CrossRef]

25.	Chin, C.L.; Collins, C.M.; Li, S.; Dardzinski, B.J.; Smith, M.B. BirdcageBuilder: Design of specified-geometry birdcage coils with desired current pattern and resonant frequency. *Concepts Magn. Reson.* **2002**, *15*, 156–163. [CrossRef] [PubMed]

26.	ANSYS Electronics Desktop. Available online: http://www.ansys.com/products/electronics/ansys-electronics-desktop (accessed on 20 December 2019).

Article

Electromagnetic Scattering from Surfaces with Curved Wedges Using the Method of Auxiliary Sources (MAS)

Vissarion G. Iatropoulos, Minodora-Tatiani Anastasiadou and Hristos T. Anastassiu *

Department of Informatics, Computer and Communications Engineering, International Hellenic University, Serres Campus, End of Magnisias Str., GR-62124 Serres, Greece; viss.iatr@gmail.com (V.G.I.); minodoratatiana@hotmail.com (M.-T.A.)
* Correspondence: hristosa@teiser.gr

Received: 17 February 2020; Accepted: 23 March 2020; Published: 27 March 2020

Abstract: The method of auxiliary sources (MAS) is utilized in the analysis of Transverse Magnetic (TM) plane wave scattering from infinite, conducting, or dielectric cylinders, including curved wedges. The latter are defined as intersections of circular arcs. The artificial surface, including the auxiliary sources, is shaped in various patterns to study the effect of its form on the MAS accuracy. In juxaposition with the standard, conformal shape, several deformations are tested, where the auxiliary sources are forced to approach the tip of the wedge. It is shown that such a procedure significantly improves the accuracy of the numerical results. Comparisons of schemes are presented, and the optimal auxiliary source location is proposed.

Keywords: method of auxiliary sources (MAS); electromagnetic scattering; wedge; numerical methods; accuracy

---

1. Introduction

Computational Electromagnetics techniques have traditionally been utilized in the mathematical modeling of problems related to radiation and the propagation of electromagnetic waves. Standard applications involve antennas, diffraction, scattering, waveguides, propagation in complex environments, etc., while boundary surfaces are typically either perfectly conducting or dielectric. Recently, important technological advances have been based on the electromagnetic properties of Near-Zero Index (NZI) materials [1,2], plasmonics [3], metasurfaces [4], graphene [5], nanoparticles [6], etc.

Analytical and asymptotic methods were originally the only mathematical tool to perform relevant calculations, naturally being restricted to canonical geometries. However, computers demonstrated explosive progress over the last few decades, thus rendering numerical techniques indispensable in extracting approximate results of controllable accuracy for arbitrary geometries. In general terms, numerical techniques discretize either integral [7,8] or differential equations [9,10] involved in the mathematical simulation of physical problems. A variant of differential equation techniques is based on the construction of equivalent circuit or transmission line models [11–13]. Each group of the aforementioned techniques is characterized by advantages and disadvantages, and therefore their suitability and efficiency depend on the particular configuration. For example, in scattering problems, where the Radar Cross Section (RCS) of a target must be calculated, differential equation methods should ideally discretize the entire space all the way to infinity, which is obviously impossible. Mesh truncation is a complicated task that needs to be resolved in order to obtain accurate results. On the contrary, integral equation methods are usually more complicated in terms of calculations, and the computational cost may be exceedingly high due to dense matrices involved in the resulting linear system of equations. However, field behavior at infinity is automatically incorporated into the Green's function, and therefore modeling the entire space is easily tractable. Integral equation solutions and their procedure optimization is an important research topic, which recently triggered some especially unconventional approaches [14].

The method of auxiliary sources (MAS) [15] is an integral equation technique that has successfully been used in computational physics, including a wide range of electromagnetic radiation and scattering applications [16]. MAS is superficially similar to the point matching version of the method of moments (MoM) [7]; however, the auxiliary current sources are located at a distance from, and not right on the surface boundaries. Moreover, the MAS basis functions set, used in the field expansions, has been proven to be complete [17], which is not always easy to prove in MoM, therefore the method is mathematically rigorous. Finally, unlike MoM, MAS is free of singularity complications, it does not require any time-consuming numerical integrations and is much easier to implement algorithmically. A quantitative assessment of MAS performance is given in [18], where a detailed comparison is carried out between the computational costs of MAS and MoM. It is demonstrated that for the same discretization density, MAS is always more efficient than MoM, but even if a higher number of unknowns is required for MAS, the latter is still less computationally intensive than MoM under certain conditions.

Although MAS has been invoked in several problems with various configurations and material properties, further research is necessary to determine the optimal source location for arbitrary geometry layouts. Particular difficulties arise when the outer boundary of the scatterer contains wedges, i.e., when the analytical expression of the boundary curve is not differentiable. In that case, the solution accuracy deteriorates because the boundary condition close to the wedge tip is hard to satisfy sufficiently well. To apply MAS to such configurations, a set of auxiliary sources (ASs) is positioned on a fictitious surface, which is generally conformal to the physical boundary, except in the vicinity of the tips. In the areas surrounding the wedges, the ASs are densely packed, simultaneously approaching the tips, to account for the edge effects, as suggested in [19]. Similar strategies were employed in the case of a scattering problem associated with coated perfectly electric conducting (PEC) surfaces including wedges [20], where the surface was modeled via the standard impedance boundary condition (SIBC) [21].

While this deformation of the auxiliary surface has proven efficient for straight wedges, especially when the latter form right angles, no evidence is known from the literature about its applicability to arbitrarily shaped wedges. The aim of this paper is to investigate whether MAS accuracy improves through this deformation, when the wedge is shaped as an intersection of circular arcs with non-coincident centers. Moreover, as an improvement over the heuristic approaches in [19,20], explicit algorithms should be developed for the definition of the deformed auxiliary surface. The scatterer is thus defined as an infinite cylinder, either conducting or dielectric, with an eye-shaped cross-section. The auxiliary surface is generally retained as conformal to the scatterer's boundary, except in the neighborhood of the wedge tips, where various deformation schemes are employed and accuracy comparisons are drawn.

The format of this paper, which is an extension of [22], including additional data on perfectly conducting case, lossless, and lossy dielectrics, is as follows: Section 2 briefly presents the mathematical formulation of MAS for 2D scatterers, illuminated by a transverse magnetic (TM) polarized plane wave. Furthermore, in Section 3 various algorithms are proposed for the deformation of the auxiliary surfaces close to the wedge tips. Section 4 includes a variety of data for the eye-shaped scatterer and checks the satisfaction of the boundary condition. Finally, the method is summarized and discussed, whereas useful conclusions are drawn.

A $+j\omega t$ time variation convention is assumed and suppressed throughout the paper.

2. MAS for Eye-Shaped Scatterers

We assume an infinitely long cylinder with a cross section that resembles an eye (Figure 1). We investigate two separate cases: perfect electric conductor (PEC) or linear- homogeneous-isotropic dielectric. The geometry of the scatterer, depicted by a solid line, comprises two circular arcs with identical radii equal to ρ, but with different centers. In particular, the Cartesian coordinates of the upper arc are given by (1):

$$x_u = \rho \cos \varphi, \ y_u = \rho \sin \varphi - d,$$ (1)

whereas those of the lower arc are given by (2):

$$x_l = \rho \cos \varphi, \ y_l = \rho \sin \varphi + d,$$ (2)

where φ is the azimuth angle and $\pm d$ is the vertical displacement of each arc center, taken equal to the arc apothem (see Figure 2). Obviously, φ does not span the entire $[0, 2\pi)$ interval, but it is limited by the arc width itself, which is given by (3):

$$\varphi_{arc} = 2\arccos\frac{d}{\rho}.$$ (3)

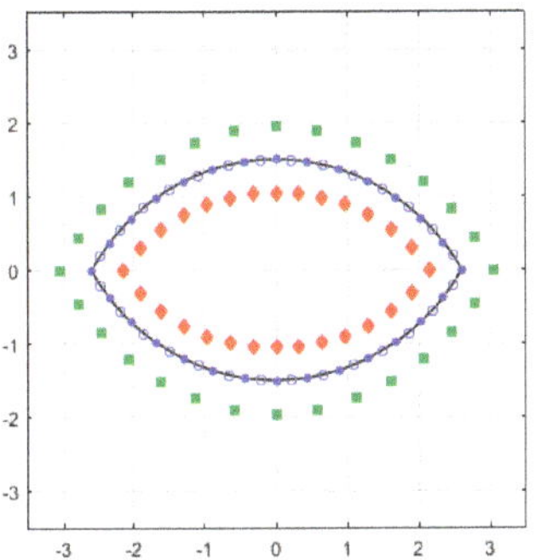

Figure 1. Geometry of the scatterer (solid line), including inner auxiliary sources (ASs) (diamonds) and outer auxiliary sources (ASs) (squares). The dots stand for collocation points (CPs) and circles for midpoints (MPs).

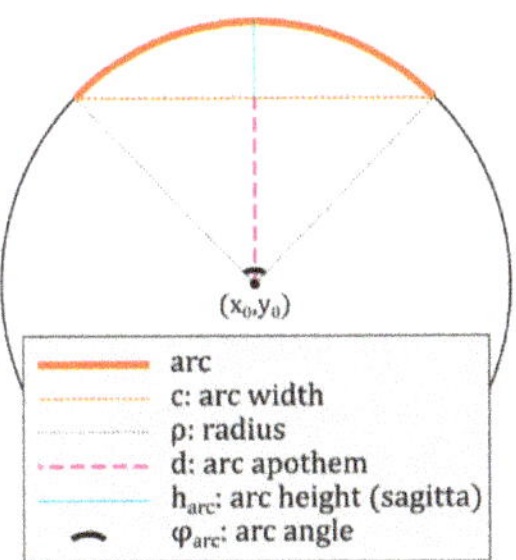

Figure 2. Construction of the geometry.

The scatterer is illuminated by a TM plane wave impinging from an azimuth angle equal to φ_{inc}. Therefore, the incident electric field E_{inc} is given by (4):

$$E_{inc}(x, y) = E_0 \exp\{jk_0(x \cos \varphi_{inc} + y \sin \varphi_{inc})\}\hat{z},$$ (4)

where E_0 is the amplitude of the incident electric field, k_0 is the free space wavenumber, and a hat denotes a unit vector along the corresponding direction. The incident magnetic field H_{inc} is given by (5):

$$H_{inc}(x,y) = -\frac{E_0}{\zeta_0}(\sin \varphi_{inc}\hat{x} - \cos \varphi_{inc}\hat{y}) \exp\{jk_0(x \cos \varphi_{inc} + y \sin \varphi_{inc})\},\tag{5}$$

where ζ_0 is the free space intrinsic impedance. To solve the scattering problem via MAS, two sets of ASs are generally defined, each one of multitude N, as shown in Figure 1. In the PEC case, only the inner set is used; the outer set is necessary only in the dielectric configuration. In standard MAS formulation, both inner and outer auxiliary surfaces are conformal to the scatterer boundary. The electric field due to the nth inner AS, located at point r_n and radiating in the outer space, is as follows:

$$E_{sn}(r) = \hat{z}E_nH_0^{(2)}(k_0|r - r_n|),\tag{6}$$

where E_n is the corresponding unknown weight, $(n = 1, 2, \ldots, N)$, and $H_0^{(2)}$ is the Hankel function of the zero order and second kind (2D Green's function). The corresponding magnetic field of the nth auxiliary source is obviously proportional to the curl of (6), given explicitly in [20]. Similar expressions hold for the outer ASs, radiating in the inner space of the dielectric scatterer, except for k_0 and ζ_0, which have to be replaced by k and ζ, respectively, corresponding to the scatterer's dielectric properties. The total scattered E field is expressed as the superposition of the fields in (6) and the H field accordingly. By applying the boundary conditions for both fields at the N collocation points (CPs) (x_m, y_m) $(m = 1, 2, \ldots, N)$ of the scattering boundary (blue solid dots in Figure 1), we cast a linear system of equations as in (7):

$$[Z]\{I\} = \{V\},\tag{7}$$

where $\{I\}$ is the column vector of the unknown weights E_n. In the PEC case, $[Z]$ is a square matrix of size $N \times N$ with elements determined by the interaction between ASs and CPs, and $\{V\}$ is the column vector of the incident E fields calculated at the CPs. In the dielectric case, the $[Z]$ interaction matrix is of size $2N \times 2N$ and $\{V\}$ is the column vector of both the incident E and H fields calculated at the CPs.

3. Improvement of the Auxiliary Surface Layout

As mentioned in [19,20], MAS becomes less accurate when the auxiliary surfaces are conformal to boundaries encompassing wedges. In particular, satisfaction of the boundary condition at midpoints (MPs) (see Figure 1) close to the tips is no longer sufficiently good. To overcome this complication, the auxiliary surface, defined by radius ρ_{aux} and using expressions analogous to (1-3), may be deformed so that ASs not only approach the tips closely, but become denser in the tip neighborhood as well. ASs may approach CPs following several patterns. In this work, two basic patterns were tested. Let M be the number of ASs to be moved. Let ρ_m be the polar radius of the mth AS $(m = 1, 2, \ldots, M)$, and let g be the maximum polar radius difference between the mth AS and the mth CP. Finally, let s be the proximity factor, defined in $[0,1]$, so that 0 corresponds to no approach and 1 corresponds to the maximum approach (resulting in coincident ASs and CPs). Then, the schemes proposed for the auxiliary surface deformation are defined in (8) as follows:

$$\rho'_m = \rho_m + gs\left(\frac{m}{M}\right)^v,\tag{8}$$

where $v = 1$ for simple and $v = 2$ for progressive reach. The deformation effect is graphically described in Figure 3.

Furthermore, in a way similar to [20], ASs and CPs, accordingly, should become denser close to the wedge tip. Again, there is no unique way to accomplish this. In this work, the scheme implemented multiplies the polar angle φ_m of the mth AS location by a factor D_m, $0 < D_{start} \leq D_m \leq 1$, where D_{start} is user-defined. For example, in the first quadrant of the 'eye', D_m is defined as being close to 0 for

ASs near the wedge tip, and close to 1 for ASs close to the vertical axis. For progressive densification, the scheme proposed is: $\varphi'_m = \varphi_m D_m^2$. Moreover, additional ASs may be superimposed to the already existing ones close to the tips if necessary. The combined effect of the tip approach and densification is depicted in Figure 4.

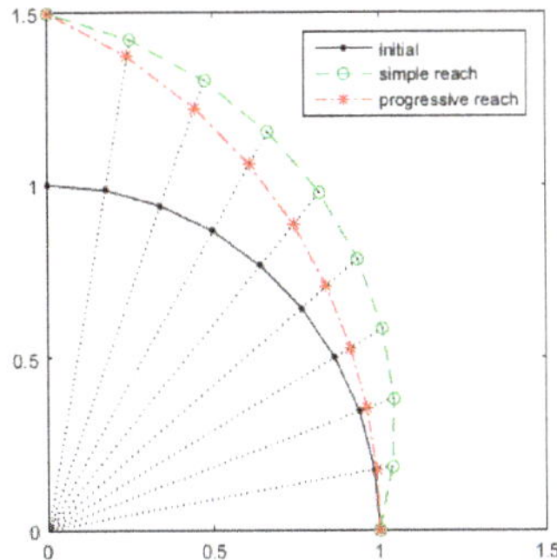

Figure 3. Polar radius decrease from 1.5 to 1 according to the proposed schemes.

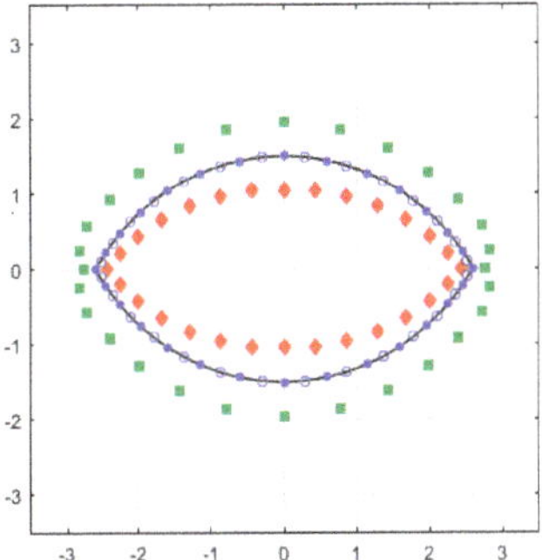

Figure 4. Deformation of the auxiliary surface, combined with densification in the vicinity of the tips: all inner ASs are allowed to approach the CPs, whereas only 1/4 of the outer ASs are allowed to do so.

4. Results

4.1. PEC Scatterer

To test the efficiency of the method, a PEC scatterer is initially defined by radius $\rho = 3\lambda$, arc displacement $d = 0.5\rho = 1.5\lambda$, incidence angle $\varphi_{inc} = 0$, auxiliary surface radius $\rho_{aux} = \rho - 0.075\lambda$, and originally 136 CPs and therefore 136 inner ASs. The solution to the problem without any deformation yields the results of Figure 5a, which depicts the quantified error in the generic boundary condition (BC) of the E field along the boundary stretch, i.e.,:

$$\Delta E_{bc} = \frac{\left| \hat{n} \times (E_{in} - E_{out}) \right|}{|E_{inc}|_{max}}, \tag{9}$$

where in (9) $\hat{n}$ is the normal unit vector on the boundary, pointing outwards, and E_{in}, E_{out} are the electric fields just inside and just outside the scatterer, respectively, the former obviously vanishing for the PEC case.

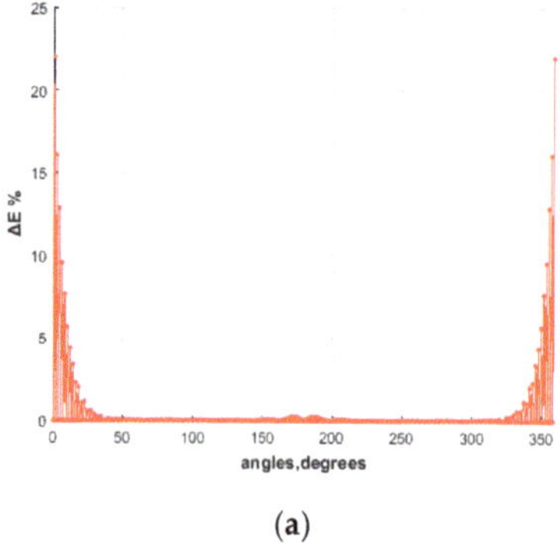 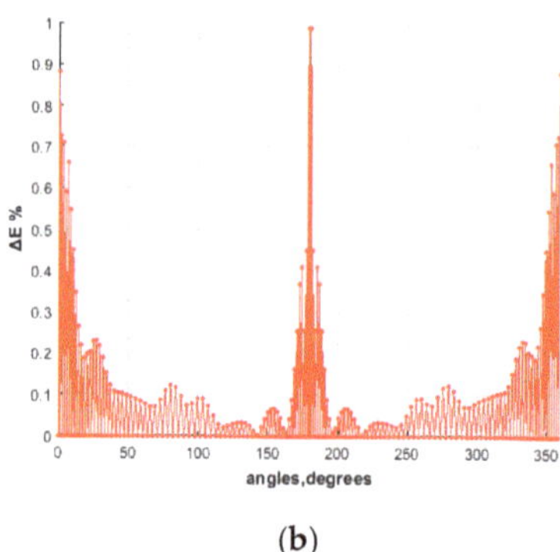

(a) **(b)**

Figure 5. Plots of *E* field boundary condition error for a perfectly electric conducting (PEC) scatterer with initial $\rho_{aux} = \rho - 0.075\lambda$ using: (**a**) the standard, conformal auxiliary surface; and (**b**) the deformed auxiliary surface.

To improve the satisfaction of the BC, the deformation scheme proposed above was implemented. After several trials, the following parameters were finally invoked: All ASs were displaced as well as densified and the proximity factor was set equal to $s = 0.88$, whereas $D_{start} = 0.9$. No extra ASs were added since their presence proved to achieve only incremental improvement. The results are displayed in Figure 5b, which shows that the overall *E* field error at MPs was substantially reduced.

It is worth noting that the proper placement of the auxiliary surface is of great importance in order to avoid caustics exclusion by the auxiliary surface or any potential resonance effects, as presented in [23]. To illustrate the possibly detrimental influence on the BC error, specifically, in this case, keeping the multitude of CPs consistent (and thus the multitude of ASs consistent) but moving the auxiliary surface further away from the scattering boundary ($\rho_{aux} = \rho - 0.15\lambda$) produced the results in Figure 6a,b, before and after applying the deformation scheme, respectively. Severe complications are revealed in Figure 6a, while it may be deduced from Figure 6b that utilizing an improved auxiliary surface layout, compared to a conformal one, may be much less prone to accuracy degradation, presumably because caustics are still retained within the auxiliary surface.

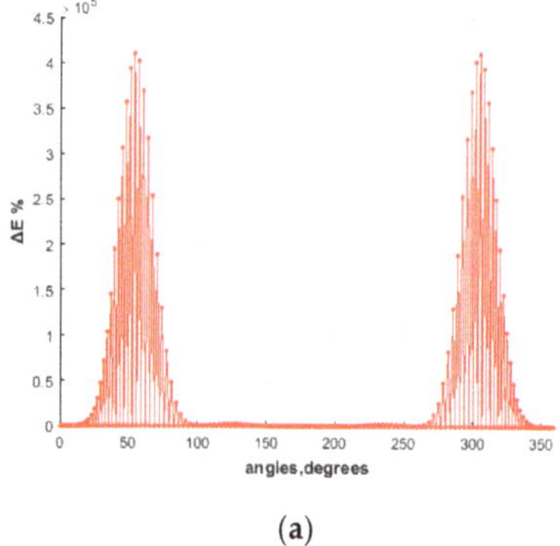 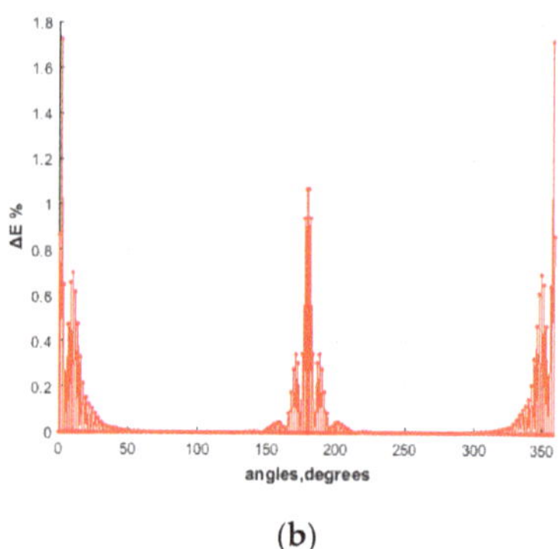

(a) **(b)**

Figure 6. Plots of *E* field boundary condition error for a PEC scatterer with initial $\rho_{aux} = \rho - 0.15\lambda$ using: (**a**) the standard, conformal auxiliary surface; and (**b**) the deformed auxiliary surface.

4.2. Lossless dielectric scatterer

Next, a lossless dielectric scatterer was considered, with radius $\rho = 3\lambda$, arc displacement $d = 0.5\rho = 1.5\lambda$, relative permittivity $\varepsilon_r = 2.56$, incidence angle $\varphi_{inc} = 0$, inner and outer auxiliary surface radius $\rho_{aux\ in} = \rho - 0.3\lambda$ and $\rho_{aux\ out} = \rho + 0.45\lambda$, and, originally, 160 CPs (hence 160 inner and 160 outer ASs). Like in the PEC case, the efficiency of MAS was initially tested using a conformal auxiliary surface and the extracted results of the BC error are presented in Figure 7a, where the upper subplot depicts the quantified error for the E field (ΔE %) and the lower one for the H field (ΔH %), both being relatively significant near the wedge tips.

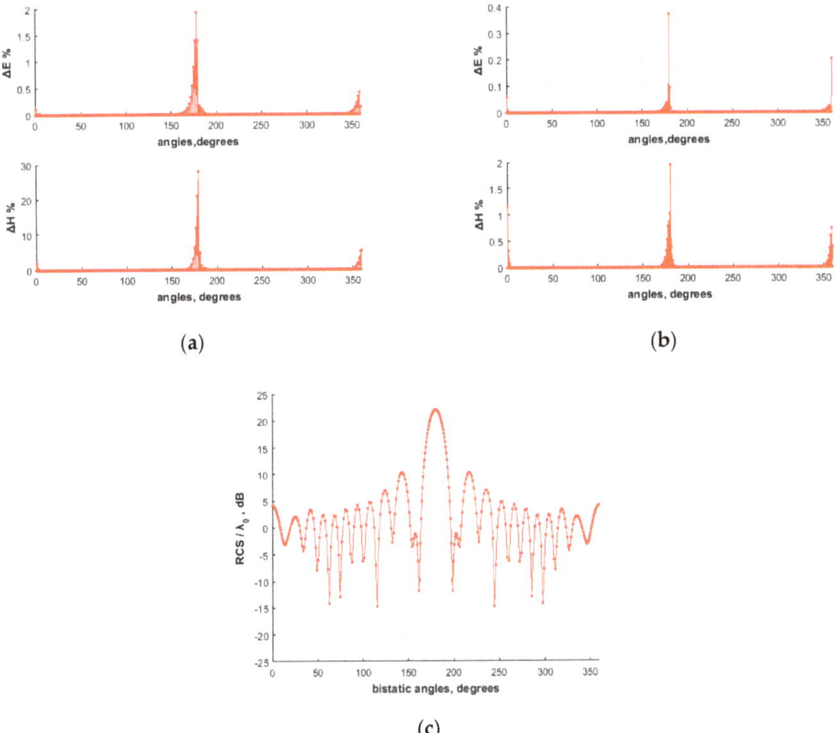

Figure 7. Results for lossless dielectric scatterer: (**a**) E and H field boundary the condition error by employing a conformal auxiliary surface; (**b**) E and H field boundary the condition error by employing a deformed auxiliary surface; (**c**) the Radar Cross Section (RCS) acquired by utilizing the auxiliary surface deformation scheme.

Afterwards, the improvement of the BC was examined by employing the aforementioned deformation scheme with the following parameters, obtained through multiple trials: the portion of ASs to be displaced was 1/5 for the inner and 1/8 for the outer ones, with a proximity factor of $s = 0.75$ and $s = 0.65$, respectively, while all of them were densified with $D_{start} = 0.80$. The addition of extra ASs offered no further refinement and was skipped once more. Figure 7b depicts the considerably decreased quantified BC error, and the corresponding radar cross section (RCS) is displayed in Figure 7c.

4.3. Lossy Dielectric Scatterer

Finally, a lossy dielectric scatterer of radius $\rho = 5\lambda$, arc displacement $d = 0.5\rho = 2.5\lambda$, relative permittivity $\varepsilon_r = 2.56 - 0.102\ j$, incidence angle $\varphi_{inc} = 0$, inner and outer auxiliary surface radius $\rho_{aux\ in} = \rho - 0.5\lambda$ and $\rho_{aux\ out} = \rho + 0.75\lambda$, and originally 180 CPs (hence 180 inner and 180 outer ASs) was analyzed. Similar BC tests were conducted for conformal auxiliary surface, once again resulting in a relatively sizable BC error in the neighborhood of the wedge tips (Figure 8a).

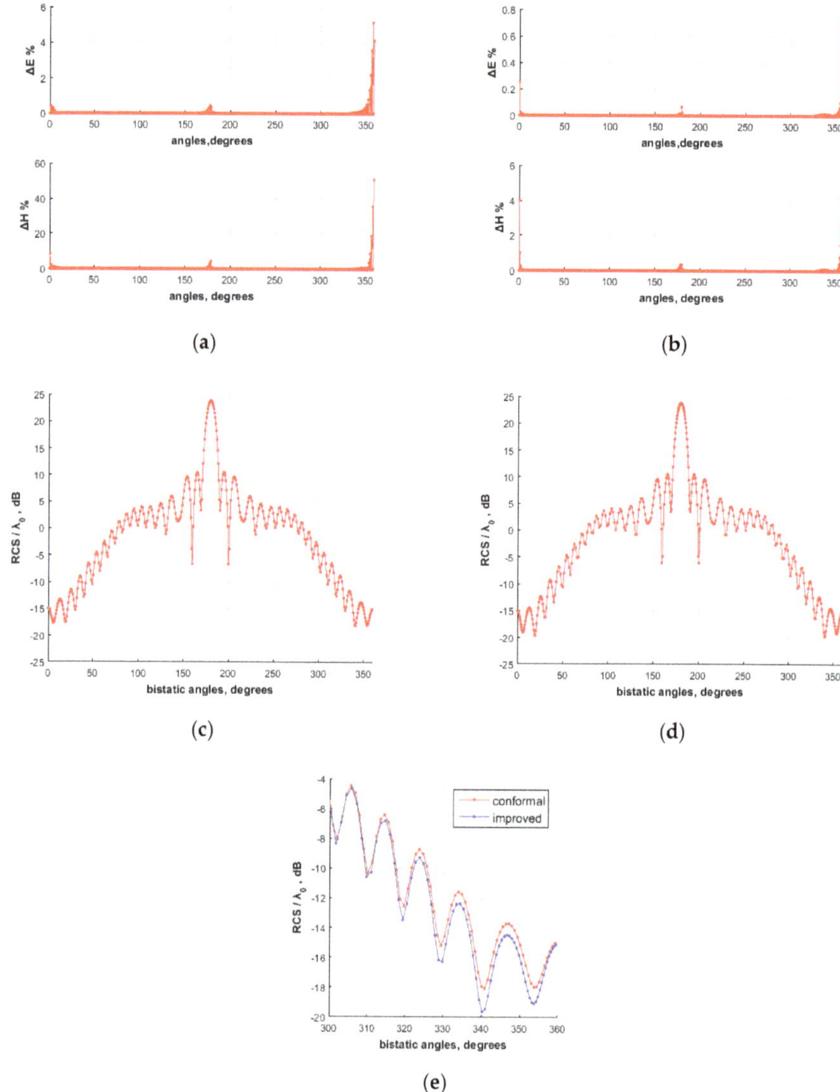

Figure 8. Results for lossy dielectric scatterer: (**a**) error in the generic boundary condition (BC) of the *E* and *H* field via conformal auxiliary surface placement; (**b**) error in the generic BC of the *E* and *H* field via improved auxiliary surface placement; (**c**) computed RCS for conformal auxiliary surfaces; (**d**) computed RCS for improved auxiliary surfaces; (**e**) RCS compared for both auxiliary surface placements (conformal shown in red, improved shown in blue), focused in angles near the backscattering direction.

The deformation scheme proposed above was also implemented in order to enhance the satisfaction of the boundary condition. The parameters found to produce the lowest BC error, shown in Figure 8b, are as follows: 1/5 of the inner ASs and 1/8 of the outer ones were displaced, with a proximity factor set equal to $s = 0.75$ and $s = 0.65$, respectively, whereas all of them were included in the densification with $D_{start} = 0.80$. Adding extra ASs did not prove to be beneficial and, therefore, was not employed.

The maximum E field error at MPs was reduced from $5.13 * 10^{-2}$ to $7.27 * 10^{-3}$ and the maximum H field error from $5.11 * 10^{-1}$ to $4.14 * 10^{-2}$.

The computed RCSs, using conformal and deformed auxiliary surface layouts, are depicted in Figure 8c,d, respectively. Although not visually discernible in these plots due to their wide dynamic range, comparing their values at angles close to the scatterer's wedges exhibits an observable improvement, e.g., from 300° to 360° in Figure 8e, in the azimuthal direction where ΔE and ΔH obtain their greatest values.

It is worth mentioning that placing ASs too close to the wedge tip is not advisable. Indeed, an AS right on the wedge tip causes severe ill-conditioning of the linear system matrix, whereas proximity factor s $= 0.99$ for both inner and outer ASs lead to inadequate error reduction, namely $3.16 * 10^{-2}$ for the E field and $4.43 * 10^{-1}$ for the H field.

Furthermore, the symmetry of the geometry should be accompanied by symmetry in the ASs/CPs. As a counterexample, the same lossy dielectric scatterer was experimentally analyzed via unequal CP densities in the two arcs, namely 45 for the upper and 90 for the lower one, leading to excessively high errors and completely wrong RCS plots, as demonstrated in Figure 9.

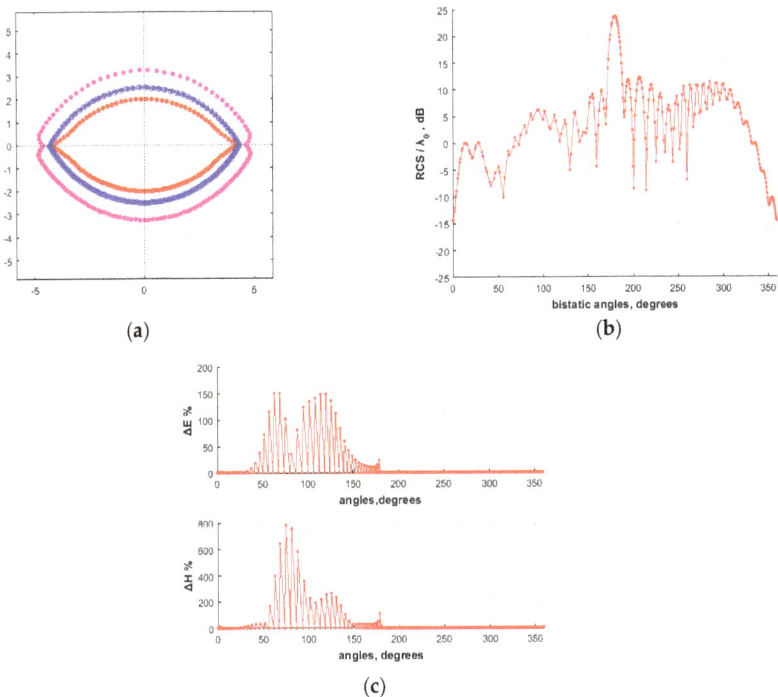

Figure 9. (a) Symmetric geometry with non-symmetric auxiliary points allocation, (b) RCS, and (c) ΔE and ΔH error.

4.4. Investigation of the Solution Behavior for Various Geometry Modifications

To check the robustness of the method described above, several modifications of the geometry were carried out and the respective results were extracted. In all cases below, lossy dielectric material with relative permittivity $\varepsilon_r = 2.56 - 0.102\,j$ is considered, whereas the incidence angle is always $\varphi_{inc} = 0$ with an arc radius of $\rho = 3\lambda$.

First, a scatterer that deviates only slightly from a circle was studied, i.e., a layout with small arc displacement $d = 0.1\rho = 0.3\lambda$, followed by a scatterer with a moderate deviation from a circle, i.e., a layout with mediocre arc displacement $d = 0.5\rho = 1.5\lambda$, and, finally, a scatterer with a significant distortion with respect to a circle, which, therefore, encompasses sharp wedges, i.e., a layout with large arc displacement $d = 0.7\rho = 2.1\lambda$. All three geometries and their computed results are depicted in Figures 10–12, respectively, while their MAS parameters and the quantified error in the generic boundary condition of the electromagnetic field are presented in Table 1.

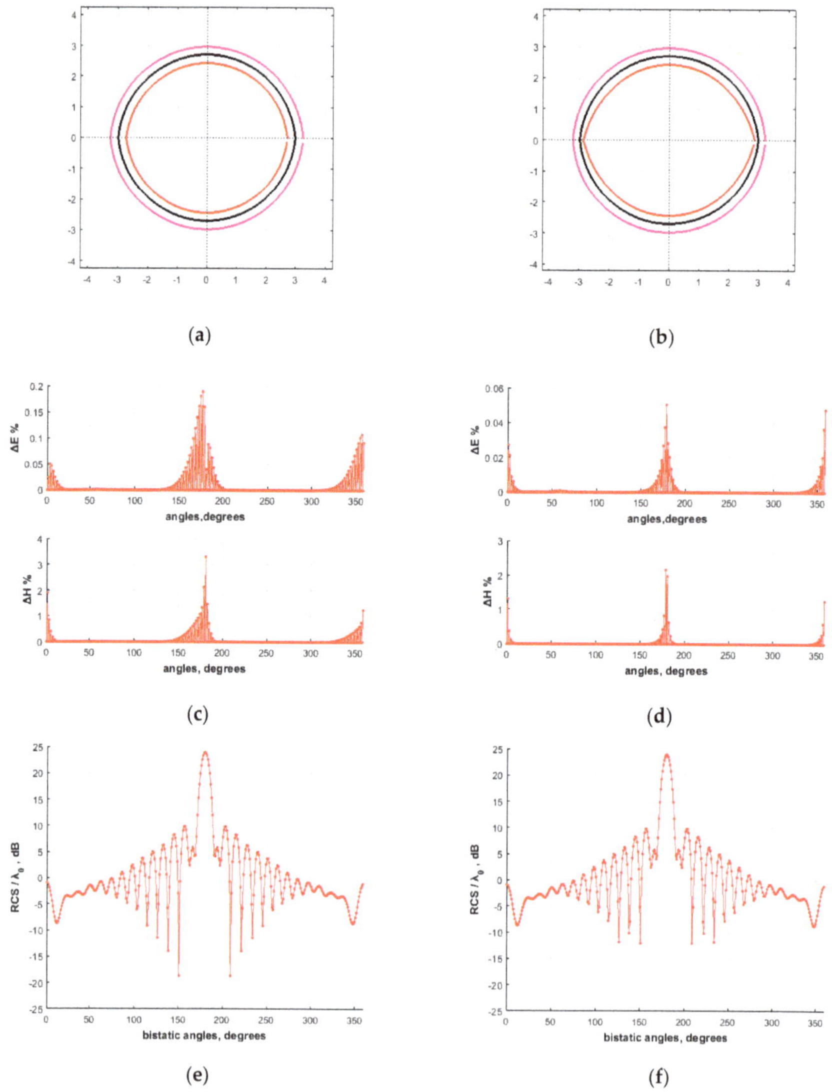

Figure 10. Small arc displacement $d = 0.1\rho = 0.3\lambda$: (**a**) geometry and conformal auxiliary surfaces; (**b**) geometry and improved auxiliary surfaces; (**c**) conformal ΔE and ΔH error; (**d**) improved ΔE and ΔH error; (**e**) conformal RCS; and (**f**) improved RCS.

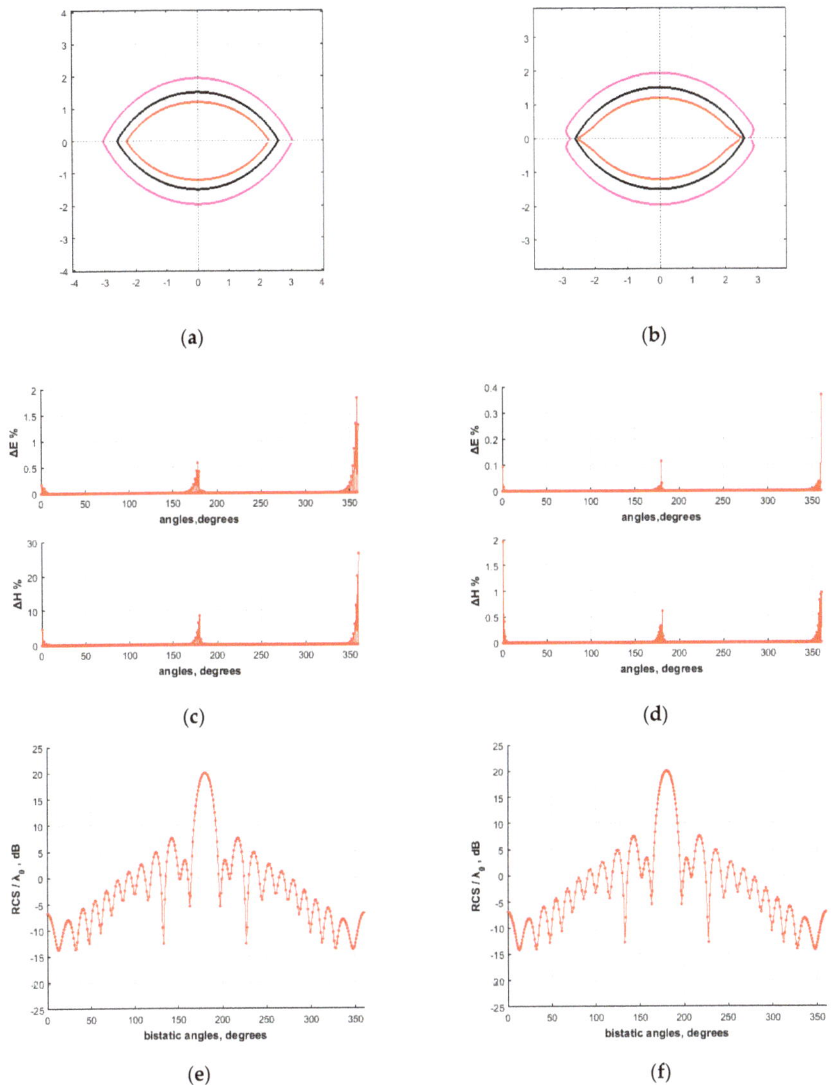

Figure 11. Mediocre arc displacement $d = 0.5\rho = 1.5\lambda$: (**a**) geometry and conformal auxiliary surfaces; (**b**) geometry and improved auxiliary surfaces; (**c**) conformal ΔE and ΔH error; (**d**) improved ΔE and ΔH error; (**e**) conformal RCS; and (**f**) improved RCS.

As a verification of the broad applicability of the technique, an asymmetric scatterer was also analyzed. Namely, two unequal circular arcs were connected, forming a "sorrowful" eye. The geometry parameters were chosen as follows: radius $\rho = 4\lambda$ and arc displacement $d = 0.5\rho = 2\lambda$ for the upper arc, radius $\rho = 5\lambda$ and arc displacement $d = 0.5\rho = 2.5\lambda$ for the lower arc, originally 160 CPs (hence 160 inner and 160 outer ASs), 1/2.5 of the inner ASs and 1/4 of the outer ones were displaced, with proximity factor set equal to $s = 0.6$ and $s = 0.6$, respectively. No densification was implemented and no extra ASs were added. The maximum E field error at MPs was reduced from $4.19 * 10^{-2}$ to $1.57 * 10^{-2}$ and the maximum H field error from $4.36 * 10^{-1}$ to $1.38 * 10^{-1}$. The results are depicted in Figure 13.

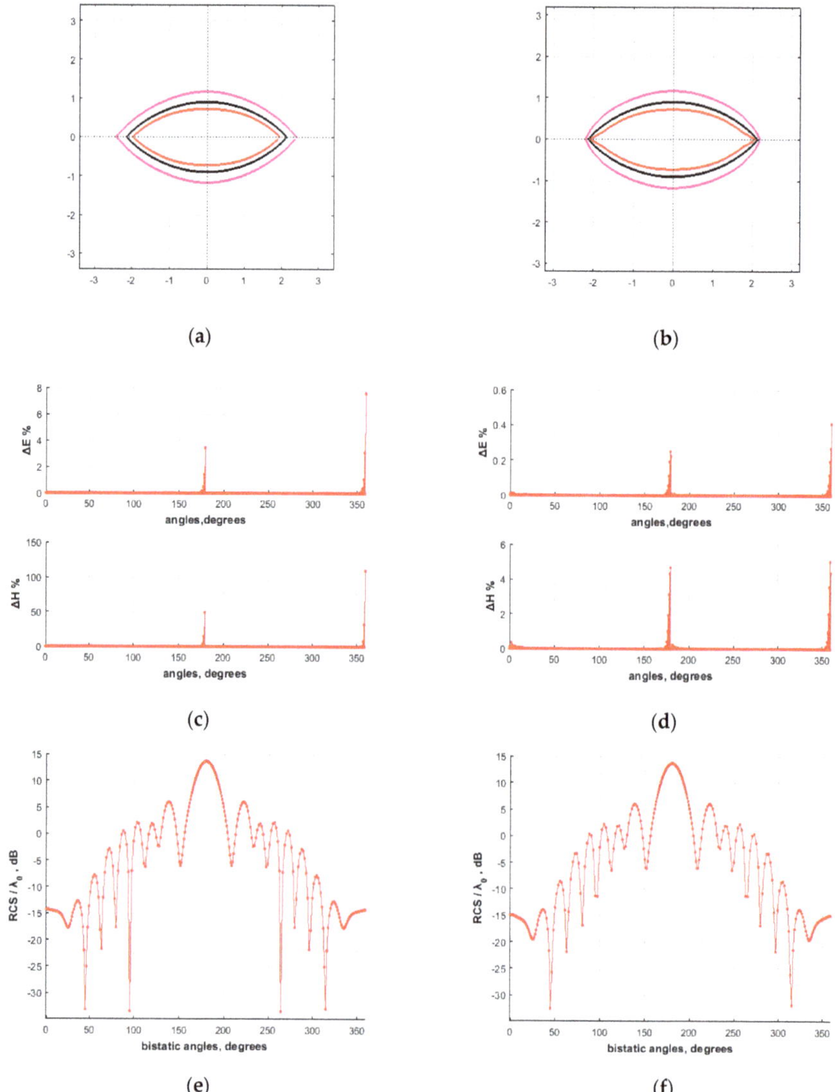

Figure 12. Large arc displacement $d = 0.7\rho = 2.1\lambda$: (**a**) geometry and conformal auxiliary surfaces; (**b**) geometry and improved auxiliary surfaces; (**c**) conformal ΔE and ΔH error; (**d**) improved ΔE and ΔH error; (**e**) conformal RCS; and (**f**) improved RCS.

Judging from the outcome of the aforementioned tests, the method proposed herein is very robust, producing reliable results for blunt, acute, or non-symmetric wedges.

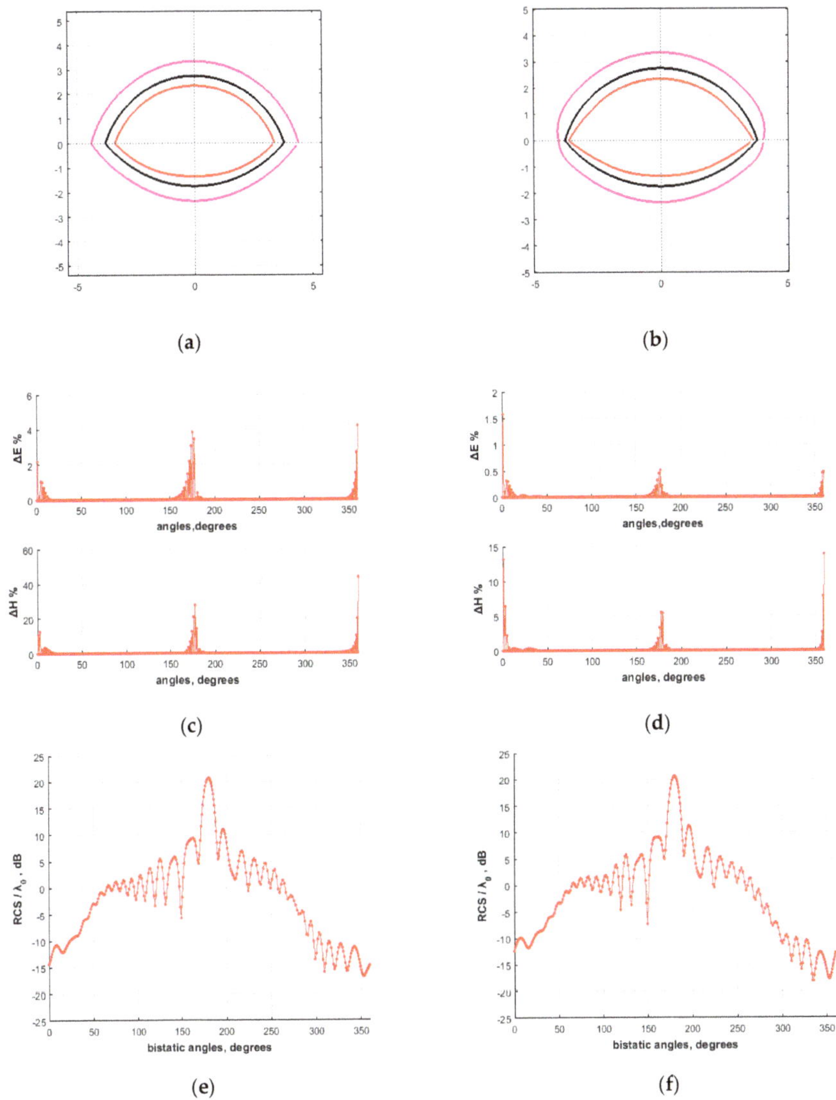

Figure 13. Non-symmetric scatterer: (**a**) geometry and conformal auxiliary surfaces; (**b**) geometry and improved auxiliary surfaces; (**c**) conformal ΔE and ΔH error; (**d**) improved ΔE and ΔH error; (**e**) conformal RCS; and (**f**) improved RCS.

Table 1. Geometry definition, method of auxiliary sources (MAS) parameters, and computed BC error of the electromagnetic field.

Arc displacement d	$0.1\rho = 0.3\lambda$	$0.5\rho = 1.5\lambda$	$0.7\rho = 2.1\lambda$
number of points (ASs, CPs)	160	160	156
displaced inner ASs fraction	1/2	1/5	1/6
displaced outer ASs fraction	1/2	1/8	1/2
inner proximity factor s	0.55	0.75	0.74

Table 1. *Cont.*

outer proximity factor s	0.15	0.65	0.8
densification factor D_{start}	0.94	0.8	0.8
maximum ΔE_{bc} [1], conformal	$1.89 * 10^{-3}$	$1.81 * 10^{-2}$	$7.58 * 10^{-2}$
maximum ΔE_{bc} [1], improved	$5.06 * 10^{-4}$	$3.68 * 10^{-3}$	$4.09 * 10^{-3}$
maximum ΔH_{bc} [1], conformal	$3.31 * 10^{-2}$	$2.64 * 10^{-1}$	1.09
maximum ΔH_{bc} [1], improved	$2.16 * 10^{-2}$	$1.96 * 10^{-2}$	$5.04 * 10^{-2}$

[1] Boundary condition error at midpoints (MPs).

5. Discussion

The applicability range of the approach presented above may span various aspects of computational electromagnetics. MAS has already been used in the literature to simulate practical problems, such as scattering by a raindrop [24], although that particular geometry is smooth, without wedges. Wedge treatment, as discussed in this paper, enhances MAS accuracy for applications such as jet engine inlet modeling [25–27], where interior blades contain sharp edges, or even further military aircraft scattering simulation, such a stealth design. Moreover, additional possible applications may include automotive modeling, for example, the functionality of a telecommunications antenna in the presence of the vehicle surface. Furthermore, the algorithm may be used in the design of absorbers and cloaking materials [28–30], or generally wave manipulation [31] when the geometry layout contains wedges.

Although the numerical results presented were very good, they are limited to wedges formed by intersecting circular arcs only. It is anticipated that analogous techniques would be equally efficient in more general cases (i.e., arbitrarily shaped curved wedges); however, this is yet to be proven in future research. Likewise, other issues to be investigated include TE (Transverse Electric) incidence, the three-dimensional counterpart of this problem, unconventional materials, as well as a robust methodology to choose the optimal deformation parameters.

Furthermore, by studying various geometries, proposing schemes for wedge treatment and obtaining solutions by choosing alternative configurations for the incident wave, material properties, etc., an extensive data set can be acquired. The latter, a combination of input variables and their respective exported results may be used as training data for building a mathematical model via Machine Learning algorithms. Machine Learning, the usage of computational methods in order to produce/"learn" information directly from data without relying on explicit instructions, improves its accuracy and performance proportionally to the amount of available training data. A larger data set is, therefore, required to find hidden patterns or intrinsic properties that can be utilized in the procedure of determining the optimal settings for the scattering problem and, hopefully, automating it.

6. Conclusions

The method of auxiliary sources (MAS) was applied to scattering from a PEC or dielectric cylinder whose geometry contains curved wedges, formed by intersections of circular arcs. The BC error in the vicinity of the wedge tips is significant for standard, conformal auxiliary surfaces due to field singularities. To overcome such a complication, the deformation of the auxiliary sources was proposed. When close to the tips, both inner and outer (when applicable) ASs approached the CPs, and their distribution was also forced to become denser. The BC error was shown to decrease significantly for both the E and the H fields, resulting in more accurate RCS results.

Author Contributions: Formal analysis, V.G.I., M.-T.A. and H.T.A.; Funding acquisition, H.T.A.; Methodology, V.G.I., M.-T.A. and H.T.A.; Project administration, H.T.A.; Software, V.G.I. and M.-T.A.; Supervision, H.T.A.; Validation, V.G.I. and M.-T.A.; Writing—Original draft, H.T.A.; Writing—Review & editing, V.G.I. and M.-T.A. All authors have read and agreed to the published version of the manuscript.

Funding: The authors wish to acknowledge financial support by the Special Account for Research Funds of the International Hellenic University (former Technological Educational Institute of Central Macedonia), Greece, under grant SAT/IE/93/6/28-2-2019/50135.

Conflicts of Interest: The authors declare no conflict of interest.

References

1. La Spada, L.; Vegni, L. Near-zero-index wires. *Opt. Express* **2017**, *25*, 23699–23708. [CrossRef] [PubMed]
2. Pacheco-Peña, V.; Engheta, N.; Kuznetsov, S.; Gentselev, A.; Beruete, M. All-metallic epsilon-near-zero graded-index converging lens at terahertz frequencies. In Proceedings of the 12th European Conference on Antennas and Propagation (EuCAP 2018), London, UK, 9–13 April 2018; pp. 320–324.
3. Greybush, N.J.; Pacheco-Peña, V.; Engheta, N.; Murray, C.B.; Kagan, C.R. Plasmonic optical and chiroptical response of self-assembled au nanorod equilateral trimers. *ACS Nano* **2019**, *13*, 1617–1624. [CrossRef] [PubMed]
4. La Spada, L.; Spooner, C.; Haq, S.; Hao, Y. Curvilinear metasurfaces for surface wave manipulation. *Sci. Rep.* **2019**, *9*, 3107. [CrossRef] [PubMed]
5. Lee, I.H.; Yoo, D.; Avouris, P.; Low, T.; Oh, S.H. Graphene acoustic plasmon resonator for ultrasensitive infrared spectroscopy. *Nat. Nanotechnol.* **2019**, *14*, 313–319. [CrossRef] [PubMed]
6. La Spada, L.; Vegni, L. Electromagnetic nanoparticles for sensing and medical diagnostic applications. *Materials* **2018**, *11*, 603. [CrossRef]
7. Harrington, R.F. *Field Computation by Moment Methods*; Mc Millan: New York, NY, USA, 1968.
8. Vegni, L.; Cichetti, R.; Capece, P. Spectral dyadic Green's function formulation for planar integrated structures. *IEEE Trans. Antennas Propag.* **1988**, *36*, 1057–1065. [CrossRef]
9. Volakis, J.L.; Chatterjee, A.; Kempel, L.C. *Finite Element Method Electromagnetics: Antennas, Microwave Circuits, and Scattering Applications*; Wiley-IEEE Press: Hoboken, NJ, USA, 1998.
10. Taflove, A.; Hagness, S.C. *Computational Electrodynamics: The Finite-Difference Time-Domain Method*, 3rd ed.; Artech House: Norwood, MA, USA, 2005.
11. Christopoulos, C. *The Transmission-Line Modeling Method: TLM*; IEEE Press: Piscataway, NJ, USA, 1995.
12. La Spada, L. Metasurfaces for advanced sensing and diagnostics. *Sensors* **2019**, *19*, 355. [CrossRef]
13. Pacheco-Peña, V.; Beruete, M.; Rodríguez-Ulibarri, P.; Engheta, N. On the performance of an ENZ-based sensor using transmission line theory and effective medium approach. *New J. Phys.* **2019**, *21*, 043056. [CrossRef]
14. Estakhri, N.M.; Edwards, B.; Engheta, N. Inverse-designed metastructures that solve equations. *Science* **2019**, *363*, 1333–1338. [CrossRef]
15. Popovidi-Zaridze, R.S.; Tsverikmazashvili, Z.S. Numerical study of a diffraction problem by a modified method of non-orthogonal series. *Zurnal. Vichislit. Mat. Mat Fiz.* **1977**, *17*, 384–393.
16. Kaklamani, D.I.; Anastassiu, H.T. Aspects of the method of auxiliary sources (MAS) in computational electromagnetics. *IEEE Antennas Propag. Mag.* **2002**, *44*, 48–64. [CrossRef]
17. Kupradze, V.D. On the approximate solution of problems of mathematical physics. *Russ. Math. Surv.* **1967**, *22*, 58–108. [CrossRef]
18. Avdikos, G.K.; Anastassiu, H.T. Computational cost estimations and comparisons for three methods of applied electromagnetics (MoM, MAS and MMAS). *IEEE Antennas Propag. Mag.* **2005**, *47*, 121–129. [CrossRef]
19. Eisler, S.; Leviatan, Y. Analysis of electromagnetic scattering from metallic and penetrable cylinders with edges using a multifilament current model. *IEE Proc. H (Microw. Antennas Propag.)* **1989**, *136*, 431–438. [CrossRef]
20. Anastassiu, H.T.; Kaklamani, D.I.; Economou, D.P.; Breinbjerg, O. Electromagnetic scattering analysis of coated conductors with edges using the method of auxiliary sources (MAS) in conjunction with the standard impedance boundary condition (SIBC). *IEEE Trans. Antennas Propag.* **2002**, *50*, 59–66. [CrossRef]
21. Senior, T.B.A.; Volakis, J.L. *Approximate Boundary Conditions in Electromagnetics*; The Institution of Engineering and Technology: London, UK, 1995.
22. Iatropoulos, V.G.; Anastasiadou, M.T.; Anastassiu, H.T. Application of the method of auxiliary sources (MAS) to the analysis of electromagnetic scattering from dielectric surfaces with curved wedges. In Proceedings of the 13th International Conference on Communications, Electromagnetics and Medical Applications (CEMA'18), Sofia, Bulgaria, 18–20 October 2018; pp. 12–16.

23. Anastassiu, H.T.; Lymperopoulos, D.G.; Kaklamani, D.I. Accuracy analysis and optimization of the method of auxiliary sources (MAS) for scattering by a circular cylinder. *IEEE Trans. Antennas Propag.* **2004**, *52*, 1541–1547. [CrossRef]

24. Kanellopoulos, S.A.; Panagopoulos, A.D.; Kanellopoulos, J.D. Analysis of the electromagnetic scattering by a raindrop using the method of auxiliary sources. In Proceedings of the IEEE Antennas and Propagation Society Symposium, Monterey, CA, USA, 20–25 June 2004; pp. 4176–4179.

25. Anastassiu, H.T. A review of electromagnetic scattering analysis for inlets, cavities and open ducts. *IEEE Antennas Propag. Mag.* **2003**, *45*, 27–40. [CrossRef]

26. Fromentin-Denozière, B.; Simon, J.; Tzoulis, A.; Weinmann, F.; Anastassiu, H.T.; Escot Bocanegra, D.; Poyatos Martínez, D.; Fernández Recio, R.; Zdunek, A. Comparative study of miscellaneous methods applied to a benchmark, inlet scattering problem. *IET Radar Sonar Navig.* **2014**, *9*, 342–354. [CrossRef]

27. Anastassiu, H.T.; Petrović, V.V.; Kaklamani, D.I.; Uzunoglu, N.K. Analysis of electromagnetic scattering from jet engine inlets using the method of auxiliary sources (MAS). In Proceedings of the COMCON 7, Seventh Annual International Conference on Advances in Communications and Control, Athens, Greece, 28 June–2 July 1999; pp. 217–226.

28. La Spada, L.; Vegni, L. Metamaterial-based wideband electromagnetic wave absorber. *Opt. Express* **2016**, *24*, 5763–5772. [CrossRef]

29. La Spada, L.; Haq, S.; Hao, Y. Modeling and design for electromagnetic surface wave devices. *Radio Sci.* **2017**, *52*, 1049–1057. [CrossRef]

30. McManus, T.M.; La Spada, L.; Hao, Y. Isotropic and anisotropic surface wave cloaking techniques. *J. Opt.* **2016**, *18*, 044005. [CrossRef]

31. Qin, F.; Ding, L.; Zhang, L.; Monticone, F.; Chum, C.C.; Deng, J.; Mei, S.; Li, Y.; Teng, J.; Hong, M.; et al. Hybrid bilayer plasmonic metasurface efficiently manipulates visible light. *Sci. Adv.* **2016**, *2*, e1501168. [CrossRef] [PubMed]

Article

Optimisation of Energy Transfer in Reluctance Coil Guns: Application to Soccer Ball Launchers

Valentin Gies [1,*], **Thierry Soriano** [2], **Sebastian Marzetti** [1], **Valentin Barchasz** [3], **Herve Barthelemy** [1], **Herve Glotin** [4] **and Vincent Hugel** [2]

[1] IM2NP—UMR 7334, CNRS, Université de Toulon, 83130 La Garde, France; smarzetti705@etud.univ-tln.fr (S.M.); herve.barthelemy@univ-tln.fr (H.B.)

[2] COSMER—EA 7398, CNRS, Université de Toulon, 83130 La Garde, France; thierry.soriano@univ-tln.fr (T.S.); vincent.hugel@univ-tln.fr (V.H.)

[3] SMIoT, Université de Toulon, 83130 La Garde, France; valentin.barchasz@univ-tln.fr

[4] LIS—UMR 7020, CNRS, Université de Toulon, 83130 La Garde, France; glotin@univ-tln.fr

*   Correspondence: gies@univ-tln.fr; Tel.: +33-628357685

Received: 13 February 2020; Accepted: 22 April 2020; Published: 30 April 2020

Abstract: Reluctance coil guns are electromagnetic launchers having a good ratio of energy transmitted to actuator volume, making them a good choice for propelling objects with a limited actuator space. In this paper, we focus on an application, which is launching real size soccer balls with a size constrained robot. As the size of the actuator cannot be increased, kicking strength can only be improved by enhancing electrical to mechanical energy conversion, compared to existing systems. For this, we propose to modify its inner structure, splitting the coil and the energy storage capacitor into several ones, and triggering the coils successively for propagating the magnetic force in order to improve efficiency. This article first presents a model of reluctance electromagnetic coil guns using a coupled electromagnetic, electrical and mechanical models. Four different coil gun structures are then simulated, concluding that splitting the kicking coil into two half size ones is the best trade-off for optimizing energy transfer, while maintaining an acceptable system complexity and controllability. This optimization results in robust enhancement and leads to an increase by 104% of the energy conversion compared to a reference launcher used. This result has been validated experimentally on our RoboCup robots. This paper also proves that splitting the coil into a higher number of coils is not an interesting trade-off. Beyond results on the chosen case study, this paper presents an optimization technique based on mixed mechanic, electric and electromagnetic modelling that can be applied to any reluctance coil gun.

Keywords: coil gun; reluctance; electromagnetic launcher; mechatronics; electronics; mechanics; simulation; RoboCup

---

1. Introduction

Propelling projectiles with a controlled speed and trajectory is a technological challenge having lots of applications ranging from kicking soccer balls to launching rockets or satellites, including testing military ammunitions. In this introduction, we focus on propulsion techniques having strong accelerations, so that an important speed can be obtained in a short distance. Common propulsion techniques of this type are among the following ones:

*   **Chemical propulsion**: mainly used for propelling weapons or rockets; chemical propulsion uses the product of a chemical explosive or expanding reaction to push out a projectile [1]. The benefit of this propulsion is to have a high density of energy stored in a small size as a chemical product leading to strong accelerations, without the need of being connected to a power supply. Its drawback is that the propulsion is a single shot one due to the chemical reaction.

- **Mechanical propulsion**: there are many types of mechanical propulsion systems. Among the solutions allowing a strong acceleration of the payload are the inertial launchers. They are mostly using an energy storage in heavy high speed rotating mechanical parts such as iron cylinders. These parts are accelerated slowly by a standard motor and part of the stored energy is transferred in a very short amount of type to a projectile by friction. These systems are very simple but their size is much more important than the size of chemical or electromagnetic launchers for a given propelling strength. An example is presented later in this paper.
- **Rail Gun propulsion**: a rail gun is composed of a pair of conductive parallel rails connected to a direct current (DC) power supply. The electrical circuit is a closed sliding conductor where a payload is placed. Once current flows through the rails, a Lorentz force is created, accelerating the payload to launch it. This propulsion technique is very efficient, and output speed can be higher than using a conventional chemical propulsion as shown in [1], for a launching structure having the same overall size. Compared with chemical propulsion, this solution can be far less expensive than chemical propulsion for launching limited size payloads such as small satellites [2]. However, the presence of a mechanical contact between the rails and the payload propeller can lead to several issues reducing the energy transfer, such as friction losses [3] and plasma phenomena at very high speeds [2,4].
- **Coil Gun propulsion**: a coil gun is an electromagnetic launcher (EML) converting electricity into kinetic energy using coils [5,6]. There are two types of coil guns. The first one is based on induction to accelerate a conductive non-magnetic projectile using eddy currents induced in a conductive/non-magnetic moving rod inside a magnetic field created by a fixed coil [7]. This solution has an important drawback due to magnetic losses leading to heat generation and controllability loss. The second one is based on accelerating a magnetic projectile by minimizing the reluctance between the projectile and a magnetic field generated by a current flowing through a fixed coil [8]. This type of coil gun, also named reluctance accelerator is simpler to drive than induction coil guns and is very compact.

1.1. Comparison of Existing Soccer Ball Launching Systems

In this article, we focus on a case study having strong constraints: a real size soccer ball kicking system embedded in robots participating in the RoboCup (the autonomous robot soccer World Cup) in Middle Size League (Figure 1). This competition puts strong constraints on the robot size and weight, requiring to choose the launcher having the highest ratio of energy transmitted to launcher volume (Figure 2).

The balls used for the competition are real soccer balls (diameter 22 cm) having a weight equal to 450 g. Size constraints on the robots are a maximum width and length $L = W = 52$ cm and a maximum height $H = 80$ cm. Considering these robots are all using omnidirectional propulsion with 3 or 4 wheels, the space for embedding the kicking system is very small and cannot exceed a length of 30 cm and a width of 20 cm, as shown in Figure 3. Moreover, its height must be limited because the mass centre of the robot must be as low as possible in order to allow high accelerations.

A comparison of existing soccer ball launchers is first proposed in this section. Chemical propulsion has not been considered because they are dangerous and not reusable. Moreover, Ref.[1] shows that the size of a chemical launching system is equivalent to an electromagnetic one. The only advantage is that it does not need any power supply, but it is not a problem in our case considering that the robot has one.

The reference for understanding ball kicking systems is the human. Professional soccer players shots can reach 130 km·h^{-1} = 36 m·s^{-1}, corresponding to an kinetic energy equal to $E_K = 290$ J. As shown in Figure 4, the surface swept by the leg during a kick is important and approximately equal to one third of the surface of a circle having a radius $R = 80$ cm. The mass of a soccer player leg is approximately equal to $m = 20$ kg.

Figure 1. RoboCup 2019—Sydney—Australia.

Figure 2. RoboCup RCT robot 2020.

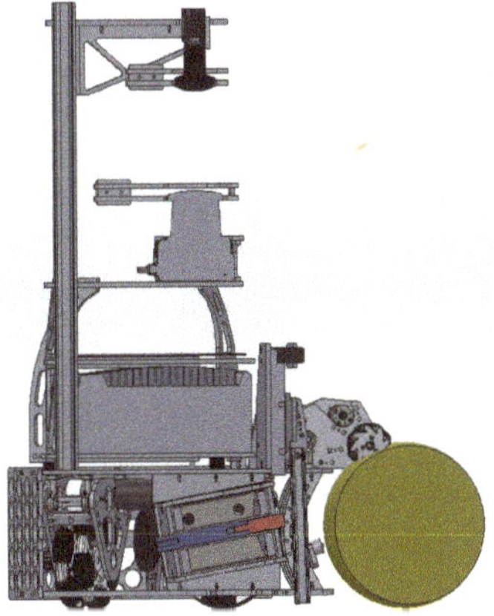

Figure 3. Cut view of RCT robot 2020.

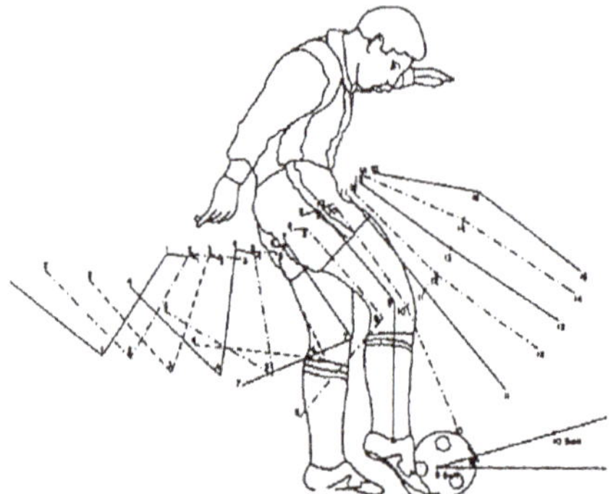

Figure 4. Human kicking sequence.

Mechanical propulsion is also one of the most commonly used methods for propelling a soccer ball. A commercial system, shown in Figure 5, is able to launch soccer balls at a maximum speed of 105 km·h^{-1} = 29 m·s^{-1} corresponding to an kinetic energy equal to E_K = 190 J, using two 10 kg cylinders coated with rubber. The propulsion part (cylinders and motors) of the system weighs 15 kg and its dimensions are W = 65 cm and L = 25 cm and H = 25 cm.

Figure 5. Mechanical inertial rotating launcher.

Another mechanical system that can be used for propelling a soccer ball is a robotic leg powered by a motor as shown in the kicking from Adidas [9] or as described in [10]. These solutions, using a robotic arm, have multiple degrees of freedom [10] or a set of rotary and linear spring-loaded actuators [9]. The Adidas solution shown in Figure 6 is composed of a 0.6 m robotic thigh rotating at 85 RPM and 0.6 m shank rotating at a maximum speed of 165 RPM, leading to a maximum ball speed of 21 m·s^{-1} corresponding to an kinetic energy equal to E_K = 100 J. However, the whole system is heavy (more than 50 kg) and its dimensions are important as it sweeps a 1.2 m radius cylinder.

These mechanical systems are interesting for simulating a football player leg [9,10] and for training humans in real conditions. However, embedding them in a RoboCup robot is very difficult. This is probably the reason why the RoboCup research community is mainly focused on electromagnetic launchers, and more especially on variable reluctance coil guns. Authors in [11] introduced a design of a variable reluctance coil gun (Figure 7) that is used in the RoboCup robots of the 2019 World Champion team. This actuator's dimensions are L = 30 cm, W = 9 cm and H = 9 cm, its weight is 4.5 kg and ball maximum speed can reach 11.4 m·s^{-1}, corresponding to a kinetic energy equal to E_K = 29 J. Output speed is smaller than mechanical design, but volume has been divided by factor 20 compared with an inertial rotating launcher.

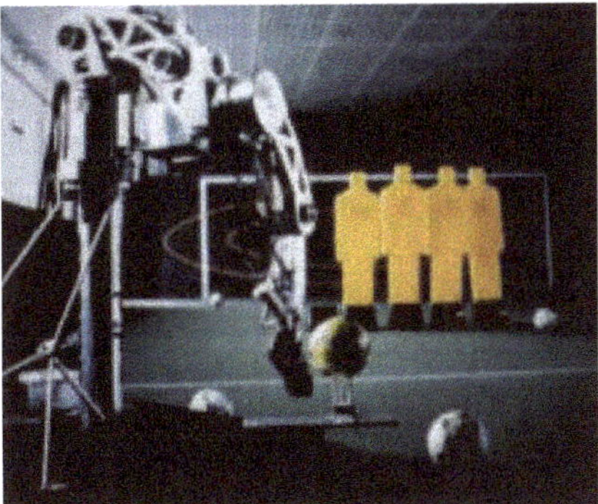

Figure 6. Mechanical robot leg.

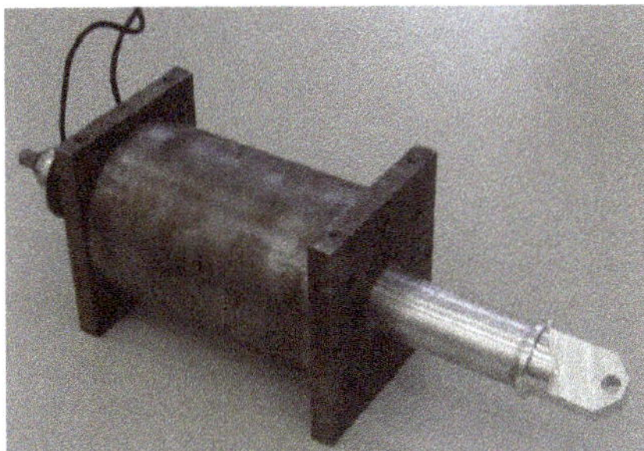

Figure 7. Tech United Reluctance Coil Gun.

Table 1 shows a comparison between existing ball launchers, including humans. These solutions are very different. This comparison is done considering the weight and size for each system, as it is a strong constraint in our case study. It is important to note that the energy transferred to the ball is close to the maximum value for all systems, except for the robot leg. This one is based on an industrial actuator able to carry heavy loads, and largely oversized for launching a soccer balls in terms of torque and power. Because it is not used at full power, its ratio of energy transmitted to launcher volume is very low compared to other solutions. However, this solution takes too much space due to the rotation of the leg and is not relevant for a small size launcher.

Table 1. Existing ball launchers comparison.

Launcher	Length (cm)	Width (cm)	Height (cm)	Volume (cm^3)	Weight (kg)	Ball Speed (m·s^{-1})	Ball Energy (J)	$\frac{Energy}{Volume}$ (J·dm^{-3})
Soccer player leg	160	20	80	133×10^3	20	36	290	**2.18**
Rotating inertial launcher	25	65	25	40×10^3	25	29	190	**4.75**
Robot arm [9]	240	240	30	1360×10^3	50	21	100	**0.07**
Reluctance coil gun [11]	30	9	9	2.4×10^3	4.5	11.4	29	**12.08**

In conclusion of this section, the most relevant launching systems in terms of energy transferred to the ball for a given actuator volume are reluctance coil guns, with a ratio of energy transmitted to launcher volume better than rotating inertial launchers by a factor 2.5, and better than most humans by a factor 5.

Since these electromagnetic coil guns seem to be the most promising solution for launching balls, this paper will only focus on improving that solution in order to maximize this energy transfer without changing the volume and the weight of the actuator.

1.2. Reluctance Coil Guns: A Ball Launcher That Can Be Optimized

Even if reluctance coil guns are a relevant solution for kicking soccer balls efficiently, it is important to note that they are not very efficient in terms of energy conversion. In [11], electrical energy for the coil gun is stored in a capacitor having a capacitance value C = 4700 μF under 425 V. Stored electrical energy is equal to:

$$E_C = \frac{1}{2} C U_C^2 = 424\,\text{J}$$

Consequently, the ratio of ball kinetic energy to the input electrical one is only 7%, and the ratio of the overall mechanical transmitted energy (including the kinetic energies of the iron rod, the lever and the ball as explained later) to the input electrical one is 12%. However, the energy necessary for kicking like human soccer players is already stored in the capacitor. This means that if a robot's kick is 10 times less powerful than a human one, it is not an issue related to available energy, but it is a problem of inefficiency of energy transfer in reluctance coil guns.

Optimizing this energy transfer can be done in two main ways without changing the size and the weight of the launcher. The first one is to adjust the initial position of the plunger, and the length of its non-magnetic extension. [12] shows that energy transmission can be increased by 70% using this technique compared to the reference case presented in [11]. This optimization is interesting because nothing is changed on the coil gun structure and size; it is only an optimization of initial conditions and a plunger parameter adjustment.

A second way of improving the energy transfer of a coil gun is to modify its inner structure by splitting the coil and the energy storage capacitor into several ones [7,13], without changing the overall quantity of coil copper and the overall capacitance value. Instead of sending an energy pulse to a single coil, a sequence of smaller energy pulses will be sent to the different coils propagating the magnetic force along the coil as the plunger enters it. The number of coils and the triggering sequence are the parameters to be optimized.

This paper focuses on this second method for optimizing the energy transfer in a reluctance coil gun. It is divided into three sections:

- Section 2 recalls the principles of coil guns.
- Section 3 describes 4 mechatronic coupled models of reluctance coil guns. All these coil guns use the same coil copper quantity and have the same overall electrical energy storage capacitance, but they have, respectively, one, two, three and four coils. The electromagnetic part of each model has been implemented using *FEMM 4.1*, a finite element electromagnetic simulation tool, and Matlab Simulink is used for modelling the electrical and mechanical parts.
- Section 4 presents results, which are discussed in order to conclude on the most relevant coil structure for maximizing the ball speed and the energy transfer of the reluctance coil gun, while maintaining a high level of robustness.

2. Principles of Coil Guns

2.1. Physical Concept

Magnetic field in a looped circuit composed of magnetic material and air gap tends to be maximized when a current is applied. Hopkinson law used in magnetic circuits tells that: $NI = R\Phi$, with:

- I: current in the coil (A)
- N: number of turns of the coil
- Φ: flux (Wb)
- R: reluctance (H^{-1})

Increasing the magnetic field is similar to an increase of Φ. For a given current and number of coil turns, this increase can be done by reducing the reluctance R of the magnetic circuit. Its expression is

$$R = \frac{1}{\mu_0 \mu_r} \frac{l}{S}, \text{with:}$$

- $\mu_0 = 4\pi10^{-7} \ H{\cdot}m^{-1}$: permeability of vacuum
- μ_r: relative magnetic permeability
- S: cross-sectional area of the circuit (m^2)
- l: length of the magnetic circuit (m)

The way to reduce reluctance is to minimize the air gap in the magnetic circuit, replacing it by a portion of the iron plunger (having a relative magnetic permeability equal to 5000 or more). Thus, under a strong coil current, the plunger will be propelled in order to reduce the air gap, with an important force dependent on the coil current and the number of coil turns. Figure 8 shows a three-stage coil gun. In this case, coil 1 is powered first, then coil 2, then coil 3. This is the principle of a multi-stage variable reluctance actuator.

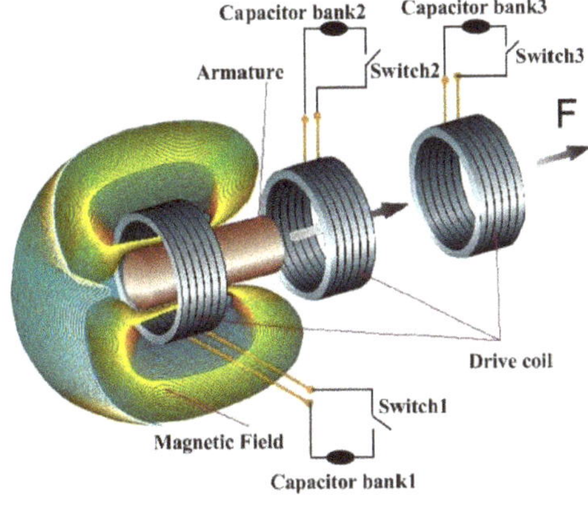

Figure 8. Three Coils Electromagnetic Launcher Principle [14].

Our case study focuses on coil gun implementations having one, two, three or four coils, with a fixed overall size and quantity of copper, as shown in Figure 9. It is important to note that between each coil, an iron plate has been placed in order to close the magnetic circuit around each coil.

Figure 9. Coil gun configurations with 1, 2, 3 and 4 coils sharing the same quantity of copper.

In our case study, each coil is powered by an identical capacitor. These capacitors have an global overall capacitance equal to (4700 µF). This capacitance is split into *n* smaller equal ones, where *n* is the number of coils. Each capacitor can be discharged, one at a time, in its corresponding coil producing a strong current which generates a magnetic force. The iron rod, mobile part of the magnetic circuit slides in a stainless steel tube in order to reduce the air gap of the magnetic circuit. This iron rod is attracted and accelerated as long as the air gap can be minimized. It is slowed down if the plunger goes to far and the air gap increases again. To avoid that, the duration of the current pulse in each coil has to be limited in time.

Discharge from the capacitor to the coil inductor can be described by a second order *RLC* differential equation. This equation has non-constant coefficients because the value of the inductor highly depends on the value of the current in the coil and on the plunger position in the sliding tube. This mixed non-linear model combining electrical and mechanical inputs will be presented in the following sections.

2.2. Electromagnetic Theory and Simulation Software

The model of an electromagnetic actuator has to take into account many non-linearities such as:

1. The saturation of magnetic materials under high currents [15] as shown in Figure 10.
2. The impact of the plunger position leading to change locally the relative magnetic permittivity by a factor 5000 or more.

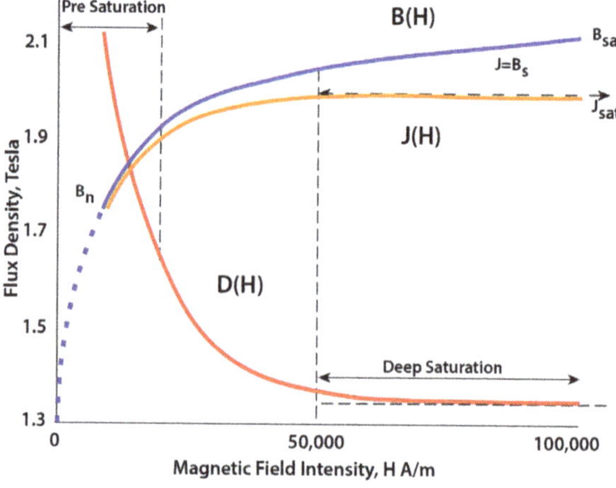

Figure 10. Example of magnetic field saturation [15].

Considering this, it is difffcult to find a theoretical solution to calculate the strength of the force applied to the rod. For taking non-linearities into account, the model used is a finite elements one obtained using an open-source simulation tool called *FEMM 4.2*. It has been developed by D.C. Meeker. This tool calculates force and inductance values under different conditions [16]. More precisely, magnetic field $\overrightarrow{B}$ and potential vector $\overrightarrow{A}$ are calculated everywhere using a successive approximation finite element solver on an axisymmetric model with a spherical boundary as shown in Figure 11.

The mesh used for this computation is determined using an heuristic approach having the following characteristics: a maximum allowable mesh size is then computed as 1% of the length of the diagonal of the bounding box of any region, leading to generate a default mesh with about 4200 elements in an empty square region as shown in Figure 12. Fine meshing is also forced in all corners and a five-degree default discretization is used for arc segments.

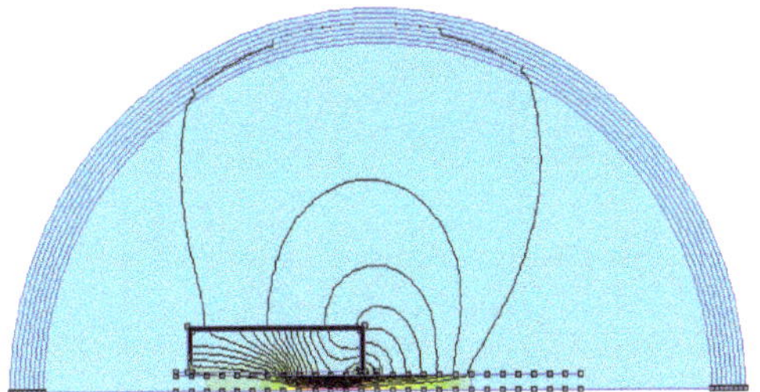

Figure 11. *FEMM 4.2* model: flux density.

Figure 12. *FEMM 4.2* mesh with its boundary.

Evaluating force and inductance value which are integral values on the mesh is also done by *FEMM 4.2* on pre-defined specific parts of the system such as the iron plunger or the inductance of the coil. Computation needs approximately 5 s on a standard *Intel Core I7* processor.

In order to compute force and inductance for all combinations of currents and plunger position, a *LUA* script is used in *FEMM 4.2*. Simulations have been done for 30 different positions of the plunger and 6 different currents for each coil: 0 A, 40 A, 80 A, 120 A, 160 A and 200 A. Simulation times are as follows:

- One coil: 180 combinations—15 min
- Two coils: 1080 combinations—90 min
- Three coils: 6480 combinations—9 h
- Four coils: 38,880 combinations—54 h

Considering the high computational cost of the electromagnetic simulations, our study has been limited to four coils, but we will show later that it is not necessary to go further.

2.3. Electrical Model

At each stage of the coil gun, a $\dfrac{4700}{n}$ µF, 450 V capacitor C is discharged in the inductance L using a controlled switch based on a MOSFET Transistor as shown in Figure 13. In this LC circuit, resistor R must be considered because its value is not negligible at all due to the important number of loops in the coil. $R = \dfrac{\rho L}{S}$ can be evaluated or measured, where ρ is the resistivity of copper, L the total length of the coil wire and S the surface of a wire section. This leads to the differential Equation (1) where L is not constant, but depends on the plunger position and on the coil current. Considering that, Equation (1) must be solved by numerical simulation.

$$\frac{d^2U_C}{dt^2} + \frac{R}{L}\frac{dU_C}{dt} + \frac{1}{LC}U_C = 0 \tag{1}$$

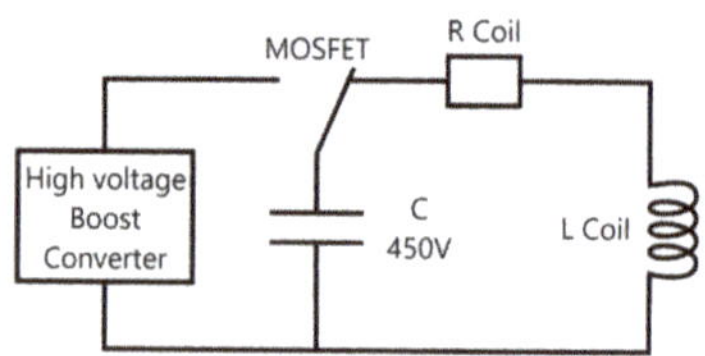

Figure 13. Electric circuit.

As shown in Figure 14, *FEMM 4.2* simulations shows that L inductance varies by a factor 20 from $L = 13$ mH to $L = 253$ mH in a 1 coil kicking system.

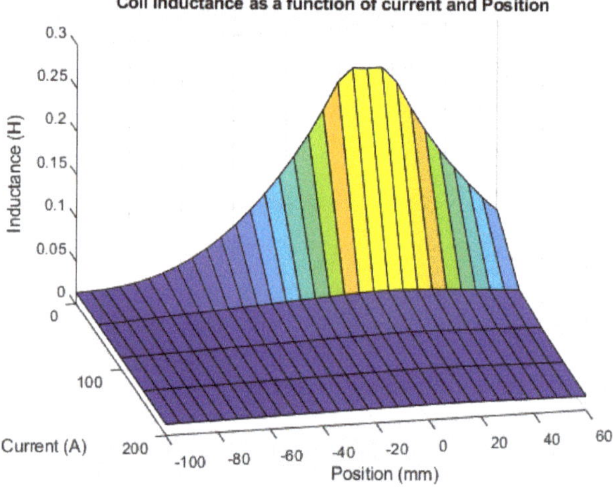

Figure 14. Variation of the inductance value depending on the plunger position and the coil current.

The inductance value depends on the plunger position and the saturation of the magnetic circuit due to the current in the coils. In Figure 14, discrete values of the inductance are calculated for positions of the plunger (Figure 15) varying from $x = -100$ mm to 60 mm by increment of 5 mm, and for coil currents varying from 1 A to 200 A by increment of 50 A.

It is important to note that:

- Inductance value L increases as the plunger enters the coil, is maximized when the plunger centre is aligned with the coil centre, and decreases after that. This is because the magnetic field is well guided when the plunger is inside the coil with a low air gap.
- The inductance value is highly dependent on the coil current. For a current $I = 1$ A, L varies, depending on plunger position, from $L = 13$ mH to $L = 253$ mH whereas for a current $I = 100$ A, L varies only from $L = 13$ mH to $L = 24$ mH. There is an important difference between maximal values because at a low current, magnetic material is not saturated, leading to a high inductance value. In contrast, there is no difference between minimal values because when the plunger is outside the coil, the air gap is so important that it leads to a huge reluctance in the air gap part which prevents saturation of the magnetic circuit.

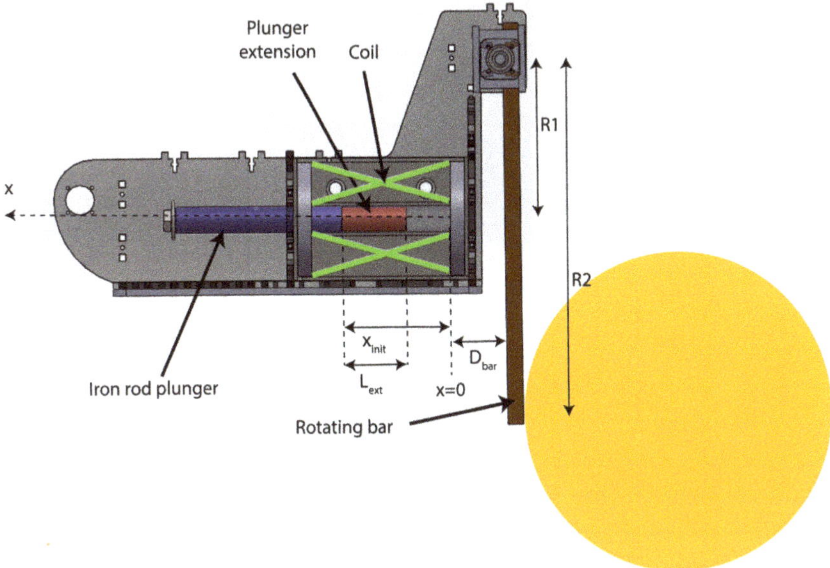

Figure 15. RoboCup reluctance coil gun kicking system.

2.4. Mechanical Model

The mechanical model has been defined following the coil gun structure described in [11]. However, our kicking system is very similar in terms of dimensions and weight to that one which serves as a reference design for most RoboCup teams. Consequently, this model will be reusable for other teams, and can be adapted for other use cases by changing the content of some blocks according to the chosen mechanical design. However, changing parameters of the coil requires to simulate again the electromagnetic part, as force depends on coil gun geometry, current and plunger position.

As for the electrical model, an analytical calculation of the force is not possible. The finite elements model using *FEMM 4.2* shows that force varies in our case study from -1900 N to 19,000 N for the same coil depending on the plunger position and on the coil current (Figure 16). Discrete values of the force are calculated for positions of the plunger (Figure 15) varying from $x = -100$ mm to 60 mm by increment of 5 mm, and for coil currents varying from 1 A to 200 A by increments of 50 A.

It is important to note that:

- Force F is almost linear with coil current in any situation.
- Force F is highly dependent on the plunger position. For a current $I = 100$ A, F varies, depending on plunger position, from $F = -954$ N to $F = 954$ N, with $F = 0$ N when the plunger is exactly aligned with the centre of the coil. This is because magnetic flux tends to be maximized in a magnetic circuit, leading to reduce the air gap. Consequently, as shown in Figure 16, the magnetic force on the plunger is symmetrical around the point where the centre of the plunger is aligned with the centre of the coil. Thus, it is important to stop powering the coil as soon as the plunger has crossed the coil.
- Force is very small if the plunger is outside the coil. This is normal considering the important length of the air gap replacing the plunger for looping back the magnetic circuit. We can also note that when the plunger is at the centre of the coil, force is null for any value of the current because magnetic flux can not be maximized.

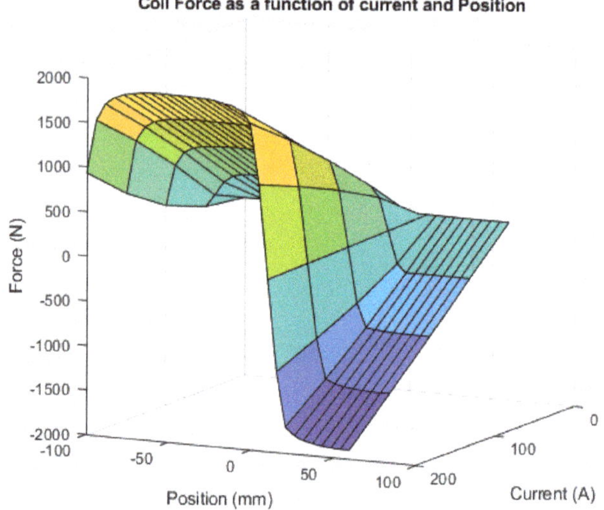

Figure 16. Variation of the force on the plunger depending on its position and the coil current.

3. Mixed Electrical and Mechanical Model of the Reluctance Coil Gun

Implementation has been done using Matlab Simulink for this mechatronic model. Simulated inductance and magnetic force values using *FEMM 4.2* are implemented in look-up tables interpolating data in order to have a force and inductance value for any position of the plunger and any coil current. Figure 17 shows the model of a single coil electromagnetic launcher used for simulations. Figure 18 shows the model of a four-coil electromagnetic launcher used for simulations. The mechanical simulation part is unique in both models, whereas the electrical part is replicated by the number of coils present in the electromagnetic launcher.

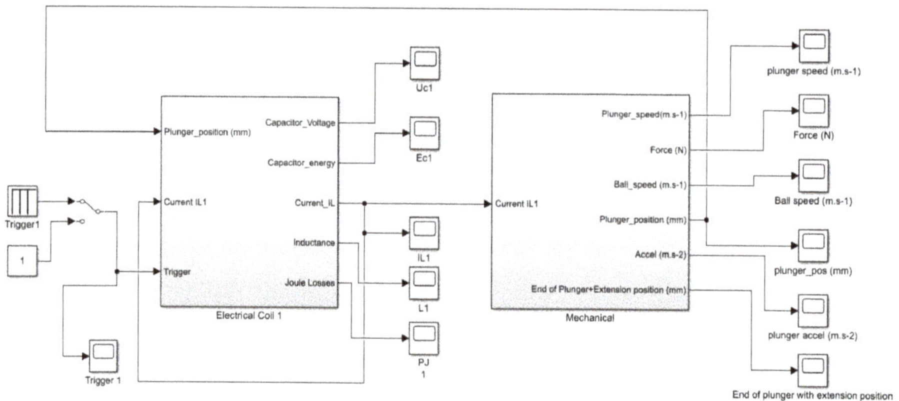

Figure 17. Mechatronic model of a 1 coil electromagnetic launcher.

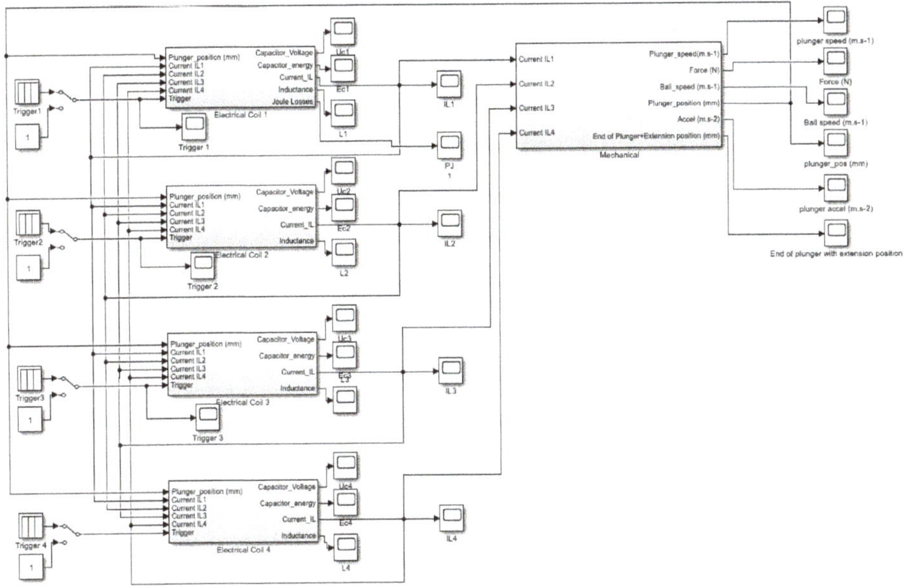

Figure 18. Mechatronic model of a 4 coils electromagnetic launcher.

3.1. Electrical Model

A model of the electrical part of the first coil of the electromagnetic launcher is described in Figure 19 in the case of a single coil launcher, and in Figure 20 in the case of a four-coil launcher. Electrical differential Equations (1) are implemented using discrete blocks because coefficients of the equation are not constant due to the dependence of inductance L on the current and position of the plunger.

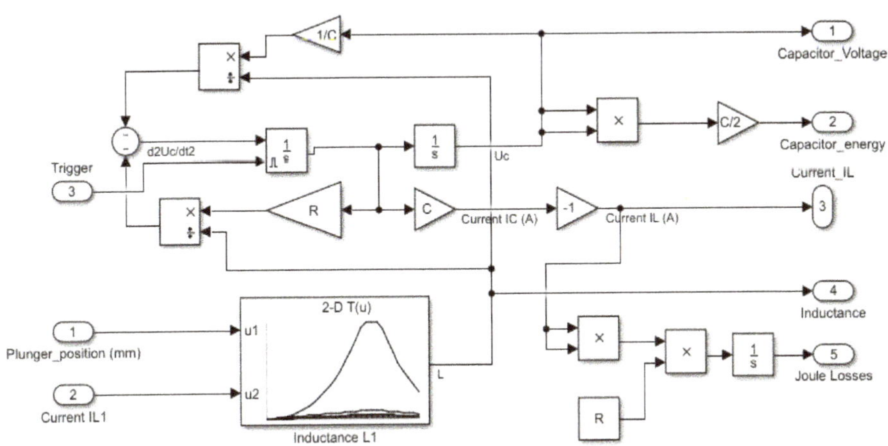

Figure 19. Electrical model of the first coil of an electromagnetic launcher with 1 coil.

As shown in Figures 19 and 20, the only difference between both models is the number of inputs of the inductance look-up table (LUT) block interpolating linearly the value of L using the simulations performed with *FEMM 4.2*. In the case of a four-coil EML, there are five inputs: plunger position, and the currents on each of the four coils.

It is important to note that there is a trigger input in each block. This input corresponds to an electronic trigger supplied by a pilot board and used for piloting the MOS transistor commutating the capacitor on the coil. There is one trigger input per coil, so that it is possible to drive them independently.

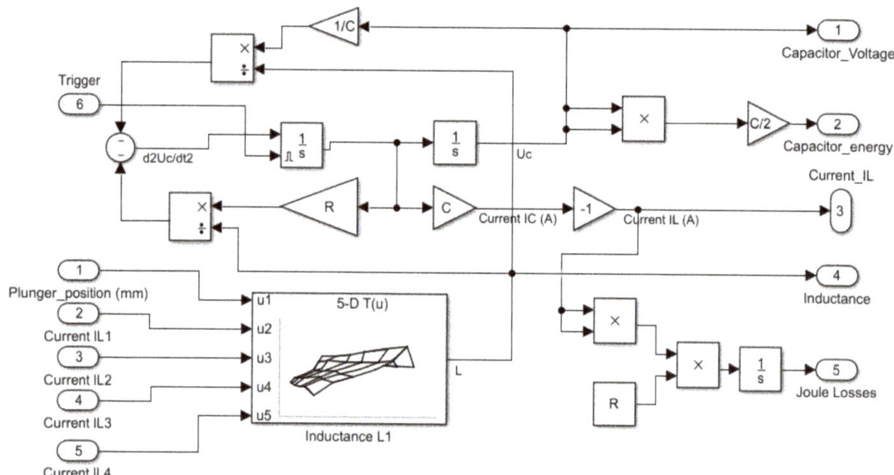

Figure 20. Electrical model of the first coil of an electromagnetic launcher with 4 coils.

3.2. Mechanical Model

The reluctance coil gun mechanical system shown in Figure 15 has been described in a former work [12]. This model takes into account the transmission of the movement from the plunger to the ball through an aluminium lever. Movement can be split into three phases as shown in Figure 21:

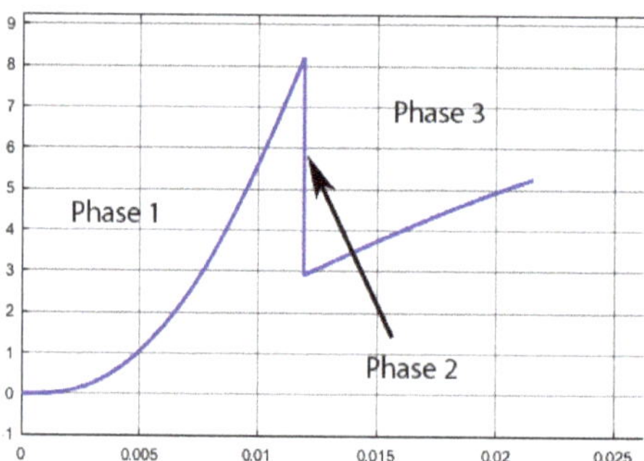

Figure 21. Plunger speed in m·s^{-1} over the time in s.

- Phase 1: Acceleration of the plunger without contact on the lever, due to the magnetic force as shown in Equation (2) without contact on the lever. Force $F_{magneto}$ depends non-linearly on plunger position and current I. A look-up table (LUT) interpolates linearly the value of F using the simulations performed with *FEMM 4.2*.

$$m_p \ddot{x} = F_{Magneto}(x, I) \tag{2}$$

- Phase 2: Impact on the lever corresponding to an elastic shock when plunger hits it at a distance R_1 from its rotation centre. Kinetic energy is conserved as described in Equation (3).

$$\frac{1}{2} m_p \dot{x}_{Init}^2 = \frac{1}{2} m_p \dot{x}_{Final}^2 + \frac{1}{2} m_B \frac{R_2^2}{R_1^2} \dot{x}_{Final}^2 + \frac{1}{2} J_{Lever} \frac{\dot{x}_{Final}^2}{R_1^2} \tag{3}$$

where J_{Lever} is the inertial moment of the lever, R_1 and R_2 the distances between the lever axis and respectively the plunger impact point and the ball impact point as shown in Figure 15. This leads to a plunger speed just after the shock equal to the $\dot{x}_{final}$ given in Equation (4).

$$\dot{x}_{Final} = \sqrt{\frac{m_p}{m_p + m_B \dfrac{R_2^2}{R_1^2} + \dfrac{J_{Lever}}{R_1^2}}} \dot{x}_{Init} \tag{4}$$

- Phase 3: the plunger is accelerated in contact with the lever, which is also in contact with the ball. This means that the lever applies a force on the plunger in subtraction of the magnetic force as shown in Equation (5). This force is an inertial one due to the acceleration of the ball and the lever as shown in Equation (6). It is important to note that theoretically, the speeds of the ball, lever and plunger are not equal after the shock, but in reality they are due to the elastic deformation of the ball as shown in the slow motion picture in Figure 22.

$$m_p \ddot{x} = F_{Magneto}(x, I) - F_{Lever} \tag{5}$$

where

$$F_{Lever} = \frac{J_{Lever} + m_B R_2^2}{R_1} \ddot{\theta} \tag{6}$$

with:

$$J_{Lever} = \frac{m_{Lever} R_2^2}{3}$$

For small θ angles, $\ddot{\theta} \simeq \dfrac{\ddot{x}}{R_1}$, this leads to :

$$\frac{m_p R_1^2 + J_{Lever} + m_B R_2^2}{R_1^2} \ddot{x} = F_{Magneto}(x, I) \tag{7}$$

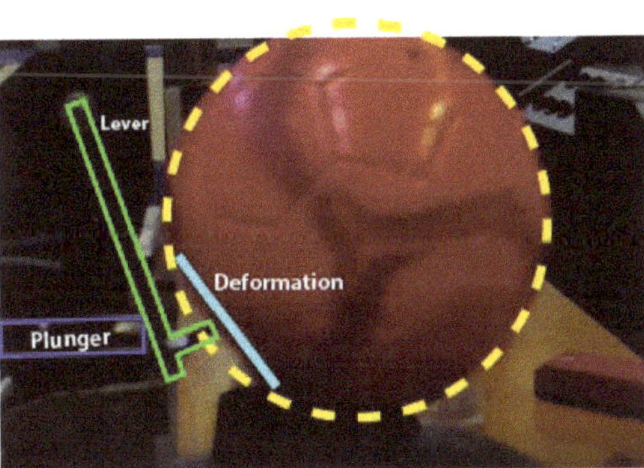

Figure 22. Ball deformation after phase 2.

Implementation of this three-phase mechanical model has been done using Matlab Simulink. Figure 23 shows the mechanical part model of a 1 coil electromagnetic launcher, whereas Figure 24 shows the mechanical part model of a 4 coils electromagnetic launcher. It is important to note that, as shown in Figure 22, plunger, lever and ball are in contact after the shock. This is due to the softness of the ball, and because the ball is close to be in contact with the lever before the impact. Thus, the hypothesis of a perfect elastic shock is almost verified except for a transitional short period of less than one millisecond after the shock of the rod on the lever.

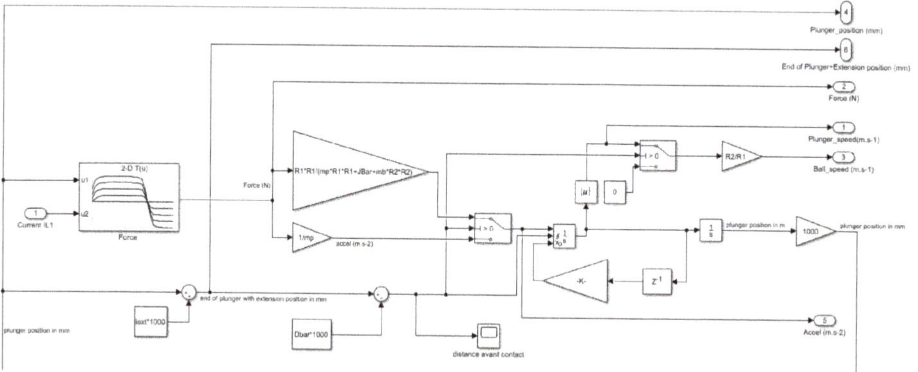

Figure 23. Mechanical part model of a 1 coil electromagnetic launcher.

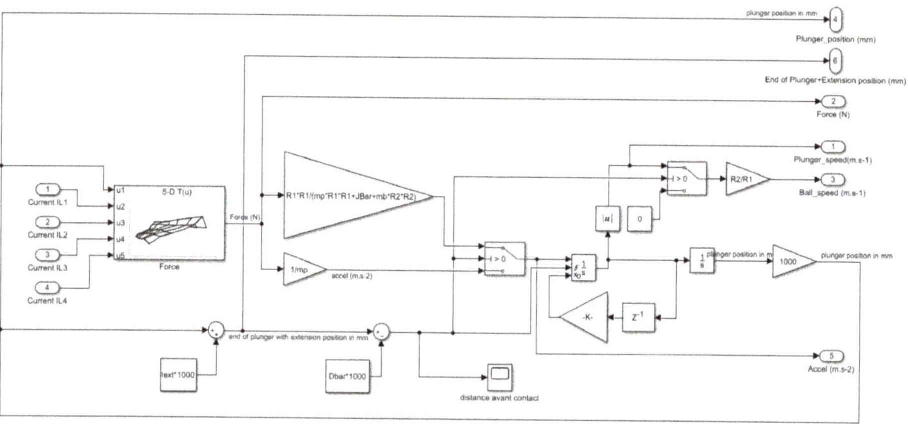

Figure 24. Mechanical part model of a 4 coils electromagnetic launcher.

However, in order to understand more accurately what is going on during this transition, impact of the plunger on the lever and impact of the lever on the soft ball will be modelled in another work (for example using MSC ADAMS software), but this is out of the scope of this paper.

As shown in Figures 23 and 24, the only difference between both models is the number of inputs of the force look-up table (LUT) block interpolating linearly the value of F using the simulations performed with *FEMM 4.2*. In the case of a four-coil EML, there are five inputs: plunger position, and the currents on each of the four coils.

4. Reluctance Coil Gun Simulations

4.1. Hypothesis

The reluctance coil gun previously described and used in robots at the RoboCup has been simulated using Matlab Simulink. In this study, we focus on optimizing the inner structure of the coil gun and especially we aim at finding the optimal number of coils and the optimal instant and duration of triggering for each coil in a sequence.

Optimality is not only focused on the ball speed which must be as important as possible, but also on the reliability and robustness of the triggering system, which can be very sensitive to a small change in the triggering delay when several coils (especially 3 or 4) are used.

This last point is important because we have decided of not adding an observer of the plunger position in the system such as a set of infrared light barriers. This choice has been done considering the mechanical difficulties for inserting sensors inside the coil gun structure, and the issues about robustness it can raise due to the huge impacts and vibrations on the EML structure. Instead of that, a simple and robust open loop driving has been chosen, each coil being commutated during a fixed time and with a fixed delay from the start of the sequence. Considering that the initial conditions of the plunger position are always the same ones (this is true because the plunger is returned to its initial position by an elastic restoring force), the ball speed has been measured to be almost the same at each shooting sequence.

This paper does not focus on optimizing the initial position x_{init} of the plunger, and the length L_{ext} of the non-magnetic extension of the plunger, as done in [12]. In this study, we started using the results presented in [12]. However, in a final step, a fine optimization has been done for getting the best possible solution on both parameters.

4.2. Model Parameters

In order to compare results with other previous studies, the kicking system simulated is identical to the *Tech United Team* one described in [11]. However, geometry of our coil gun is very similar to this reference design. Parameters of the model are as follows:

- Distance from lever axis to plunger touch point: $R_1 = 13$ cm
- Distance from lever axis to ball touch point: $R_2 = 24$ cm
- Coils number (each coil as been chosen identical): $nb_{Coils} = 1, 2, 3, 4$
- Coils length (for each coil): $L_{Coil} = 11.5/nb_{Coils}$ cm
- Coils number of turns (for each coil): $N_{Coil} = 1000/nb_{Coils}$ turns
- Coils resistance (for each coil): $2.5/nb_{Coils}\Omega$
- Capacitors number: $nb_{Capacitors} = nb_{Coils}$
- Capacitors value (for each capacitor): $4700\ \mu F/nb_{Coils}$
- Capacitors charge voltage: 425 V
- Plunger iron rod diameter: $D_{Plunger} = 25$ mm
- Plunger iron rod length: $L_{Plunger} = 11.5$ cm
- Plunger iron rod mass: $m_{Plunger} = 690$ g
- Plunger extension diameter: $D_{Ext} = 18$ mm
- Plunger extension length: L_{Ext} cm
- Plunger extension mass: $m_{Ext} = 0.68 * L_{Ext}$ (in m)
- Distance from coil to lever: $D_{Lever} = 4$ cm
- Vertical lever mass: $m_{Lever} = 80$ g
- Ball mass: $m_{Ball} = 450$ g

4.3. Simulations

Simulations have been done using coil guns having one, two, three and four coils. As explained before, the overall quantity of copper and the global number of coil turns is a constant, as is the sum of the capacitors value. Delays and durations of each trigger pulses for each coil have been optimized manually in order to maximize the ball speed.

Results of simulation for an EML having one coil are presented in Figure 25. $Coil_1$ is triggered during 25 ms. Optimal initial position of the plunger is -92 mm. Ball speed reaches 13.45 m·s^{-1}.

Results of simulation for an EML having two coils are presented in Figure 26. $Coil_1$ is triggered during 15 ms, $Coil_2$ is triggered during 14 ms with a delay of 10 ms. Optimal initial position of the plunger is -82 mm and optimal plunger extension length is 104 mm. Ball speed reaches 16 m·s^{-1}.

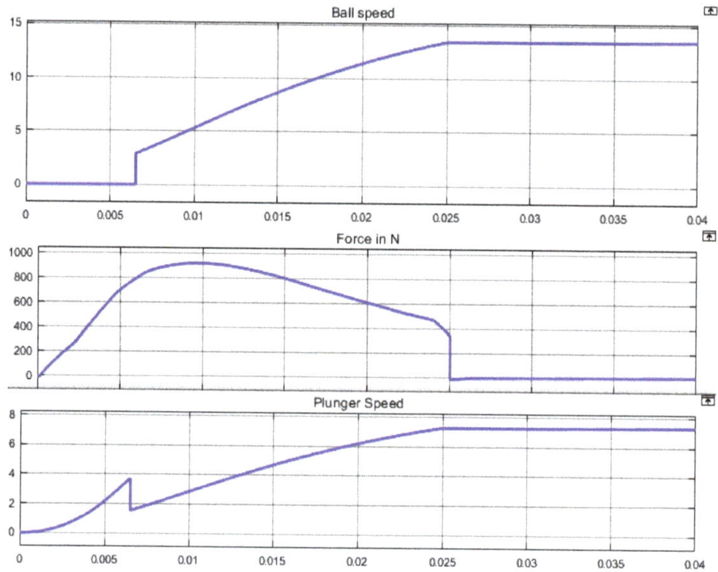

Figure 25. Simulation of one coil EML used in an optimal way.

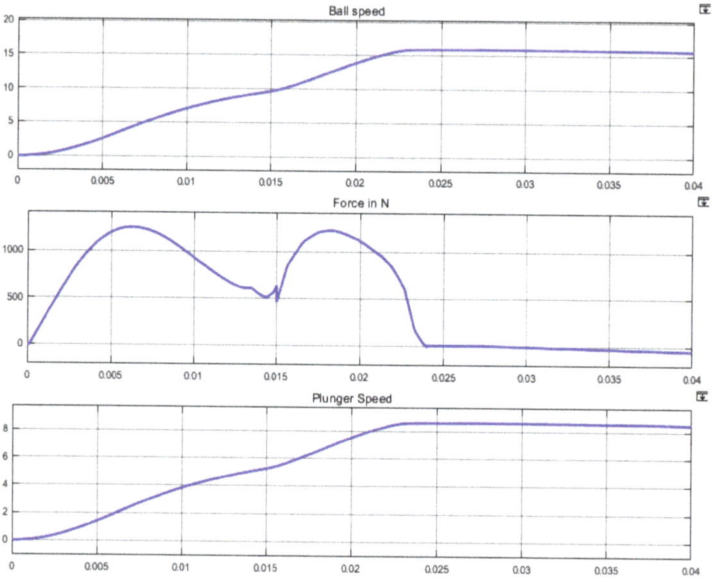

Figure 26. Simulation of a two-coil EML used in an optimal way.

Results of simulation for an EML having three coils are presented in Figure 27. $Coil_1$ is triggered during 10 ms, $Coil_2$ is triggered during 15 ms with a delay of 7 ms, $Coil_3$ is triggered during 12 ms with a delay of 11 ms. Optimal initial position of the plunger is -82 mm and optimal plunger extension length is 90 mm. Ball speed reaches 16.1 m·s^{-1}.

Results of simulation for an EML having 4 coils are presented in Figure 28. $Coil_1$ is triggered during 10 ms, $Coil_2$ is triggered during 10 ms with a delay of 7 ms, $Coil_3$ is triggered during 12 ms with a delay of 11 ms, $Coil_4$ is triggered during 14 ms with a delay of 15.5 ms. Optimal initial position of the plunger is −92 mm and optimal plunger extension length is 104 mm. Ball speed reaches 16.4 m·s^{-1}.

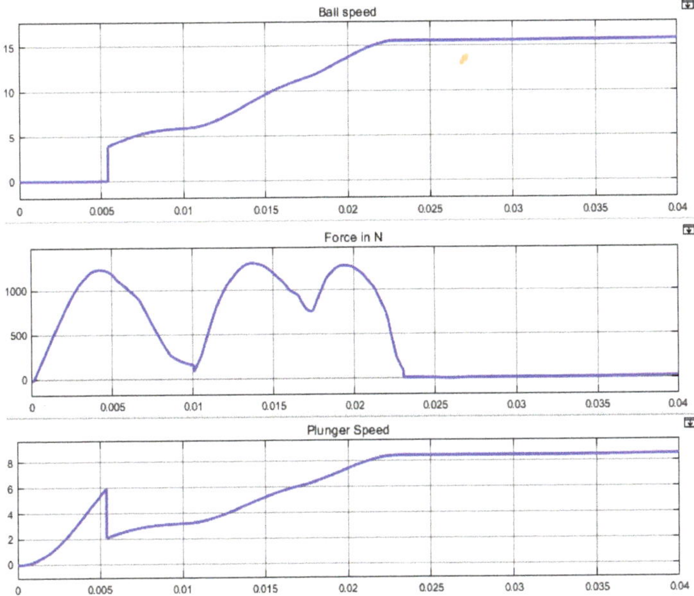

Figure 27. Simulation of a two-coil EML used in an optimal way.

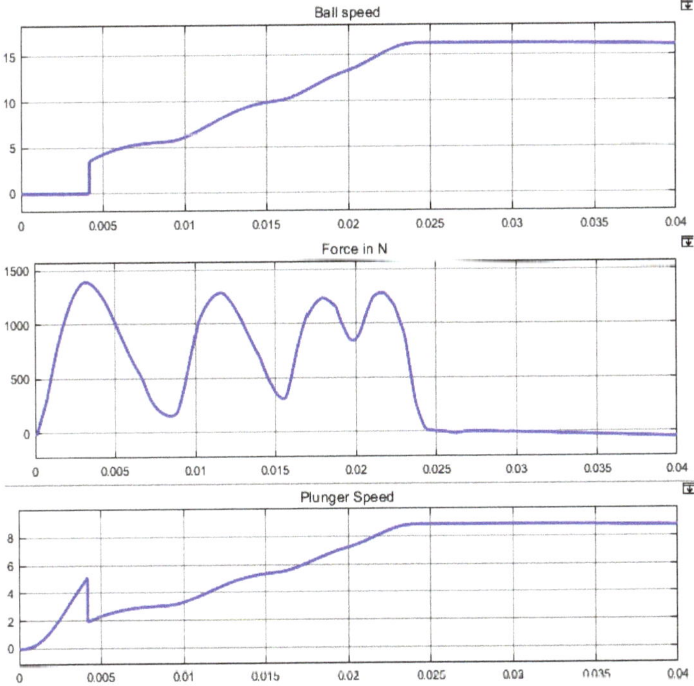

Figure 28. Simulation of a four-coil EML used in an optimal way.

5. Results and Discussion

5.1. Discussion on Simulation Results

5.1.1. Impact of the Coil Gun Structure

As presented in the results section, increasing the number of coils allows to transfer more power to the ball as shown in Table 2:

Table 2. Optimal ball speed depending on the number of coils in the EML.

Number of Coils	1	2	3	4
Optimized Ball Speed (m·s^{-1})	13.5	16	16.1	16.3
Kicking Range (m)	18.6	26.1	26.4	27

Increasing the number of coils from one to two allows to increase the speed by 18%, corresponding to an energy transfer optimization of 40%. However, increasing the number of coils from two to three or four allows to increase the speed by only respectively 0.6% and 1.8%, corresponding to an energy transfer optimization of respectively 1.2% and 3.6%. This result is not intuitive and is important. Consequently, considering the impact of adding a coil to the EML in terms of mechanical and electrical integration, the two-coil EML seems to be the best configuration in our case.

5.1.2. Comparison between the Reference Case and the Chosen Configuration

Comparing the reference situation described in [11], optimization of the number of coils of the EML, the initial position and the extension length leads to increase the ball speed by 42%, from 11.2 m·s^{-1} to 16 m·s^{-1}. This corresponds to an energy transfer improved by 104% compared to the reference situation, without complicating the coil gun structure too much.

Shooting range, which is defined as the distance of the first rebound in case of a 45° kick, without considering air friction, increases from 12.5 m in the reference situation described in [11] to 26 m in the optimal configuration.

5.1.3. Impact of a Variation of the Coil Triggering Instants

Having a powerful coil gun is important, but its behaviour robustness is also a key factor. This is especially true for the triggering instants which are important parameters for optimizing power transmission and have to be tuned carefully.

In the case of a two-coil EML, simulations show that the second coil optimal triggering instant is 10 ms after the first coil. Table 3 shows that a variation of this triggering instant of ±1 ms has a very limited impact on the ball speed, which is only reduced by less than 2%, leading to a good robustness of this system.

Table 3. Impact of a variation of the second coil triggering instant in the case of a 2 coils EML.

Triggering Instant (ms)	7	8	9	10	11	12	13
Optimized Ball Speed (m·s^{-1})	14.4	15.2	15.8	16.1	16	15.8	15.6

In the case of a four-coil EML, the fourth coil optimal triggering instant is 15.5 ms after the first coil. Table 4 shows that a variation of this triggering instant of ±1 ms has an important impact on the ball speed which is reduced by 11% or more. Consequently, the EML using four coils is far less robust than the EML using two coils in terms of sensitivity to the coil-triggering instants.

Table 4. Impact of a variation of the fourth coil triggering instant in the case of a four-coil EML.

Triggering Instant (ms)	13.5	14.5	15.5	16.5	17.5
Optimized Ball Speed (m·s^{-1})	13.3	14.6	16.4	14.3	13.9

5.2. Experimental Results

In order to validate the simulation results, experiments have been done using a two-coil reluctance coil gun corresponding to the optimal configuration (as explained in Section 5.1). This launcher is shown in Figure 29. It is part of our new RoboCup robot presented in Figure 30.

Figure 29. 2 coils optimized reluctance coil gun.

Figure 30. RoboCup robot for testing the optimized coil gun.

A custom four-channel coil gun driver has been designed but is not in the scope of this paper. It includes four MOSFET for capacitor commutation with a 160 A current peak under 450 V on each coil. A safety system for dissipating energy stored in the capacitors when the system is switched off or stopped has been implemented in this driver, justifying the aluminium ventilated power heat sink that can be seen on the right board of Figure 31. For triggering the coils in a very accurate time sequence (and for controlling DC motors and low level sensors), a micro-controller board has been designed and can be seen on the left side of Figure 31. In our test corresponding to the chosen optimal case, the second coil has been triggered exactly 10 ms after the first one.

Figure 31. Four-channel reluctance coil gun driver (on the right).

Ball speed measurements have been done using a high speed camera on 20 successive tests. Average measured ball velocity is equal to $V_{Ball} = 15.5\text{ m·s}^{-1}$. This is consistent with the theoretical value (16 m·s^{-1}). Error is only 3.1% and dispersion is low ($\sigma = 0.2$ m·s^{-1}). These results show that the simulation model used in this paper is accurate, despite many strongly non-linear effects, and that the structure of a reluctance coil gun can be optimized very efficiently without changing the amount of copper used and the size of the actuator.

6. Conclusions

In this paper, a method for optimizing the structure of a reluctance coil gun has been proposed. Kicking real soccer balls used at the RoboCup in the Middle Size League has been chosen as a case study. After having presented the principles of coil guns, a mechatronic model coupling mechanic, electromagnetic and electric ones has been proposed and implemented. Simulation results have been explained and discussed so that this optimization method can be easily reproduced in another application.

Results show that the output speed of the non-magnetic object propelled by the EML highly depends on the structure of the coil gun, the sequence for triggering it, the initial position of the iron plunger and the size of its non-magnetic extension. Among the results of this paper, we show that

- Using a two-coil EML is 104% energetically more efficient than the reference situation of an existing coil gun [11], without adding too much mechanical, electrical and algorithmic complexity to the EML. As shown in Table 5, it is also ten times more efficient than a human considering the ball energy to launcher volume ratio.
- Having a high number of coils is not necessary for optimizing the energy transfer. In our case, having two coils in the EML is an excellent trade-off between energy transfer optimization and system complexity.
- Robustness in terms of sensitivity to the coil triggering instants decreases with the number of coils.

Table 5. Ball launchers comparison including optimized launcher.

Launcher	Length (cm)	Width (cm)	Height (cm)	Volume (cm^3)	Weight (kg)	Ball Speed (m·s^{-1})	Ball Energy (J)	$\frac{Energy}{Volume}$ (J·dm^{-3})
Soccer player leg	160	20	80	133×10^3	20	36	290	**2.18**
Rotating inertial launcher	25	65	25	40×10^3	25	29	190	**4.75**
Robot arm [9]	240	240	30	1360×10^3	50	21	100	**0.07**
Reluctance coil gun [11]	30	9	9	2.4×10^3	4.5	11.4	29	**12.08**
Optimized coil gun	30	9	9	2.4×10^3	4.5	16.4	60.5	**25.21**

Author Contributions: Writing—original draft preparation, V.G.; writing—review and editing, T.S., S.M., H.G. and V.H.; visualization, V.B. and H.B. All authors have read and agreed to the published version of the manuscript.

Funding: The APC was funded by [the Conseil Départemental du Var, France].

Acknowledgments: The authors would like to thank Pôle INPS of Toulon University and the Embedded Electronics Technology Platform of Toulon University named SMIoT (Scientific Microsystems for Internet of Things) [17] for its support in the achievement of this work, and the Conseil Départemental du Var, France, for funding this project.

Conflicts of Interest: The authors declare no conflict of interest.

References

1. Tzeng, J.T.; Schmidt, E.M. Comparison of Electromagnetic and Conventional Launchers Based on Mauser 30-mm MK 30-2 Barrels. *IEEE Trans. Plasma Sci.* **2011**, *39*, 149–152. [CrossRef]

2. McNab, I. Progress on Hypervelocity Railgun Research for Launch to Space. *IEEE Trans. Magn.* **2009**, *45*, 381–388. [CrossRef]

3. Cao, B.; Guo, W.; Ge, X.; Sun, X.; Li, M.; Su, Z.; Fan, W.; Li, J. Analysis of Rail Erosion Damage during Electromagnetic Launch. *IEEE Trans. Plasma Sci.* **2017**, *45*, 1263–1268. [CrossRef]

4. Wetz, D.; Stefani, F.; McNab, I. Experimental Results on a 7-m-Long Plasma-Driven Electromagnetic Launcher. *IEEE Trans. Plasma Sci.* **2011**, *39*, 180–185. [CrossRef]

5. Orbach, Y.; Oren, M.; Golan, A.; Einat, M. Reluctance Launcher Coil-Gun Simulations and Experiment. *IEEE Trans. Plasma Sci.* **2019**, *47*, 1358–1363. [CrossRef]

6. Meessen, K.J.; Paulides, J.J.H.; Lomonova, E.A. Analysis and design of a slotless tubular permanent magnet actuator for high acceleration applications. *J. Appl. Phys.* **2009**, *105*, 07F110. [CrossRef]

7. Bencheikh, Y.; Ouazir, Y.; Ibtiouen, R. Analysis of capacitively driven electromagnetic coil guns. In Proceedings of the XIX International Conference on Electrical Machines—ICEM 2010, Rome, Italy, 6–8 September 2010; pp. 1–5.

8. Abdo, T.M.; Elrefai, A.L.; Adly, A.A.; Mahgoub, O.A. Performance analysis of coil gun electromagnetic launcher using a finite element coupled model. In Proceedings of the 2016 Eighteenth International Middle East Power Systems Conference (MEPCON), Cairo, Egypt, 27–29 December 2016; pp. 506–511. [CrossRef]

9. Schempf, H.; Kraeuter, C.; Blackwell, M. Roboleg: A robotic soccer-ball kicking leg. In Proceedings of the 1995 IEEE International Conference on Robotics and Automation, Nagoya, Japan, 21–27 May 1995; Volume 2, pp. 1314–1318. [CrossRef]

10. Vahidi, M.; Moosavian, S. Dynamics of a 9-DoF robotic leg for a football simulator. In Proceedings of the 2015 3rd RSI International Conference on Robotics and Mechatronics (ICROM), Tehran, Iran, 7–9 October 2015; pp. 314–319. [CrossRef]

11. Meessen, K.J.; Paulides, J.J.H.; Lomonova, E.A. A football kicking high speed actuator for a mobile robotic application. In Proceedings of the IECON 2010—36th Annual Conference on IEEE Industrial Electronics Society, Glendale, AZ, USA, 7–10 November 2010; pp. 1659–1664. [CrossRef]

12. Gies, V.; Soriano, T. Modeling and Optimization of an Indirect Coil Gun for Launching Non-Magnetic Projectiles. *Actuators* **2019**, *8*, 39. [CrossRef]

13. Williamson, S.; Horne, C.D.; Haugh, D.C. Design of pulsed coil guns. *IEEE Trans. Magn.* **1995**, *31*, 516–521. [CrossRef]

14. Kang, Y. A high power-density, high efficiency front-end converter for capacitor charging applications. In Proceedings of the Applied Power Electronics Conference and Exposition (APEC 2005), Austin, TX, USA, 6–10 March 2005; Volume 2, pp. 1258–1264. [CrossRef]

15. Rao, D.K.; Kuptsov, V. Effective Use of Magnetization Data in the Design of Electric Machines with Overfluxed Regions. *IEEE Trans. Magn.* **2015**, *51*, 1–9. [CrossRef]

16. Lequesne, B.P. Finite-element analysis of a constant-force solenoid for fluid flow control. *IEEE Trans. Ind. Appl.* **1988**, *24*, 574–581. [CrossRef]

17. SMIOT: Scientific Microsystems for the Internet of Things. 2019. Available online: http://www.smiot.fr (accessed on 2 March 2020).

 *applied sciences*

Article

Distribution of Magnetic Field in 400 kV Double-Circuit Transmission Lines

Ramūnas Deltuva * and Robertas Lukočius

Faculty of Electrical and Electronics Engineering, Kaunas University of Technology,
LT-51367 Kaunas, Lithuania; robertas.lukocius@ktu.lt
*   Correspondence: ramunas.deltuva@gmail.com; Tel.: +370-612-98623

Received: 24 March 2020; Accepted: 1 May 2020; Published: 8 May 2020

Abstract: A high-voltage AC double-circuit 400 kV overhead power transmission line runs from the city of Elk (Poland) to the city of Alytus (Lithuania). This international 400 kV power transmission line is potentially one of the strongest magnetic field-generating sources in the area. This 400 kV voltage double-circuit overhead transmission line and its surroundings were analyzed using the mathematical analytical methods of superposition and reflections. This research paper includes the calculation of the numerical values of the magnetic field and its distribution. The research showed that the values of the magnetic field strength near the international 400 kV power transmission line exceed the threshold values permitted by relevant standards. This overhead power line is connected to the general (50 Hz) power system and generates a highly intense magnetic field. It is suggested that experimental trials should be undertaken in order to determine the maximum values of the magnetic field strength. For the purpose of mitigating these values, it is suggested that the height of the support bars should be increased or that any individual and commercial activities near the object under investigation should be restricted.

Keywords: magnetic field strength; magnetic flux density; magnetic potential; current density; power transmission line

1. Introduction

The institutions involved in electric power transmission in Lithuania and Poland have decided to implement the electricity link "LitPol Link". This link is designed to connect the power systems of the Baltic States to those in Western Europe and also contributes to the development of the general European electricity market as well as boosting the reliability of the power supply. In pursuit of increasing the capacities of electric power transmission, the decision was made to supplement the LitPol Link by installing an extra 400 kV high-voltage AC double-circuit transmission line.

Recently, there has been increasingly widespread concern over the influence of the electromagnetic field (50 Hz) on the health of residents. One of the fundamental challenges is ensuring safe conditions for those living in locations where such an electromagnetic field is present.

The recently adopted Hygiene Standard HN 104:2011 "Human protection against electromagnetic fields caused by overhead power lines" is currently valid in the European Union (EU) and states that the effective values of the magnetic field strength must never exceed exceed 32 A/m or 40 µT (magnetic flux density) in residential environments, or 16 A/m or 20 µT (magnetic flux density) within residential and public service buildings. These values should never be exceeded irrespective of the duration of a person's exposure to an electromagnetic field.

The double-circuit 400 kV power transmission line is one of the most powerful electrical installations in the entire power system; at a frequency of 50 Hz, it generates an intense magnetic field. Therefore, it is necessary to focus on the installation and safe operation of the 400 kV double-circuit overhead transmission line in order to prevent the magnetic field generated by this power transmission line from exceeding the requirements set by the European Union (EU) Hygiene Standard HN 104:2011.

2. Extra High-Voltage Double-Circuit Electric Power Transmission Line

As the strongest magnetic field in the electrical power system can be generated by the double-circuit 400 kV power transmission line, it is necessary to determine how the magnetic field is distributed in the surroundings of this overhead power transmission line.

This paper determines a three-bundled, double-circuit 400 kV AC power transmission line. Figure 1 shows a power transmission line of the low-reactance orientation type. The height of conductors shown in the figure is the maximum sag position. The lowest conductors are C_1 and C_2 at the height of 10 m above the ground surface level [1]. Each phase conductor is 0.04 m in diameter. The overhead ground wire has a diameter of 0.015 m.

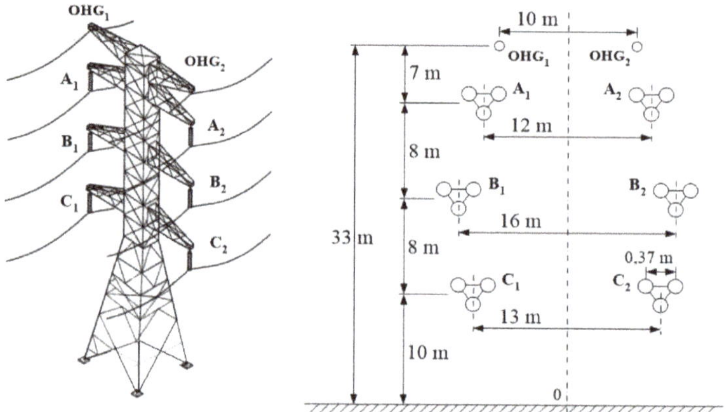

Figure 1. High-voltage double-circuit: three-bundled transmission line.

The height of the 400 kV transmission line supports is 54 m. In this particular case, the lower-phase wires (C_1 and C_2) (see Figure 1), including their insulators, should be arranged at a height of at least 26 m to the ground surface. The sag of the transmission line amounts for 16 m. Adjacent support-bars are spaced by 350 to 550 m [2].

The magnetic field generated by the 400 kV overhead transmission line can be approached as the superposition of the magnetic fields of six long, thin current leads. It is also possible to calculate this magnetic field analytically. As the current leads of the 400 kV overhead transmission line are very long and thin wires, difficult mathematical equations are derived for the purpose of calculating the magnetic field they generate.

Moreover, as regards the effect on humans and the environment, the distribution of these current leads is of utmost importance. However, typical analytical mathematical equations for the calculation of this magnetic field are also missing. Thus, digital modeling should be selected as the main method for the investigation of the magnetic field generated by the 400 kV overhead transmission line [3–6]. Besides the multitudes of advantages it possesses, modeling also enables the simple assessment of the change in the magnetic field when varying the structure and dimensions of the installation. It also allows the different irregularities and variations of the surroundings to be taken into account. Analytical mathematical equations for the calculation of the magnetic field generated by the double-circuit 400 kV overhead transmission line were derived with the aim of verifying the results of modeling [7–16].

3. Magnetic Field Generated by the Single Thin Current Lead

As the double-circuit 400 kV transmission line is comprised of six long, thin current leads, the magnetic field generated by it can be calculated using the method of superposition. We therefore analyze how the magnetic field generated by a single current lead can be calculated. When the magnetic permeability of all points in the space is the same and equal to the permeability of free space μ_0, the calculation of the magnetic field of the given symmetric current system uses the LaPlace–Poisson equations. When current is passed through the current lead with the element idl, the length is l_i if r represents a radius-vector pointing from the current element idl to the point of observation, and the electrical current is distributed within the volume of the current lead Vj, the vectors of the magnetic field strength H and of the magnetic field density B at the point of observation M are calculated as follows [17–22]:

$$H = \int_{Vj} dH; \; dH = \frac{(rJ)dV}{4\pi r^3}; \; dH = \frac{JdV}{4\pi r^2}\sin \angle r, J; \; B = \mu_r\mu_0 H. \tag{1}$$

The magnetic field generated by the current lead can be calculated using a vector potential. When the electric current flows through the lead, making a circuit l_i, and the element dl and electric current density J within the volume of the current lead Vj are known, the vector potential A is expressed as follows:

$$A = \int_{Vj} \frac{\mu_r\mu_0 J dV}{4\pi r}. \tag{2}$$

The density of the magnetic field generated by all the current elements idl of the current lead is expressed based on the superposition principle by summing up the constituents of the flux generated by individual current elements [7–15,23].

Due to axial symmetry, the magnetic field strength line of a long and thin current lead with an electric current passing through it have a circular shape, with their centers on the geometrical axis of the current lead, and they are located in planes that are perpendicular to this axis. At any point of such a circle, the directions of the magnetic field strength vector H and distance element dl match.

4. Analysis of Magnetic Field from 400 kV AC Power Transmission Line

In this challenging task regarding magnetic fields, the vectors of magnetic flux density B or magnetic field strength H are found to be functions of coordinates when the dependency of the current density vector J on coordinates is known. As the magnetic field strength or magnetic flux density is to be found in areas in which there are no currents, LaPlace's equation is solved for the scalar magnetic potential. Such a magnetic field will also be of a potential nature [7–15,23].

For the purpose of solving this task, the method of reflections is used. The magnetic field of the conductor system in the double-circuit 400 kV overhead transmission line is examined when the conductor system is comprised of long, thin, round cylindrical conductors running in parallel and with flat conductive surfaces with the symmetrical current systems i_{i1} ($i_1 = A_1, B_1, C_1$) and i_{i2} ($i_2 = A_2, B_2, C_2$) (see Figure 2). The radii of current leads $r_i \ll h_i$ are significantly lower than the distance from the ground surface to the current leads. The sag of the double-circuit overhead power line is not taken into consideration.

Following the method of reflections, the magnetic field in which the distribution of currents over the flat surface of the conductor is known is replaced with the magnetic field of currents i_{in} and that of their reflections-i_{in}. It is assumed that the distances from the conductors i_{in} to their respective reflections-i_{in} are equal; i.e., y_{in} ($i = A, B, C$) ($n = 1, 2$).

With the values of the currents $i_{A1}, i_{B1}, i_{C1}, i_{A2}, i_{B2}, i_{C2}$ being known, the values of reflections of these loads-$i^*_{A1}, -i^*_{B1}, -i^*_{C1}, -i^*_{A2}, -i^*_{B2}, -i^*_{C2}$ are assumed to have the opposite signs.

In order to determine how the values of currents change in a single period of sinusoidal quantity, calculations are made by varying the angle of the phase current at steps of 10^0; i.e., $\triangle wt = 10^0$.

Instantaneous phase current values of a symmetrical current system are interconnected, as follows:

$$\begin{cases} i_A = I_m \, \sin(wt), \\ i_B = I_m \, \sin(wt\text{-}120^0), \\ i_C = I_m \, \sin(wt\text{+}120^0); \end{cases} \tag{3}$$

where I_m is the maximum amplitude value of the phase current, in A.

The 400 kV double-circuit power transmission line has a maximum effective linear current $I_1 = I_f = 2500$ A; thus, the value of the amplitude phase current I_{mf} is as follows: $I_m = I_1 \cdot \sqrt{2} = 3540$ A.

Using Equation (3), instantaneous current values are calculated at steps of $\triangle wt = 10^0$.

$$H_{in} = \frac{\pm i_{in}}{2\pi r_{in}}; \tag{4}$$

where i_{in} $(i = A, B, C)$ $(n = 1, 2)$ shows the phase current flowing through the conductor, which is calculated with Equation (3), A, r_{in} $(i = A, B, C)$ $(n = 1, 2)$ are the lengths of distances from phases and their reflections to the reference point M, in m., which are calculated from Figure 2, and H_{in} is the magnetic field strength, in A/m.

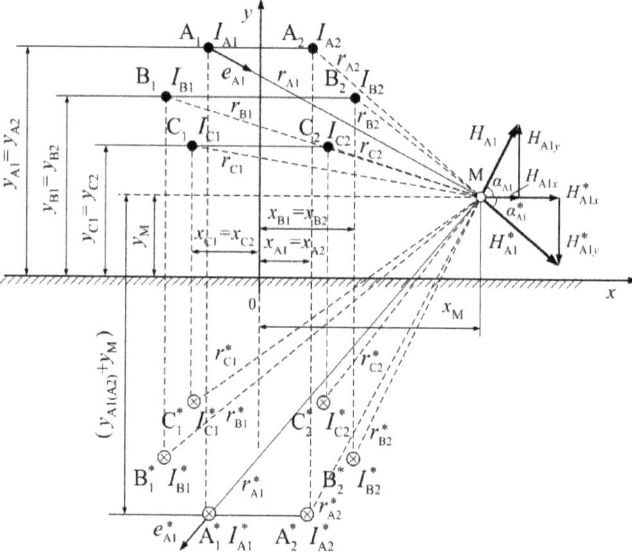

Figure 2. Components of the magnetic field vectors at the point M.

The magnetic field strength generated by the double-circuit three-phase 400 kV overhead transmission line in the plane in Figure 2 is calculated as follows:

$$H_M = \sqrt{H_{Minx}^2 + H_{Miny}^2}; \tag{5}$$

where H_{Minx} and H_{Miny} are the constituents of the magnetic field strength vectors with respect to x and y axes.

Next, we must examine how the values of H vary with the measurement point M moving between the phase conductors of the power line. The location of point M with respect to the ground is $y_0 = y_M$.

The vectors of the magnetic field strength generated at point M are H_{in} and H_{in}^* (i = A, B, C) (n = 1, 2). After distributing the directions of x and y axes, the following is obtained:

$$H_x = H_{A1x} + H_{B1x} + H_{C1x} + H_{A2x} + H_{B2x} + H_{C2x} + H_{A1x}^* + H_{B1x}^* + H_{C1x}^* + H_{A2x}^* + H_{B2x}^* + H_{C2x}^*, \quad (6)$$

$$H_y = H_{A1y} + H_{B1y} + H_{C1y} + H_{A2y} + H_{B2y} + H_{C2y} + H_{A1y}^* + H_{B1y}^* + H_{C1y}^* + H_{A2y}^* + H_{B2y}^* + H_{C2y}^*. \quad (7)$$

Assuming that constituents of the magnetic field strength vectors at the observation point M are as follows,

$$E_{Minx} = \sqrt{\frac{1}{T} \sum_{i=1}^{n} E_{Minx\,max}^2}, \quad (8)$$

$$E_{Miny} = \sqrt{\frac{1}{T} \sum_{i=1}^{n} E_{Miny\,max}^2}, \quad (9)$$

the effective value of magnetic field strength at observation point M H_M is calculated using Equation (5). One millionth of a Tesla (1 µT) corresponds to 0.8 A/m.

As initial phases of conductors in the double-circuit symmetric 400 kV overhead power line are different by 120^0, the total magnetic field will be a rotating one, and for the magnetic field vector, at any observation point M, the time course will define an ellipse in a general way. The normal value of the rotating magnetic field strength is assumed to be the effective value of the sinusoid, the amplitude of which is equal to the semi-major axis of the ellipse orbited by the strength vector at the given point.

To express respective constituents of the magnetic field strength (see Figure 2), the following markings are introduced: coordinate y of point M, y_M; height of the conductors above the ground surface, $y_{A1} = y_{A2}$, $y_{B1} = y_{B2}$, $y_{C1} = y_{C2}$. The phase reflections A_1^*, B_1^*, C_1^*, A_2^*, B_2^*, C_2^* are situated at the same distances above the ground surface in the direction of the y axis. The distances from the conductors to the point M in the direction of the x axis are as follows:

1. $x_{A1M} = x_{A1} + x_M$ is the distance from phase A_1 to point M;
2. $x_{B1M} = x_{B1} + x_M$ is the distance from phase B_1 to point M;
3. $x_{C1M} = x_{C1} + x_M$ is the distance from phase C_1 to point M;
4. $x_{A2M} = x_M - x_{A2}$ is the distance from phase A_2 to point M;
5. $x_{B2M} = x_M - x_{B2}$ is the distance from phase B_2 to point M;
6. $x_{C2M} = x_M - x_{C2}$ is the distance from phase C_2 to point M.

The conductors' phase reflections A_1^*, B_1^*, C_1^*, A_2^*, B_2^*, C_2^* are situated at the same distances in the direction of the x axis.

Figure 2 shows that magnetic field strength vector H_{in} is distributed into two vectors as follows: vector H_{iny}, which varies following the law of sines; and vector H_{inx}, which varies following the law of cosines. The magnetic field strength reflection vector H_{in}^* is distributed analogically. The numerical values of the constituents H_{inx}, H_{inx}^*, H_{iny}, and H_{iny}^* (i = A, B, C) (n = 1, 2) are found as follows:

$$H_{inx} = H_{in} \cos \alpha_{in}, \quad (10)$$

$$H_{inx}^* = H_{in}^* \cos \alpha_{in}^*, \quad (11)$$

$$H_{iny} = H_{in} \sin \alpha_{in}, \quad (12)$$

$$H_{iny}^* = H_{in}^* \sin \alpha_{in}^*. \quad (13)$$

The values of the angles α_{in} and α_{in}^* (i = A, B, C) (n = 1, 2) are used to consider the positions of the conductor phases and their reflections of the double-circuit 400 kV overhead power line as well as the position of the reference point M with respect to the ground surface and their distances. The values of these trigonometric functions are found from Figure 2. In the general case, these angles α_{in}, α_{in}^* are calculated in the following sequence:

1. The angle α_{in} or α_{in}^{*} is deducted from the conductor phase under investigation or its reflection's magnetic field vector H_{in} and measurement point M's position with respect to the ground surface.

2. The arctan function of the angle α_{in} or α_{in}^{*} (i = A, B, C) (n = 1, 2) is calculated as follows:

$$\alpha_{in} = \arctan\frac{y_{in} - y_{M}}{x_{inM}}, \tag{14}$$

$$\alpha_{in}^{*} = \arctan\frac{y_{in} + y_{M}}{x_{inM}}. \tag{15}$$

3. The functions $\cos\alpha_{in}$ and $\sin\alpha_{in}$ of angles α_{in} or α_{in}^{*} are calculated with the sign "$\pm$" depending on the trigonometric function $\cos\alpha_{in}$ and $\sin\alpha_{in}$ quarter in which the phase load electric field vector H_{in} is calculated.

For the purpose of the mathematical calculation of the magnetic field strength, a 400 kV double-circuit three-phase AC conductor system was selected, which was laid out vertically starting from the ground surface as follows: 1. $C_1(C_2)$ is at a height of 10 m; 2. $C_1(C_2)$ and $B_1(B_2)$ are at a height of 18 m; 3. $B_1(B_2)$ and $A_1(A_2)$ are at a height of 26 m.

The following are the horizontal distances between the different link phases of the conductor: 1. 12 m between A_1 and A_2; 2. 16 m between B_1 and B_2; 3. 13 m between C_1 and C_2.

Measurement point M is located 1.5 m vertically from the ground surface. In order for the obtained results of the mathematical calculations to be as precise as possible, 15 total positions of the measurement point M were selected, which were spaced by 5 m and located at 1.5 m vertically from the ground surface (see Figure 3).

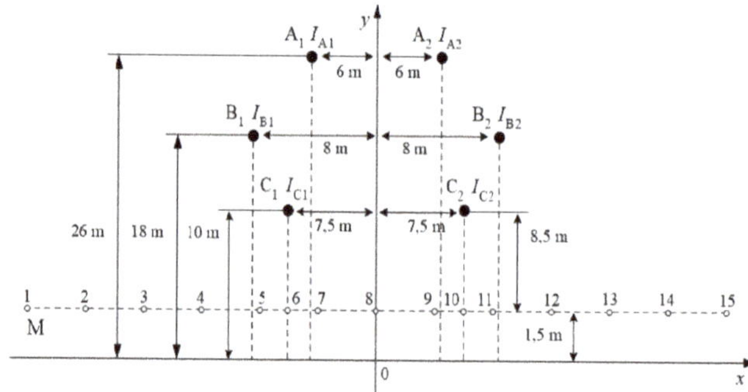

Figure 3. Positions of the measurement point M.

In this paper, extra high-voltage, double-circuit power transmission lines are studied; all six groups with six types of conductor transmission line transpositions are included in the case of long-distance distribution. For 400 kV AC lines, there are six types of first group transposition. There are 36 different transposition layouts possible of the phases in 400 kV double-circuit overhead power lines. For these six different groups of phase conductor transpositions, changes in their transposition sequences influence the results of mathematical calculations [1,24]. The six groups of transposition are as follows (see Figure 1):

Group 1 ($A_1 - C_2$, $B_1 - B_2$, $C_1 - A_2$);
Group 2 ($B_1 - C_2$, $A_1 - B_2$, $C_1 - A_2$);
Group 3 ($A_1 - C_2$, $C_1 - B_2$, $B_1 - A_2$);
Group 4 ($C_1 - C_2$, $A_1 - B_2$, $B_1 - A_2$);
Group 5 ($B_1 - C_2$, $C_1 - B_2$, $A_1 - A_2$);
Group 6 ($C_1 - C_2$, $B_1 - B_2$, $A_1 - A_2$).

5. Discussion

The results obtained from mathematical calculations of the magnetic field in the selected locations of point M_i are presented mathematically in Table 1. The mathematical calculation results in Table 1 suggest that the magnetic field strength achieves its maximum values at points M_6 and M_{11}, and these values are grouped into six categories with six types of conductor transposition.

Table 1. The mathematical calculation of the effective values of the magnetic field strength at 1.5 m above the ground.

Point No.	Distance [m]	Group 1 [A/m]	Group 2 [A/m]	Group 3 [A/m]	Group 4 [A/m]	Group 5 [A/m]	Group 6 [A/m]
M_1	0	2.63	2.85	3.14	2.76	2.76	2.95
M_2	5	4.51	4.66	5.23	4.75	4.94	5.42
M_3	10	8.84	9.12	9.41	8.74	9.79	9.6
M_4	15	18.18	19.67	19.95	19.48	21.38	22.33
M_5	20	35.72	37.05	35.63	40.38	40.85	41.8
M_6	22	39.48	41.33	38.48	46.08	44.65	47.5
M_7	25	32.9	35.15	31.83	43.7	41.33	45.6
M_8	30	18.8	18.53	9.5	35.63	30.88	38.0
M_9	35	37.13	31.35	31.83	43.7	41.33	45.6
M_{10}	38	42.3	38.0	38.48	46.08	44.65	47.5
M_{11}	40	37.6	34.2	35.63	40.38	40.85	41.8
M_{12}	45	19.72	18.81	19.95	19.49	21.38	22.33
M_{13}	50	9.21	8.74	9.41	8.74	9.79	9.6
M_{14}	55	4.8	4.37	5.23	4.75	4.94	5.42
M_{15}	60	3.01	2.47	3.14	2.76	2.76	2.95

These points of observation are situated near phases B_1 and B_2 and between phases C_1 and C_2 of the double-circuit 400 kV overhead power line conductor system. The observation point M_8 is at the very centre of the 400 kV overhead power line (see Figure 3). At this particular point, the magnetic field strength values are lower.

These points of observation are situated near phases B_1 and B_2 and between phases C_1 and C_2 of the double-circuit 400 kV overhead power line conductor system. The observation point M_8 is at the very centre of the 400 kV overhead power line (see Figure 3). At this particular point, the magnetic field strength values are also lower.

The obtained analytical results were also verified through a simulation using the software package COMSOL Multiphysics 3.5. The model simulation additionally assessed the marginal and ambient conditions [6]. Results obtained from a finite element method (FEM) simulation of the magnetic field in the selected locations of point M_i are presented mathematically in Table 2. The FEM simulation results in Table 2 suggest that the magnetic field strength reaches its maximum values at points M_6 and M_{11}, and these values are also grouped into six categories with six types of conductor transposition.

The results of the simulation (see the upper curve of Figure 4) were found to be close to the analytical results; consequently, the proposed methodology can be used to investigate the relevant magnetic field. The assessment of our findings and the obtained results lead us to the following suggestions: the mean error between the analytical findings and simulation results is below 4%. The methodology described here allows the calculation of the magnetic field strength at any point under the three-phase double-circle power transmission line.

The graph in Figure 4 shows that as the measurement point M_i gets farther from the outside phases of the double-circuit 400 kV overhead power line conductor system, the magnetic field strength decreases proportionally. However, at the distance of 12 m from phases B_1 and B_2, the magnetic field strength still exceeds 16 A/m (i.e., B is more than 20 µT).

The assessment of the calculation results revealed that values of the magnetic field strength exceed the numerical values established in the EU Hygiene Standard HN 104:2011. The regulation on the hygienic norms of the European Union and of the Republic of Lithuania states that the numerical values of the electromagnetic field parameters of electric power transmission lines in residential and public buildings—as well as in residential areas—should never exceed the permissible values provided for in EU HN 104:2011.

Table 2. The finite element method (FEM) simulation of the effective values of the magnetic field strength at 1.5 m above the ground.

Point No.	Distance [m]	Group 1 [A/m]	Group 2 [A/m]	Group 3 [A/m]	Group 4 [A/m]	Group 5 [A/m]	Group 6 [A/m]
M_1	0	2.8	3.0	3.3	2.9	2.9	3.1
M_2	5	4.8	4.9	5.5	5.0	5.2	5.7
M_3	10	9.4	9.6	9.9	9.2	10.3	10.1
M_4	15	20.0	20.7	21.0	20.5	22.5	23.5
M_5	20	38.0	39.0	37.5	42.5	43.0	44.0
M_6	22	42.0	43.5	40.5	48.5	47.0	50.0
M_7	25	35.0	37.0	33.5	46.0	43.5	48.0
M_8	30	20.0	19.5	10.0	37.5	32.5	40.0
M_9	35	39.5	33.0	33.5	46.0	43.5	48.0
M_{10}	38	45.0	40.0	40.5	48.5	47.0	50.0
M_{11}	40	40.0	36.0	37.5	42.5	43.0	44.0
M_{12}	45	21.0	19.8	21.0	20.5	22.5	23.5
M_{13}	50	9.8	9.2	9.9	9.2	10.3	10.1
M_{14}	55	5.1	4.6	5.5	5.0	5.2	5.7
M_{15}	60	3.2	2.6	3.3	2.9	2.9	3.1

When locations are identified in which the magnetic field strength values are exceeded, the regulation on the hygienic norms of the European Union and of the Republic of Lithuania obliges operators of the power transmission system, who are responsible for the power transmission lines in operation, to ensure that the permissible values of the electromagnetic field parameters provided for in EU Hygiene Standard HN 104:2011 are adhered to. If the electromagnetic field parameters are found to exceed permissible values, it is compulsory to undertake appropriate actions and to reduce the values of the electromagnetic field parameters to the levels allowed.

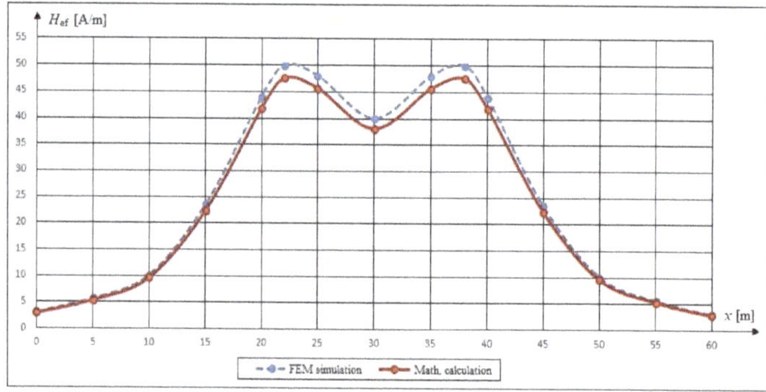

Figure 4. Distribution of the six groups of effective values of magnetic field strength at 1.5 m above the ground, obtained through calculation and FEM simulation.

With the aim of reducing the potential hazards to human health and to ensure safe living conditions in residential areas, it is suggested that experimental measurements of electromagnetic fields should be undertaken in locations or areas with the maximum exposure to electrical or magnetic fields. To mitigate the values of such magnetic fields, it is also recommended that the height of support bars should be increased or that any individual or commercial activities in their surroundings should be restricted.

6. Conclusions

The magnetic field strength of this power line was found to exceed the numerical values permitted by the Hygiene Standard HN 104:2011 HN, revealing that any person living or working in these areas was in serious danger. Any individual or commercial activities performed by the residents closer than 15 m to the double-circuit 400 kV overhead power line must be restricted.

For these reasons, the magnetic field in these locations must be reduced by installing higher support bars, and persons must wear appropriate personal protective equipment to protect them from exposure to electromagnetic field. For the purpose of more comprehensive analysis, it is recommended that experimental measurements should be made.

The distribution of the magnetic field strength at the high-voltage, double-circuit transmission line could be calculated using the methods of the magnetostatic field or by simulation using the finite element method (FEM). The difference is less than 5% between the results of the mathematical calculation and FEM simulation.

This paper has studied the magnetic field distribution resulting from all six groups with six types of long-distance distributing transposition. As a result, we can see how the impact of the calculation of the six groups' long-distance distribution transposition changes the magnetic field surrounding the transmission line.

The developed mathematical model for the calculation of the magnetic field generated by the conductors in the double-circuit 400 kV overhead power line can be used for the identification of hazardous areas and cases of electromagnetic pollution.

Author Contributions: Conceptualization, R.D.; Methodology, R.L.; Software, R.D.; Validation, R.L.; Formal analysis, R.D.; Investigation, R.D., R.L.; Resources, R.L.; Data curation, R.L.; Writing—Original draft preparation, R.L.; Writing—Review and editing, R.D., R.L.; Visualization, R.L.; Supervision, R.L. All authors have read and agreed to the published version of the manuscript.

Funding: This research received no external funding.

Conflicts of Interest: The authors declare no conflict of interest.

References

1. Safigianni, A.S.; Tsompanidou, C.G. Measurements of electric and magnetic fields due to the operation of in door power distribution substations. *IEEE Trans. Power Deliv.* **2005**, *20*, 1800–1805. [CrossRef]
2. Jin, J.-M. *Theory and Computation of Electromagnetic Fields*; IEEE Press: Champaign, II., USA, 2010; pp. 399–410.
3. Haiqi, Y.; Miaoyong, Z. Three-Dimensional Magnetohydrodynamic Calculation for Coupling Multiphase Flow in Round Billet Continuous Casting Mold with Electromagnetic Stirring. *IEEE Trans. Magn.* **2010**, *46*, 82–86. [CrossRef]
4. Velickovic, D.; Aleksic, S.; Bozic, M. Electromagnetic field in power line surroundings. *Facta Univ. Work. Living Environ. Prot.* **1996**, *1*, 69–79.
5. Dezelak, K.; Jakl, F.; Stumberger, G. Arrangements of overhead power line phase conductors obtained by differential evolution. *Electr. Power Syst. Res.* **2011**, *81*, 2164–2170. [CrossRef]
6. Deltuva, R.; Lukočius, R. Electric and magnetic field of different transpositions of overhead power line. *Arch. Electr. Eng. J. Pol. Acad. Sci.* **2017**, *66*, 595–605. [CrossRef]
7. Olsen, R.G.; Deno, D. Magnetic fields from electric power lines: theory and comparison to measurements. *IEEE Trans. Power Deliv.* **1998**, *3*, 2127–2136. [CrossRef]
8. El Dein, A.Z. Magnetic Field Calculation Under EHV Transmission Lines for More Realistic Cases. *IEEE Trans. Power Deliv.* **2009**, *24*, 2214–2222. [CrossRef]
9. Begamudre, R.D. *Extra High Voltage AC. Transmission Engineering*, 3rd ed.; Wiley Press: New York, NY, USA, 2006; pp. 172–205.

10. Brandão Faria, J.A.; Almeida, M.E. Accurate Calculation of Magnetic Field Intensity Due to Overhead Power Lines with or without Mitigation Loops with or without Capacitor Compensation. *IEEE Trans. Power Deliv.* **2007**, *22*, 951–959. [CrossRef]

11. Havas, M. Intensity of electric and magnetic fields from power lines within the business district of 60 Ontario communities. *Sci. Total Environ.* **2002**, *298*, 183–206. [CrossRef]

12. Peric, M.; Ilic, S.; Aleksic, S. Electromagnetic field analysis in vicinity of power lines. *Elektrotech. Elektron.* **2008**, *1112*, 51–56.

13. Iyyuni, G.B.; Sebo, S.A. Study of transmission line magnetic fields. In Proceedings of the Twenty-Second Annual North American, IEEE Power Symposium, Auburn, AL, USA, 15–16 October 1990; pp. 222–231.

14. Abdel-Salam, M.; Abdallah, H.; El-Mohandes, M.T.; El-Kishky, H. Calculation of magnetic fields from electric power transmission lines. *Electr. Power Syst. Res.* **1999**, *49*, 99–105. [CrossRef]

15. Ege, Y.; Kalender, O.; Nazlibilek, S. Electromagnetic Stirrer Operating in Double Axis. *IEEE Trans. Ind. Electron.* **2010**, *57*, 2444–2453. [CrossRef]

16. Nicolaou, Ch.P.; Papadakis, A.P.; Razis, P.A.; Kyriacou, G.A.; Sahalos, J.N. Measurements and predictions of electric and magnetic fields from power lines. *Electr. Power Syst. Res.* **2011**, *81*, 1107–1116. [CrossRef]

17. Li, L.; Yougang, G. Analysis of magnetic field environment near high voltage transmission lines. In Proceedings of the International Conferences on Communication Technology, Beijing, China, 22–24 October 1998; pp. S26-05-1–S26-05-5.

18. Dahab, A.A.; Amoura, F.K.; Abu-Elhaija, W.S. Comparison of magnetic field distribution of noncompact and compact parallel transmission line configurations. *IEEE Trans. Power Deliv.* **2005**, *20*, 2114–2118. [CrossRef]

19. Sadiku, M.N.O. *Numerical Techniques in Electromagnetics*; CRC Press: New York, NY, USA, 2001; pp. 743–793.

20. Stewart, J.R.; Dale, S.J.; Klein, K.W. Magnetic field reduction using high phase order lines. *IEEE Trans. Power Deliv.* **1993**, *8*, 628–636. [CrossRef]

21. Jin, J.-M. *The Finite Element Method in Electromagnetics*, 3rd ed.; Wiley-IEEE Press: New York, NY, USA, 2014; pp. 661–735.

22. Mamishev, A.V.; Nevels, R.D.; Russell, B.D. Effects of conductor sag on spatial distribution of power line magnetic field. *IEEE Trans. Power Deliv.* **1996**, *11*, 1571–1576. [CrossRef]

23. Chari, M.V.K.; Salon S.J. *Numerical Methods in Electromagnetism*; Academic Press: New York, NY, USA, 2000; pp. 1–61.

24. Hunt, R.W.; Zavalin, A.; Bhatnagar, A.; Chinnasamy, S.; Das, K.C. Electromagnetic biostimulation of living cultures for biotechnology, biofuel and bioenergy applications. *Int. J. Mol. Sci.* **2009**, *10*, 4515–4558. [CrossRef] [PubMed]

Article

3-D Integral Formulation for Thin Electromagnetic Shells Coupled with an External Circuit

Tung Le-Duc [1,*] and Gerard Meunier [2]

[1]   Department of Power Systems, School of Electrical Engineering, Hanoi University of Science and Technology, Hanoi City 100000, Vietnam
[2]   Université Grenoble Alpes, CNRS, Grenoble INP, G2Elab, F-38000 Grenoble, France; gerard.meunier@g2elab.grenoble-inp.fr
*   Correspondence: tung.leduc1@hust.edu.vn

Received: 11 May 2020; Accepted: 19 June 2020; Published: 22 June 2020

Abstract: The aim of this article is to present a hybrid integral formulation for modelling structures made by conductors and thin electromagnetic shell models. Based on the principle of shell elements, the proposed method provides a solution to various problems without meshing the air regions, and at the same time helps to take care of the skin effect. By integrating the system of circuit equations, the method presented in this paper can also model the conductor structures. In addition, the equations describing the interaction between the conductors and the thin shell are also developed. Finally, the formulation is validated via an axisymmetric finite element method and the obtained results are compared with those implemented from another shell formulation.

Keywords: electromagnetic modelling; integral formulation; skin effect; thin shell approach; mutual inductance; finite element method; partial element equivalent circuit method

---

1. Introduction

All of the electromagnetic phenomena occurring in the electrical systems are described by Maxwell's equations, together with the constitutive material laws. It is a set of partial differential equations associated with the relationships of electromagnetic field ($\mathbf{E}$, $\mathbf{H}$) distributing into space and varying into time, the distribution of currents and charges ($\mathbf{J}$, ρ) and material properties (μ, σ) [1,2].

In order to compute the electromagnetic field (i.e., solutions to Maxwell's equations), an analytical method or a numerical method can be used. For these devices with simple geometries, a correct analytical solution can be identified. However, in some general cases, especially for the complex structure of electrical devices, the numerical methods are used as a sole solution. The numeric methods applied in modelling of electromagnetic fields can be divided into two categories: finite methods like FEM (Finite Element Method), FVM (Finite Volume Method) and the numerical integration methods such as BEM (Boundary Elements method), MoM (Methods of Moments), PEEC method (Partial Element Equivalent Circuit) [3–13]. The choice of an appropriate method totally depends on the physical phenomena that need modelling such as the physical phenomena in high or low frequencies, with or without magnetic material, considering the effect of inductance or capacitance, external excitation source. However, there is no universal, optimal method for these problems, and the choice of the most suitable method depends on the nature of the electrical devices and their operation range.

Generally, the electromagnetic modelling of structures including thin shell model and thin wires is a complex problem in the fields of electrical engineering. Its geometry is characterized by a high ratio of the length and the thickness. Thus, the use of a volume mesh leads to a large number of elements and/that makes this method expensive and time consuming to apply to practical devices. Furthermore, due to the skin effect, when the skin depth δ is thinner than the thickness e, the size of the studied meshes must be smaller than that of the skin depth. This leads to the difficulties when the thickness of some parts of the structures is very small in comparison with the overall size of the actual devices.

The shell element formulations have recently been developed by many authors in order to cope with difficulties in thin plate models, e.g., with the boundary element method (BEM) [3], with the finite element method (FEM) [5] and with the integral methods [11,14–16]. In [11], we presented a coupling method between the Partial Element Equivalent Circuit (PEEC) and an integro-differential method to model the devices that include thin electromagnetic shells and complicated conductor systems. However, this coupling formulation cannot be applied to problems when the skin depth is low in comparison with the thickness of the thin shell. In [14], the authors presented a hybrid of volume and surface integral formulations for the eddy current solution of the conductive regions with arbitrary geometry. However, this formulation cannot be used for the magnetic material and as in [11], the skin effect is not yet considered. In [16], we developed a general shell element formulation for modelling of thin magnetic and conductive regions. The thin shell is modelled by an integral method to avoid meshing the air region. As in [3,5,15], the field variation through the thickness is considered. Based on a simple discretization of the shell averaged surface, the number of unknowns is greatly reduced. However, this formulation is not suitable for model systems with complex conductive structures.

In this paper, a hybrid integral formulation is proposed to allow the modelling of an inhomogeneous structure constituted by conductors and thin magnetic and conductive shells in the general case ($\delta > e$ or $\delta \approx e$ or $\delta < e$). The modelling of thin magnetic and conductive shell regions is determined thanks to the integral formulation in [15,16]. A method that allows the modelling of the contributions of the inductors fed with the alternating currents will be presented in Section 3. The coupling formulations for integrating the interaction between conductors and thin shells material will be fully developed in Section 4. Finally, two numerical examples are presented in Section 5. The results obtained from this formulation are compared with those obtained from the FEM. Strong and weak points of our formulation are also analyzed.

2. Thin Shell Equation

A thin electromagnetic shell (Ω) with thickness e, conductivity σ and linear permeability μ_r in this study is illustrated in Figure 1, where Γ_1 and Γ_2 are the surface boundaries of the shell with the air region.

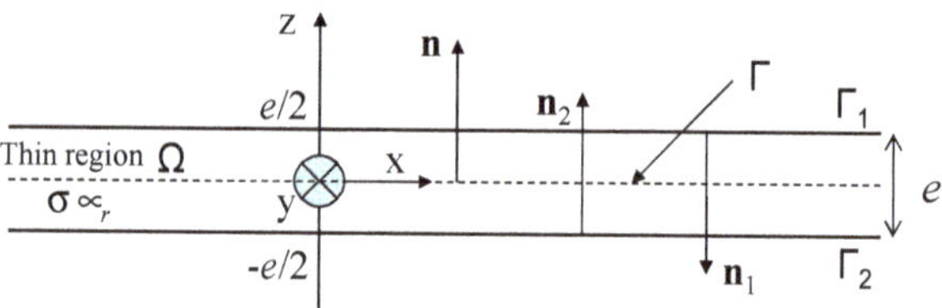

Figure 1. Thin region and associated notations [16].

The electromagnetic behavior for the side "1" of the shell regions is represented by [15,16]:

$$\int_{\Gamma_1} \mathbf{grad}_s w \cdot (\alpha \mathbf{H}_{1s} - \beta \mathbf{H}_{2s}) d\Gamma + j\omega \int_{\Gamma_1} w \cdot \mathbf{B}_1 \cdot \mathbf{n}_1 d\Gamma = 0, \tag{1}$$

where δ is the skin depth associated to the shell; $a = (1 + j)/\delta$; $\alpha = a/\sigma \text{th}(ae)$, $\beta = a/\sigma \text{sh}(ae)$; w is a set of nodal surface weighting functions; $\mathbf{H}_{1s}$ and $\mathbf{H}_{2s}$ are the tangential magnetic fields on both sides Γ_1, Γ_2 and $\mathbf{n}_1$ is the normal vector of the side Γ_1.

The symmetrical equation corresponding to the other side of the shell is obtained by simply exchanging the indices "1" and "2":

$$\int_{\Gamma_2} \mathbf{grad}_s w \cdot (\alpha \mathbf{H}_{2s} - \beta \mathbf{H}_{1s}) d\Gamma + j\omega \int_{\Gamma_2} w \cdot \mathbf{B}_2 \cdot \mathbf{n}_2 d\Gamma = 0, \tag{2}$$

where $\mathbf{n}_2$ is the normal vector corresponding the side "2" of the shell.

Equations (1) and (2) are written on the averaged surface Γ of the thin shell region. Subtracting Equations (2) and (1) leads to:

$$(\alpha + \beta) \int_\Gamma \mathbf{grad}_s w (\mathbf{H}_{2s} - \mathbf{H}_{1s}) d\Gamma + j\omega \int_\Gamma w(2 \cdot \mathbf{B}_a \cdot \mathbf{n}) d\Gamma = 0, \tag{3}$$

where $\mathbf{B}_a = (\mathbf{B}_1 + \mathbf{B}_2)/2$ is defined as the averaged induction and $\mathbf{n}$ denotes the normal vector of the surface Γ (Figure 1).

The expressions related to the tangential magnetic fields on both sides and the outside of the shell are:

$$\mathbf{H}_{1s} = \mathbf{H}_{01s} - \mathbf{grad}_s \phi_1; \quad \mathbf{H}_{2s} = \mathbf{H}_{02s} - \mathbf{grad}_s \phi_2 \tag{4}$$

where $\mathbf{H}_{01s}$ and $\mathbf{H}_{02s}$ are the tangential fields generated by the inductors at side Γ_1, Γ_2 and ϕ_1, ϕ_2 denote the corresponding reduced scalar magnetic potentials.

By using (4) and assuming the small variations of $\mathbf{H}_{0s}$ through the thickness of the thin shell, Equation (3) becomes:

$$(\alpha + \beta) \int_\Gamma \mathbf{grad}_s w \cdot \mathbf{grad}_s \Delta\phi \, d\Gamma + 2j\omega \int_\Gamma w \cdot \mathbf{B}_a \cdot \mathbf{n} d\Gamma = 0, \tag{5}$$

where $\Delta\phi = \phi_1 - \phi_2$ is the scalar magnetic potential discontinuity.

Using magnetization law of the linear material, equation (5) can be rewritten as:

$$(\alpha + \beta) \int_\Gamma \mathbf{grad}_s w \cdot \mathbf{grad}_s \Delta\phi \, d\Gamma + 2j\omega \frac{\mu_0 \mu_r}{\mu_r - 1} \int_\Gamma w \cdot \mathbf{M}_a \cdot \mathbf{n} d\Gamma = 0, \tag{6}$$

where $\mathbf{M}_a = (\mathbf{M}_1 + \mathbf{M}_2)/2$ is defined as the averaged magnetization.

On the averaged surface Γ, the total magnetic field $\mathbf{H}_a$ is the sum of the inductor field $\mathbf{H}_0$, the field created by magnetization $\mathbf{H}_M$ and the field created by the eddy currents flowing in the shell $\mathbf{H}_{EC}$:

$$\mathbf{H}_a(P) = \mathbf{H}_0(P) + \mathbf{H}_M(P) + \mathbf{H}_{EC}(P) \tag{7}$$

where P is a point located in the surface region Γ.

The field $\mathbf{H}_M$ is equal to [10]:

$$\mathbf{H}_M(P) = -\mathbf{grad} \frac{1}{4\pi} \int_\Omega \frac{(\mathbf{M} \cdot \mathbf{r})}{r^3} d\Omega \tag{8}$$

where $\mathbf{r}$ is the vector between the integration point on Ω and the point P.

By using Biot and Savart law, the field $\mathbf{H}_{EC}$ can be written as:

$$\mathbf{H}_{EC}(P) = \frac{1}{4\pi} \int_\Omega \frac{\mathbf{J} \times \mathbf{r}}{r^3} d\Omega, \tag{9}$$

where $\mathbf{J}$ denotes the eddy current density of the thin shell.

The integration of the tangent magnetization through the depth of the thin shell is written as follows [3,5,16]:

$$\int_{-e/2}^{+e/2} \mathbf{M} dz = \frac{\overline{G}(\mathbf{M}_1 + \mathbf{M}_2)}{2} = \overline{G} \mathbf{M}_a, \tag{10}$$

where $\overline{G} = \mathrm{th}[(1 + j)e/2\delta]/[(1 + j)e/2\delta]$.

Using (10), Equation (8) becomes:

$$\mathbf{H_M}(P) = -\mathbf{grad}\frac{\overline{G}}{4\pi}\int_\Gamma \frac{(\mathbf{M_a} \cdot \mathbf{r})}{r^3}d\Gamma. \tag{11}$$

The equivalent shell current $\mathbf{K}$ is defined as [3]:

$$\mathbf{K} = \int_{-e/2}^{+e/2} \mathbf{J}dz = \mathbf{n} \times \mathbf{grad}\Delta\phi. \tag{12}$$

Using (12), Equation (9) becomes:

$$\mathbf{H_{EC}}(P) = \frac{1}{4\pi}\int_\Gamma \frac{\mathbf{n} \times \mathbf{grad}\Delta\phi \times \mathbf{r}}{r^3}d\Gamma. \tag{13}$$

Combining (11) with (12), Equation (7) is rewritten as:

$$\mathbf{H_a}(P) = \mathbf{H_0}(P) - \mathbf{grad}\frac{\overline{G}}{4\pi}\int_\Gamma \frac{(\mathbf{M_a} \cdot \mathbf{r})}{r^3}d\Gamma + \frac{1}{4\pi}\int_\Gamma \frac{\mathbf{n} \times \mathbf{grad}\Delta\phi \times \mathbf{r}}{r^3}d\Gamma. \tag{14}$$

Let us consider a linear magnetic law for the material $\mathbf{M_a}(P) = (\mu_r - 1)\mathbf{H_a}(P)$, then Equation (14) becomes [10,11]:

$$\frac{\mathbf{M_a}(P)}{\mu_r - 1} = \mathbf{H_0}(P) - \mathbf{grad}\frac{\overline{G}}{4\pi}\int_\Gamma \frac{(\mathbf{M_a} \cdot \mathbf{r})}{r^3}d\Gamma + \frac{1}{4\pi}\int_\Gamma \frac{\mathbf{n} \times \mathbf{grad}\Delta\phi \times \mathbf{r}}{r^3}d\Gamma, \tag{15}$$

In [16], Equations (6) and (15) have been resolved and validated by the comparison with the axisymmetric FEM and the FEM 3D with the shell elements. Let us note that in the proposed equations, the integrals are determined only by the surface numerical integration. Therefore, the 3D implementation is simple and reliable. The comparison results with different ratios (e/δ) have demonstrated the ability and the advantages of the method.

3. Conductor System Modelling

We now consider m volume conductors fed with alternating sources placed in an air region. The external electric field incident at point P can be written [7–9]:

$$\mathbf{E_{ext}}(P) = \frac{\mathbf{J_c}(P)}{\sigma} + j\omega\mathbf{A}(P) + \mathbf{grad}V(P), \tag{16}$$

where $\mathbf{A}(P)$ is the vector potential and $V(P)$ is the scalar potential.

The magnetic vector potential generated by the current density $\mathbf{J_c}$ is:

$$\mathbf{A}(P) = \sum_{k=1}^{m} \frac{\mu_0}{4\pi}\int_{\Omega_{c_k}} \frac{\mathbf{J_c}}{r}d\Omega, \tag{17}$$

where Ω_{c_k} is the volume of the conductor k; r is the distance between the integration point on Ω_{c_k} and the point P.

In the context of the quasi-stationary regime operated at the frequency range up to ten MHz, it is possible to neglect the capacitive effect. Equation (16) is written in the form:

$$\mathbf{E_{ext}}(P) = \frac{\mathbf{J_c}(P)}{\sigma} + j\omega\sum_{k=1}^{m} \frac{\mu_0}{4\pi}\int_{\Omega_{c_k}} \frac{\mathbf{J_c}}{r}d\Omega. \tag{18}$$

The PEEC method is particularly pertinent and reliable to solve this kind of problem. Let us assume that the current density in each conductor is uniform. For each conductor, Equation (18) can be associated to the electrical equivalent circuit presented by the self and mutual inductances in series with the resistances [8,9]. In most respects, what we know about this approach is difficult to the magnetic material or the magnetic conductive material. However, the mesh of the air region can be neglected and some conventional SPICEs-like or Saber circuit solvers can be employed to analyze the equivalent circuit. As a result, the behavior of several electrical devices that have electrical interconnections can be studied in only one system. With all of the advantages described above, the PEEC is well-suited for modelling real industrial devices [17–21].

Despite these clear advantages, this technique has a major drawback due to the necessity to store a fully dense matrix inherent to the use of the integral formulation in Maxwell's equation. Consequently, this approach is strongly limited in modelling large-scale electrical devices requiring substantial meshes. Let us denote N as the number of unknown then the needed memory for the fully dense matrix storage leads to O(N2) and the computational costs of a direct solver (LU-decomposition) increases in proportion to O(N3). For example, a large size modelling problem containing 50,000 unknowns requires at least 50,000 × 50,000 × 8 bytes or approximately 20 GB in the memory to store a fully dense matrix and the computational costs of finding one solution by direct solver are roughly several weeks. In order to study the characteristic of the devices in a frequency range or in a period of time by the PEEC method, the total computational costs must be multiplied with the number of frequency or the time steps. Moreover, none of the circuit solvers can handle the equivalent circuit composing of 50,000 × 50,000 basic circuit elements RLMC (resistor, inductor, partial inductor, and capacitor). Overall, in both cases the computational time tends to infinity, or it is impossible to solve. To solve this problem, the algorithms coupling different matrix compression algorithms or the model order reduction technique with integral methods have been developed with the purpose of limiting matrix storage [19,21].

In addition, the expansion of the PEEC method for more general problems has been investigated [11,18,20]. However, this method is still difficult for considering special structures such as a circular coil, thick circular coil, and thin disk coil. This drawback can be easily improved by using the semi-analytic methods developed in [22,23].

4. Coupling Thin Shell with Circuit Equation

We now consider a general system composed of *m* conductors and thin magnetic conductive shell (volume Ω and average surface Γ) (Figure 2). Let us assume that the current density in each conductor is uniform.

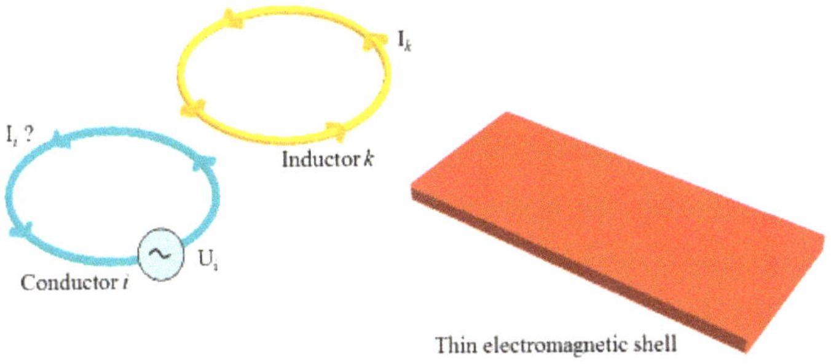

Figure 2. Representation of a general problem with conductors and thin electromagnetic shell.

4.1. Influence of the Conductor Current on the Thin Region

The term $\mathbf{H}_0$ in Equation (15) can be computed by using Biot and Savart law:

$$\mathbf{H}_0(P) = \sum_{k=1}^{m} \frac{1}{4\pi} \frac{I_k}{S_k} \int_{\Omega c_k} \frac{l_k \times \mathbf{r}}{r^3} d\Omega,$$ (19)

where I_k and S_k are respectively the current and the cross section of the conductor k; Ωc_k is the volume of conductor k; l_k is the vector unit of current direction in the conductor k and $\mathbf{r}$ is the vector between the integration point on Ωc_k and the point P where the field is expressed.

Equation (15) is then rewritten as:

$$\frac{\mathbf{M}_a(P)}{\mu_r - 1} = \sum_{k=1}^{m} \frac{1}{4\pi} \frac{I_k}{S_k} \int_{\Omega c_k} \frac{l_k \times \mathbf{r}}{r^3} d\Omega - \mathbf{grad} \frac{\overline{G}}{4\pi} \int_{\Gamma} \frac{(\mathbf{M}_a \cdot \mathbf{r})}{r^3} d\Gamma + \frac{1}{4\pi} \int_{\Gamma} \frac{\mathbf{n} \times \mathbf{grad} \Delta \phi \times \mathbf{r}}{r^3} d\Gamma.$$ (20)

4.2. Influence of Thin Shell Magnetization on the Conductor

In order to take into account the influence of the field created by the thin shell magnetization, we have to integrate the magnetic vector potential $\mathbf{A}_M$ generated by the magnetized thin shell on the conductor. This magnetic vector potential is expressed as [11,24]:

$$\mathbf{A}_M(P) = \frac{\mu_0}{4\pi} \int_{\Omega} \frac{\mathbf{M} \times \mathbf{r}}{r^3} d\Omega,$$ (21)

where $\mathbf{r}$ denotes the vector between the integration point and the point P.

Using (10), Equation (21) can be rewritten as:

$$\mathbf{A}_M(P) = \frac{\mu_0 \overline{G}}{4\pi} \int_{\Gamma} \frac{\mathbf{M}_a \times \mathbf{r}}{r^3} d\Gamma$$ (22)

4.3. Influence of Thin Shell Eddy Current on the Conductor

The eddy current $\mathbf{J}$ in the thin shell generates a magnetic potential vector on the conductor:

$$\mathbf{A}_{EC}(P) = \frac{\mu_0}{4\pi} \int_{\Omega} \frac{\mathbf{J}}{r} d\Omega,$$ (23)

where r is the distance between the integration point on Ω and the point P.

Using (12), Equation (23) can be rewritten as:

$$\mathbf{A}_{EC}(P) = \frac{\mu_0}{4\pi} \int_{\Gamma} \frac{\mathbf{n} \times \mathbf{grad} \Delta \phi}{r} d\Gamma.$$ (24)

Combining (17), (22), (23), Equation (18) is rewritten as:

$$\mathbf{E}_{ext}(P) = \frac{J(P)}{\sigma} + j\omega \sum_{k=1}^{m} \frac{\mu_0}{4\pi} \int_{\Omega c_k} \frac{J_c}{r} d\Omega + j\omega \frac{\mu_0 \overline{G}}{4\pi} \int_{\Gamma} \frac{\mathbf{M}_a \times \mathbf{r}}{r^3} d\Gamma$$
$$+ j\omega \frac{\mu_0}{4\pi} \int_{\Gamma} \frac{\mathbf{n} \times \mathbf{grad} \Delta \phi}{r} d\Gamma.$$ (25)

4.4. Final System of Equations

Equations (6), (20) and (25) are solved by using a numerical method. The most suitable way is to mesh the averaged surface Γ into n triangular elements and p nodes. Let us assume that the tangential component of the eddy current and the magnetization in each element are uniform.

The Galerkine projection method is then applied to Equation (6), and we get the following matrix system [11,15,16]:

$$[\mathbf{A}]\cdot[\mathbf{M}] + [\mathbf{B}]\cdot[\Delta\Phi] = 0, \tag{26}$$

where:

[**M**] is a complex vector of dimension $3n$; $[\Delta\Phi]$ is a complex vector of dimension p;

[**A**] is a $(p \times 3n)$ matrix expressed as follows:

$$A(i,k) = 2j\omega \frac{\mu_0\mu_r}{\mu_r - 1} \int_\Gamma w_i \cdot \mathbf{n}_k d\Gamma = 0, \tag{27}$$

[**B**] is a $(p \times p)$ sparse matrix and can be written as:

$$B(i,k) = (\alpha + \beta) \int_\Gamma \mathbf{grad}_s w_i \cdot \mathbf{grad}_s w_k d\Gamma. \tag{28}$$

A matrix system representing Equation (20) can also be determined thanks to a point matching approach at the element centroids. This linear matrix system is expressed as [10,11,16]:

$$\left(\frac{[\mathbf{I}_d]}{\mu_r - 1} + [\mathbf{F}]\right)\cdot[\mathbf{M}] - [\mathbf{C}]\cdot[\Delta\Phi] - [\mathbf{D}]\cdot[\mathbf{I}] = 0, \tag{29}$$

where $[\mathbf{I}_d]$ represents the identity matrix; [**F**] is a $(3n \times 3n)$ dense matrix; [**C**] is a $(3n \times p)$ also dense matrix; [**D**] is a $(3n \times m)$ matrix and [**I**] is a $(m \times 1)$ vector. The coefficients of these matrices are expressed as follows:

$$F(i,j) = \left[\left(\frac{\overline{G}}{4\pi}\mathbf{grad}\int_{\Gamma_j}\frac{\mathbf{u}_j\cdot\mathbf{r}_i}{r_i^3}d\Gamma\right), \mathbf{u}_i\right] \tag{30}$$

where $[,]$ is defined as the scalar product operator and $\mathbf{u}_i$ is the vector basis of the element i and $\mathbf{r}_i$ is the vector between the integration point on Γ_j to the centroid of the element i.

$$C(i,j) = \frac{1}{4\pi}\int_\Gamma \frac{\mathbf{n}_j \times \mathbf{grad}w_i \times \mathbf{r}_i}{r_i^3}d\Gamma. \tag{31}$$

$$D(i,j) = \frac{1}{4\pi}\frac{1}{S_j}\int_{\Omega c_j}\frac{\mathbf{l}_j \times \mathbf{r}_i}{r_i^3}d\Omega. \tag{32}$$

By integrating Equation (25) on each conductor, we have the link between the partial voltages of the conductors to the currents flowing in them. Thus, the voltage appearing on the conductor k is given by:

$$U_k = R_kI_k + j\omega\sum_{i=1}^m m_{ik}I_i + j\omega\sum_{i=1}^n m_{ik}^f\mathbf{M_{ai}} + j\omega\sum_{i=1}^p m_{ik}^q\cdot\phi_i, \tag{33}$$

where R_k is the resistance of the kth conductor; m_{ik} is the mutual inductance between the two conductors k, i; m_{ik}^f is defined as the mutual inductance between the magnetization $\mathbf{M_{ai}}$ of the shell element and the conductor k; m_{ik}^q is defined as the mutual inductance between the scalar magnetic potential discontinuity $\Delta\phi$ of the shell node and the conductor k. These mutual inductances are expressed as follows:

$$m_{ik} = \frac{\mu_0}{4\pi}\frac{1}{S_iS_k}\int_{\Omega c_k}\int_{\Omega c_i}\frac{\mathbf{l}_i\mathbf{l}_k}{r}d\Omega_{ci}d\Omega_{ck}, \tag{34}$$

$$m_{ik}^f = \frac{\mu_0\overline{G}}{4\pi}\frac{1}{S_k}\int_{\Omega c_k}\int_{\Gamma_i}\frac{\mathbf{u}_i \times \mathbf{r}}{r^3}d\Gamma_k d\Omega_{ck}, \tag{35}$$

$$m_{ik}^q = \frac{\mu_0}{4\pi} \frac{1}{S_k} \int_{\Omega_{c_k}} \int_{\Gamma_i} \frac{\mathbf{n}_i \times \mathbf{grad} w_i}{r} \, d\Gamma_i l_k d\Omega_{ck}, \tag{36}$$

where $\mathbf{u}_i$ is a basis vector of magnetizations; $\mathbf{n}_i$ is the normal vector of the element Γ_i and $\mathbf{l}_k$ is the vector unit of the current direction in the conductor k. Let us note that the mutual inductance terms m_{ik} can be calculated by the well-known semi-analytic formulas in [7–9,22,23]. Whereas, the terms m_{ik}^f, m_{ik}^q are computed with a standard numerical integration.

Writing Equation (33) for all conductors, we get a matrix system known as the impedance matrix system:

$$[\mathbf{U}] = [\mathbf{Z}] \cdot [\mathbf{I}] + \left[\mathbf{L}^f\right] \cdot [\mathbf{M}] + [\mathbf{L}^q] \cdot [\Delta\mathbf{\Phi}], \tag{37}$$

where $[\mathbf{Z}]$ is a $(m \times m)$ matrix; $[\mathbf{U}]$ is a $(m \times 1)$ vector; $\left[\mathbf{L}^f\right]$ is a $(m \times 3n)$ matrix and $[\mathbf{L}^q]$ is a $(m \times p)$ matrix.

Finally, the algebraic linear systems (26), (29) and (37) is considered as $p + 3n + m$ complex unknowns:

$$\begin{bmatrix} \mathbf{A} & \mathbf{B} & \mathbf{0} \\ \mathbf{MoM} & -\mathbf{C} & -\mathbf{D} \\ \mathbf{L}^f & \mathbf{L}^q & \mathbf{Z} \end{bmatrix} \times \begin{bmatrix} \mathbf{M} \\ \Delta\mathbf{\Phi} \\ \mathbf{I} \end{bmatrix} = \begin{bmatrix} \mathbf{0} \\ \mathbf{0} \\ \mathbf{U} \end{bmatrix} \tag{38}$$

where $\mathbf{MoM} = \frac{[\mathbf{I}_d]}{\mu_r - 1} + [\mathbf{F}]$ is usually called the magnetic moment method matrix.

Let us note that Equation (29) has a disadvantage when the permeability of the material is high. Indeed, in such cases, the matrix term $[\mathbf{I}_d]/(\mu_r - 1)$ becomes very small in comparison with the $[\mathbf{F}]$ matrix. This can lead to the singularity of matrix $[\mathbf{MoM}]$. It should be noted that for the linear case, the simpler formulations can be developed. In such cases, Laplace equation is applied in the whole electromagnetic thin shell regions. Formulation (30) can be determined by a surface integral equation generated by charge distributions (Coulombian approach) or current distributions (Amperian approach) located on the surface area bounding the volume Ω of the thin shell region. In such configuration, only this surface area must be discretized. Consequently, the degrees of freedom of the obtained matrix system is lower. Moreover, Newton-Raphson algorithm can be easily used for the non-linear case [25].

Besides, in order to solve the equation system (38) with a reduced number of degrees of freedom, a Kirchhoff's mesh rule is introduced. Consequently, the current and the voltage values become associated to each of the independent circuit mesh. The coupling formulation has been implemented for 3D geometry.

5. Numerical Examples

In this section, we consider two numerical examples. To valid our formulation, three numerical methods are compared. The first one is a 3D shell element FEM coupled with the circuit equations [5]. In this modelling, a special care is proposed for the mesh around the conductor to ensure accurate results. The second one is a 2D FEM formulation. This approach has already shown its good precision with a few numbers of elements and is considered as the reference. The last one is our coupling integral formulation.

5.1. Validation Through an Academic Example

In the first example, a thin magnetic conductive disk is considered with the circular conductor fed by a voltage of 1V and a frequency of 10Hz (Figure 3).

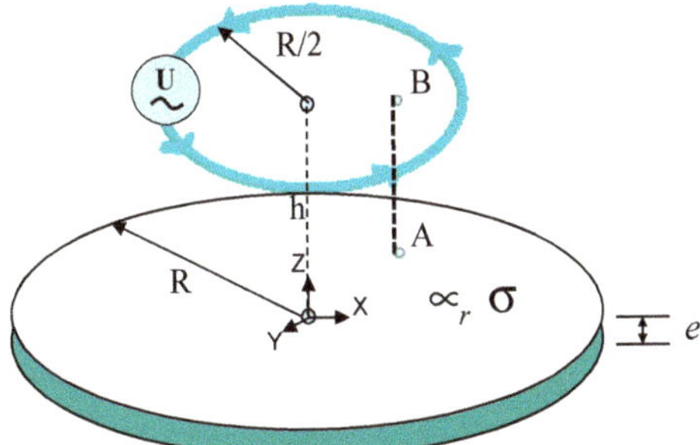

Figure 3. Thin magnetic conductive disk and conductor.

The device parameters are presented as below:

- Thin disk:

  - Electrical conductivity $\sigma_{disk} = 6 \times 10^7$ S/m
  - Permeability $\mu_r = 200$
  - $R = 1$ m, $e = 50$ mm
  - The skin depth $\delta = 1.45$ mm is smaller than the disk's thickness

- Conductor:

  - Electrical conductivity σ conductor $= 5.79 \times 107$ S/m
  - $h = R/4 = 0.25$ m

This academic example is solved by three numerical methods, where the first one is the axisymmetric FEM, the second one is a shell element formulation implemented in 3D FEM method [5] and the last one is the proposed integral method with the surface elements. In order to valid our method and to compare different approaches, we mainly focus on the computed current in the conductor and the magnetic field in the air region close to the device (calculated on the path AB, for A (0.25; 0; 0.1) and B (0.25; 0; 0.25)).

Let us note that the problem must be meshed very finely to have an accurate result with the FEM 3D (Table 1 and Figure 4) because of the high variations of the fields around the conductor. The current values greatly vary according to the number of the elements.

If the axisymmetric FEM is considered as our reference, the coupling integral method leads to an error of 0.1% (Table 1). Our method leads to more accurate results than the same shell element formulation but considering the air region treated with the 3D FEM.

Table 1. Current values in the conductor, where j is an imaginary unit.

FEM Axisymmetric		
Number of elements	15,000	30,000	55,000
Current values (A)	1905.95 − j493.12	1905.78 − j492.87	**1905.75 − j492.86**
FEM 3D with shell elements		
Number of elements	130,000	500,000	950,000
Current values (A)	1866.75 − j460.52	1841.78 − j505.48	1839.23 − j509.75
Integral method		
Number of elements	300	800	1000
Current values (A)	1908.21 − j495.44	1907.85 − j495.37	1907.76 − j495.35

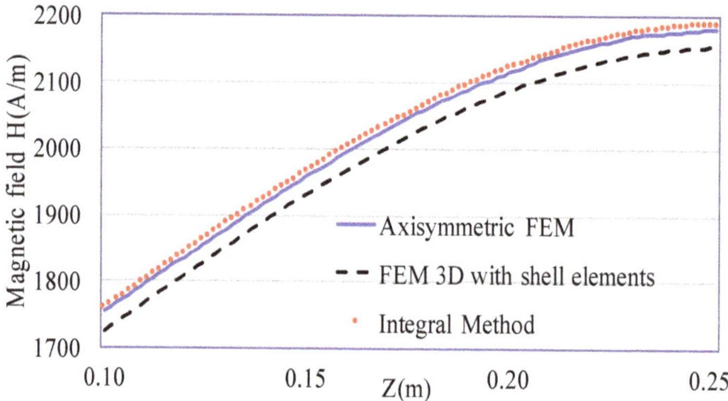

Figure 4. Magnetic field on path AB.

5.2. A Pratical Device Example

The second test problem is the modelling of a practical device proposed by the EDF (Electricité de France) [26]. The power station "Folies" is equipped with a three-phase reactance to limit short-circuit currents. In this case, the current in each reactance is 1000 A phase-shifted with 120 degrees, the frequency is 50 Hz (Figure 5).

Figure 5. Three phase reactance in the power station "Folies".

The parameters are presented as below (Figure 6):

- Outside diameter: $\emptyset e = 1.6$ m
- Inside diameter: $\emptyset i = 0.74$ m
- Winding thickness: ep = 0.5 m
- Center distance: Ent = 1.4 m

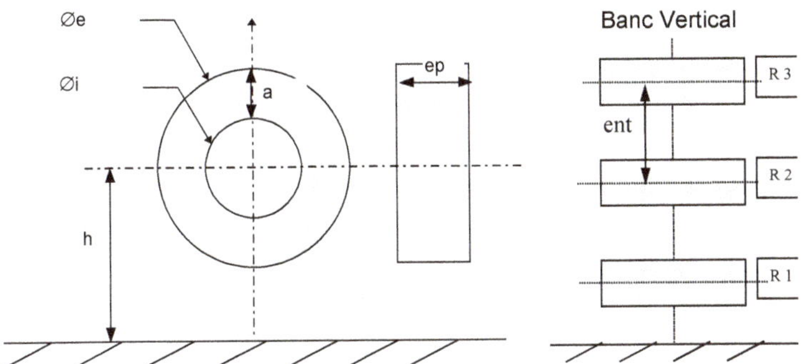

Figure 6. Parameters of three reactances.

In nominal conditions, a three-phase reactance generate a leakage induction in the neighborhoods. In order to minimize this electromagnetic disturbance, an electromagnetic shielding and a passive loop are added between the magnetic field source and the protected area (Figure 7). The device parameters are presented as below:

- Passive loop:

  - Permeability $\mu_r = 1$,
  - Electrical conductivity $\sigma = 3.03 \times 10^7$ S/m
  - Section radius $r_s = 9.25$ mm
  - $Z_{loop} = 3.85$ m

- Shielding:

  - Permeability $\mu r = 20{,}000$,
  - Electrical conductivity $\sigma = 2.2 \times 106$ S/m
  - Thickness = 3.5 mm
  - Zshielding = 5.15 m
  - The skin depth $\delta = 0.339$mm is thinner than the shielding's thickness.

The last case is tested by our integral method and the FEM 3D with shell elements. Let us note that this example is in 3-dimensional space and cannot be modelled by the 2D FEM. The current values in the passive loop and the current distribution in the shells are also compared.

The obtained results from the coupling method converge quite close to the current values presented in Table 2. Figure 8 also shows that the surface distribution of the current in the shell is quite similar to the two methods. The results achieved by the coupling integral method are very encouraging. The convergence is reached with few elements (about 1000 elements).

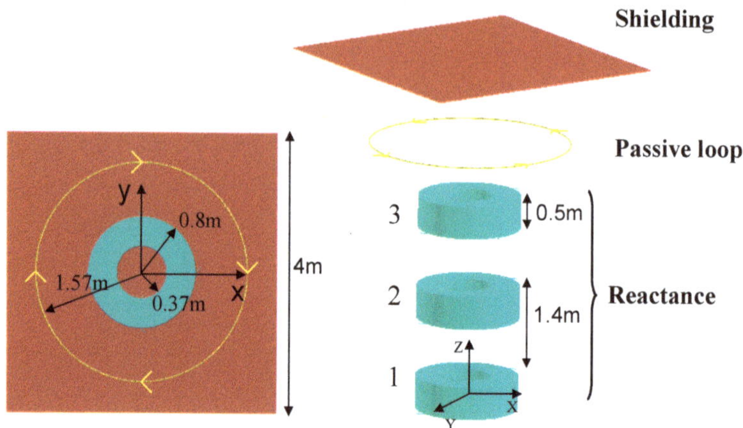

Figure 7. Geometry of the test case "Folies".

Table 2. Current values in the passive loop, where j is an imaginary unit.

FEM 3D with Shell Elements			
Number of elements	450,000	800,000	1,100,000
Current values (A)	619.14 + j1946.05	602.72 + j1878.72	606.87 + j1870.24
Integral method			
Number of elements	430	850	1020
Current values (A)	615.21 + j1742.30	619.86 + j1745.98	621.38 + j1746.60

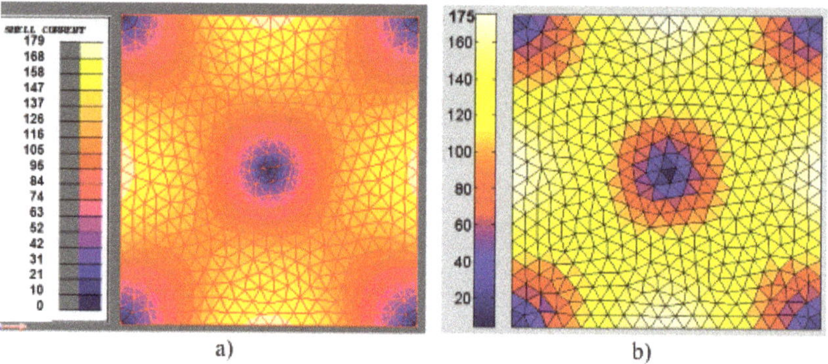

Figure 8. Surface distribution of the current (A/m) in the thin shell for the test case "Folies": (**a**) FEM 3D with shell elements; (**b**) Integral method.

However, some matrices of Equation (38) are fully populated and compression algorithms must be applied if there is a large number of elements. The coupling of the model order reduction techniques or the matrix compression algorithms with some integral methods clearly demonstrates its efficiency. For example, the coupling of a matrix compression algorithm like the FMM and the MoM or the PEEC method has reduced the computation time and the memory requirements down to more than 10 times and the compressed ratio is more than 80 percent [18,27]. Moreover, the acquired model can be reused to build a real circuit, which is easy to employ in all conventional SPICEs-like circuit solvers [28,29].

6. Conclusions

In this paper, we have presented a coupling integral method in order to model thin magnetic and conductive regions which can be coupled with an external electric circuit. This coupling enables the model of conductors with complex shapes, and various skin effects across the thickness of the thin shell are taken into account. Two numerical examples have been presented and the results highlighted the accuracy of the solution provided by our proposed method. In our suggestions, the problem dense matrix and the full memory need could be solved with the compression algorithms.

Author Contributions: Conceptualization, Methodology, Validation: T.L.-D. and G.M.; Investigation: T.L.-D.; Data Curation: T.L.-D.; Writing—Original Draft Preparation, Writing—Review & Editing: T.L.-D. and G.M; Funding Acquisition: T.L.-D. All authors have read and agreed to the published version of the manuscript.

Funding: This research was funded by Ministry of Education and Training, Vietnam, under grant number B2018-BKA-11-CtrVL.

Conflicts of Interest: The authors declare that there are no conflicts of interest regarding the publication of this paper.

Abbreviations

The following abbreviations are used in this paper:

FEM	Finite Element Method
FVM	Finite Volume Method
BEM	Boundary Elements method
MoM	Methods of Moments
PEEC	Partial Element Equivalent Circuit

References

1. Salon, S.; Chari, M.V.K. *Numerical Methods in Electromagnetism*; Academic Press: Cambridge, MA, USA, 1999.
2. Meunier, G. *The Finite Element Method for Electromagnetic Modeling*; Wiley: Hoboken, NJ, USA, 2008.
3. Krähenbühl, L.; Muller, D. Thin Layers in electrical engineering. Example of Shell Models in Analyzing Eddy-Currents by Boundary and Finite Element Methods. *IEEE Trans. Magn.* **1993**, *29*, 1450–1455. [CrossRef]
4. Igarashi, H.; Kost, A.; Honma, T. A Three-Dimensional Analysis of Magnetic Fields around a Thin Magnetic Conductive Layer Using Vector Potential. *IEEE Trans. Magn.* **1998**, *34*, 2539–2542. [CrossRef]
5. Guérin, C.; Meunier, G. 3-D Magnetic Scalar Potential Finite Element Formulation for Conducting Shells Coupled with an External Circuit. *IEEE Trans. Magn.* **2012**, *48*, 323–326. [CrossRef]
6. Dang, Q.V.; Dular, P.; Sabariego, R.V.; Krahenbuhl, L.; Geuzaine, C. Subproblem Approach for Thin Shell Dual Finite Element Formulations. *IEEE Trans. Magn.* **2012**, *48*, 407–410 [CrossRef]
7. Hoer, C.; Love, C. Exact Inductance Equations for Rectangular Conductors with Applications to More Complicated Geometries. *J. Res. Natl. Bur. Stand.-C. Eng. Instrum.* **1965**, *69C*, 127–137. [CrossRef]
8. Ruehli, A.E. Equivalent circuit models for three dimensional multiconductor systems. *IEEE Trans. Microw. Theory Tech.* **1974**, *22*, 216–221. [CrossRef]
9. Clavel, E.; Roudet, J.; Foggia, A. Electrical modeling of transformer connecting bars. *IEEE Trans. Magn.* **2002**, *38*, 1378–1382. [CrossRef]
10. Chadebec, O.; Coulomb, J.-L.; Janet, F. A review of Magnetostatic Moment Method. *IEEE Trans. Magn.* **2006**, *42*, 515–520. [CrossRef]
11. Le-Duc, T.; Chadebec, O.; Guichon, J.M.; Meunier, G. Coupling between partial element equivalent circuit method and an integro-differential approach for solving electromagnetics problems. *IET Sci. Meas. Technol.* **2012**, *6*, 394–397. [CrossRef]
12. Ishibashi, K.; Andjelic, Z.; Pusch, D. Nonlinear Eddy Current Analysis by BEM Minimum Order Formulation. *IEEE Trans. on Magn.* **2010**, *46*, 3085–3088. [CrossRef]
13. Ishibashi, K.; Andjelic, Z.; Takahashi, Y.; Takamatsu, T.; Tsuzaki, K.; Wakao, S.; Fujiwara, K.; Ishihara, Y. Ishihara. Some Treatments of Fictitious Volume Charges in Nonlinear Magnetostatic Analysis by BIE. *IEEE Trans. Magn.* **2012**, *48*, 463–466. [CrossRef]

14. Bettini, P.; Passarotto, M.; Specogna, R. Coupling Volume and Surface Integral Formulations for Eddy-Current Problems on General Meshes. *IEEE Trans. Magn.* **2018**, *54*, 1–4. [CrossRef]

15. Le-Duc, T.; Meunier, G.; Chadebec, O.; Guichon, J.M. A New Integral Formulation for Eddy Current Computation in Thin Conductive Shells. *IEEE Trans. Magn.* **2012**, *48*, 427–430. [CrossRef]

16. Le-Duc, T.; Meunier, G.; Chadebec, O.; Guichon, J.M.; Bastos, J.P.A. General integral formulation for the 3D thin shell modelling. *IEEE Trans. Magn.* **2013**, *49*, 1989–1992. [CrossRef]

17. De Camillis, L.; Ferranti, F.; Antonini, G.; Vande Ginste, D.; De Zutter, D. Parameterized Partial Element Equivalent Circuit Method for Sensitivity Analysis of Multiport Systems. *IEEE Trans. Compon. Packag. Manuf. Technol.* **2012**, *2*, 248–255. [CrossRef]

18. Meunier, G.; Guichon, J.M.; Chadebec, O.; Bannwarth, B.; Krähenbühl, L.; Guérin, C. Unstructured–PEEC Method for Thin Electromagnetic Media. *IEEE Trans. Magn.* **2020**, *56*, 1–5. [CrossRef]

19. Nguyen, T.S.; Guichon, J.M.; Chadebec, O.; Meunier, G.; Vincent, B. An independent loops search algorithm for solving inductive PEEC large problems. *Prog. Electromagn. Res. M.* **2012**, *23*, 53–63. [CrossRef]

20. Lombardi, L.; Romano, D.; Antonini, G. Efficient Numerical Computation of Full-Wave Partial Elements Modeling Magnetic Materials in the PEEC Method. *IEEE Trans. Microw. Theory Tech.* **2020**, *68*, 915–925. [CrossRef]

21. Ferranti, F.; Nakhla, M.S.; Antonini, G.; Dhaene, T.; Knockaert, L.; Ruehli, A.E. Multipoint full-wave model order reduction for delayed PEEC models with large delays. *IEEE Trans. Electromagn. Compat.* **2011**, *53*, 959–967. [CrossRef]

22. Ishida, K.; Itaya, T.; Tanaka, A.; Takehira, N. Mutual inductance of arbitrary-shaped coils using shape functions. *IET Sci. Meas. Technol.* **2019**, *13*, 1085–1091. [CrossRef]

23. Babic, S.I.; Akyel, C. New analytic-numerical solutions for the mutual inductance of two coaxial circular coils with rectangular cross section in air. *IEEE Trans. Magn.* **2006**, *42*, 1661–1669. [CrossRef]

24. Antonini, G.; Sabatini, M.; Miscione, G. PEEC modeling of linear magnetic materials. *IEEE Int. Symp. Electromagn. Compat.* **2006**, *1*, 93–98.

25. Carpentier, A.; Chadebec, O.; Galopin, N.; Meunier, G.; Bannwarth, B. Resolution of Nonlinear Magnetostatic Problems with a Volume Integral Method Using the Magnetic Scalar Potential. *IEEE Trans. Magn.* **2013**, *49*, 1685–1688. [CrossRef]

26. Abakar, A.; Meunier, G.; Coulomb, J.L.; Zgainski, F.X. 3D Modeling of Shielding Structures Made by Conductors and Thin Plates. *IEEE Trans. Magn.* **2000**, *36*, 790–794. [CrossRef]

27. Nguyen, T.S.; Guichon, J.M.; Chadebec, O.; Labie, P.; Coulomb, J.L. Ships Magnetic Anomaly Computation with Integral Equation and Fast Multipole Method. *IEEE Trans. Magn.* **2011**, *47*, 1414–1417. [CrossRef]

28. Nguyen, T.S.; Duc, T.L.; Tran, T.S.; Guichon, J.M.; Chadebec, O.; Meunier, G. Adaptive Multipoint Model Order Reduction Scheme for Large-Scale Inductive PEEC Circuits. *IEEE Trans. Electromagn. Compat.* **2017**, *59*, 1143–1151. [CrossRef]

29. Nguyen, T.S.; Le Duc, T.; Tran, S.T.; Guichon, J.M.; Chadebec, O.; Sykulski, J. Circuit realization method for reduced order inductive PEEC modeling circuits. *Compel. Int. J. Comput. Math. Electr. Electron. Eng.* **2016**, *35*, 1203–1217. [CrossRef]

 *applied sciences*

Article

Destruction of Fibroadenomas Using Photothermal Heating of Fe₃O₄ Nanoparticles: Experiments and Models

Ivan B. Yeboah [1], Selassie Wonder King Hatekah [1], Yvonne Kafui Konku-Asase [1], Abu Yaya [2] and Kwabena Kan-Dapaah [1,*]

[1] Department of Biomedical Engineering, School of Engineering Sciences, University of Ghana, P.O. Box. LG 77, Legon Accra, Ghana; iv_yeb@yahoo.com (I.B.Y.); hwk.selassie@gmail.com (S.W.K.H.); yvonnekafui@gmail.com (Y.K.K.-A.)

[2] Department of Materials Science and Engineering, School of Engineering Sciences, University of Ghana, P.O. Box. LG 77, Legon Accra, Ghana; ayaya@ug.edu.gh

* Correspondence: kkan-dapaah@ug.edu.gh; Tel.: +233-23-536-4134

Received: 23 May 2020; Accepted: 30 June 2020; Published: 24 August 2020

Abstract: Conventionally, observation (yearly breast imaging) is preferred to therapy to manage small-sized fibroadenomas because they are normally benign tumors. However, recent reports of increased cancer risk coupled with patient anxiety due to fear of malignancy motivate the need for non-aggressive interventions with minimal side-effects to destroy such tumors. Here, we describe an integrated approach composed of experiments and models for photothermal therapy for fibroadenomas destruction. We characterized the optical and structural properties and quantified the heat generation performance of Fe₃O₄ nanoparticles (NPs) by experiments. On the basis of the optical and structural results, we obtained the optical absorption coefficient of the Fe₃O₄ NPs via predictions based on the Mie scattering theory and integrated it into a computational model to predict in-vivo thermal damage profiles of NP-embedded fibroadenomas located within a multi-tissue breast model and irradiated with near-infrared 810 nm laser. In a series of temperature-controlled parametric studies, we demonstrate the feasibility of NP-mediated photothermal therapy for the destruction of small fibroadenomas and the influence of tumor size on the selection of parameters such as NP concentration, treatment duration and irradiation protocols (treatment durations and laser power). The implications of the results are then discussed for the development of an integrated strategy for a noninvasive photothermal therapy for fibroadenomas.

Keywords: magnetite nanoparticles; Mie scattering theory; near infrared laser; photothermal therapy; finite element method; bioheat transfer; diffusion approximation; Arrhenius integral; breast cancer

---

1. Introduction

Fibroadenoma is one of the commonest benign female breast diseases. Histologically, it is a well-circumscribed homogeneous biphasic solid lump with distinct imaging features made up of epithelial and stromal tissues [1]. Definitive diagnostic techniques include ultrasound, mammography, magnetic resonance imaging or stereotactic guided needle biopsy [2]. Their sizes are normally small (<2.5 cm), but can become giant juvenile tumors (>10 cm) during puberty or pregnancy [3] causing considerable pain and cosmetic deformity of the breast. Although it accounts for 25% of all breast masses in women [4], the numbers are higher in adolescents: 68% of all breast masses and 44–94% of biopsied breast lesions [5,6]. Furthermore, available data seem to suggest that incidence and recurrence rates are common in black race [7–9], who are more likely to develop breast cancer at a younger age [10].

Management of fibroadenomas can take two forms: observation and therapy. For fibroadenomas that cause pain, deform the breast, persist without any regression and are histologically complex, therapy is warranted [2]. Available options include open surgical excision as well as several modern minimally invasive probe-based thermal therapies including cryotherapy, radiofrequency ablation (RFA), microwave ablation (MWA), focused ultrasound (FUS) and laser-induced thermotherapy (LITT) [11,12]. On the other hand, observation, which involves yearly breast imaging, is usually recommended when the tumor is asymptomatic, small and not rapidly increasing in size to cause cosmetic deformity and pain. However, there are situations when patients who qualify for observation agitate due to the fear of malignancy leading to significant anxiety [12]. Furthermore, a recent study reported a 41% increase in cancer risk for women diagnosed with fibroadenomas compared to those without them [13]. Issues related to superficial skin burns, hemorrhage and hematoma, cost and complexity of technique that are associated with options stated earlier limit their use for small-sized fibroadenoma [11,14]. An ideal treatment will be one that is noninvasive with no side-effects. Recent advances in nanomedicine offer the opportunity for the design of smart strategies that can potentially overcome drawbacks with conventional techniques to reduce invasiveness and complexity.

Nanomedicine involves the use of nanomaterials—metallic and ceramic (iron-oxide) nanoparticles (NPs)—for theranostic purposes in living organisms. Photothermal therapy (PTT) is an emerging localized cancer treatment whereby NPs embedded in the tumor convert near-infrared light, which is minimally absorbed by biological tissue, to heat leading cell death. Traditionally, metallic NPs such as gold, silver, copper as well as carbon-nanotubes or graphene have been used for PTT [15]. Although several promising results have been reported in the literature for both in-vitro (cells) and in-vivo (animals), issues related to NP biocompatibility and stability have limited their progression to the clinics [15]. Unlike their metallic counterparts, ceramic NPs—Fe_3O_4 and γ-Fe_2O_3—have been used in human trials for magnetic hyperthermia treatment of brain [16] and prostate [17] cancers. Furthermore, these ceramic NPs have very recently been tested for photothermal therapy in both in-vitro and in-vivo studies. Chu et al. [18] showed that various shapes of Fe_3O_4 nanoparticles (NPs) were able to kill cancer cells and tumors in in-vitro (esophageal cancer cell) and in-vivo (mouse esophageal tumor) models, respectively. In another study, Espinosa and co-workers [19], demonstrated the ability of the iron-oxide NPs to act as magnetic and photothermal agents simultaneously—so called magnetophotothermal approach—and showed their unprecedented heating powers and remarkable heating efficiencies (up to 15-fold amplifications).

Here, we describe an integrated approach composed of experiments for NP characterization and models for optical property predictions and computational treatment planning. Our long term goal is to develop a noninvasive but highly efficacious treatment method for the destruction of fibroadenomas. The feasibility of such integrated approaches for photothermal therapies have been previously reported for different application in the literature [20,21]. We characterized the material properties and quantified the photothermal heat generation of Fe_3O_4 NPs by experimental measurements, obtained their optical absorption coefficient via experimentally guided Mie scattering theory and integrated it into a computational—finite element method (FEM)—model to predict in-vivo thermal damage of a NP-embedded tumor located in a multi-tissue breast model during irradiation by a near-infrared (NIR) 810 nm laser. Using a temperature-controlled parametric study, we explored the feasibility of NP-mediated photothermal therapy for the destruction of fibroadenomas and the influence of tumor size on parameters such as NP concentration, treatment duration and irradiation protocols (laser power and duration). The implications of the results are discussed for the development of an integrated strategy for photothermal therapy for the destruction of fibroadenomas.

2. Results

Optical and structural characterization of Fe_3O_4 NPs. Structural characterization of the Fe_3O_4 NPs—purchased commercially—were done to verify the specification provided by the manufacturer and also predict the optical absorption coefficient. X-ray diffraction spectra of the Fe_3O_4 NPs revealed the presence of peaks at 2θ = 31.5°, 35.8°, 38.35°, 42.75°, 47.2°, 54.04°, 57.24°, and 62.75° (Figure 1a). The observed peaks correspond to diffraction planes: (220), (311), (222), (400), (110), (422), (511), and (440), which have been attributable to cubic spinel phase of Fe_3O_4 (space group, *Fd-3m*, JCPDS-#19-0629). Since no other prominent phase was detected, the result implied that the NPs are essentially crystalline Fe_3O_4. Transmission electron method (TEM) confirmed the morphology of the NPs to be spherical (with agglomerations) and size distribution to be between 15 and 20 nm in diameter as indicated by the manufacturer (Figure 1b). The agglomeration revealed in the TEM image have been attributed to dipolar coupling between the NPs [22,23]. For any NP, its NIR photothermal effects are controlled by their NIR optical absorbance. UV-vis-NIR spectra of the NPs showed an extended optical absorption that slowly increased in the NIR region relative to the visible light region (Figure 1c). The absorbance intensity at 810 nm increased linearly with concentration, from 0.35 ([Fe_3O_4] = 6 mM) to 1.51 ([Fe_3O_4] = 24 mM) (Figure 1d). The absorbance band in the NIR region of UV-vis-NIR optical spectra is consistent with the results in the literature and has been attributed to multiple charge (electron) transfer [24]. Furthermore, the linear increase of absorbance for the range of concentration tested in this work has been previously reported elsewhere [19,25]. Shen and co-workers [25] showed that saturation starts occurring at high concentration (100 mM, absorbance values > 3 at 808 nm). In an effort to translate the experimentally measured photothermal heat generation capabilities of the Fe_3O_4 NPs tested in this study, we followed the flow-chart shown in Figure 1e to obtain the extinction cross sections of the MNPs, which was then used in Equation (2) to predict absorbance, A_{pred}. The validity of A_{pred} was tested by evaluating its agreement with the experimentally measured absorbance, A_{exp}, for the different concentrations of Fe_3O_4 (6, 12, 24 mM). We observed that the predictions agreed reasonably well with experiments to within 2% for all concentrations when the sample size, n, in Equation (5) was equal to 5 (see Figure 1f).

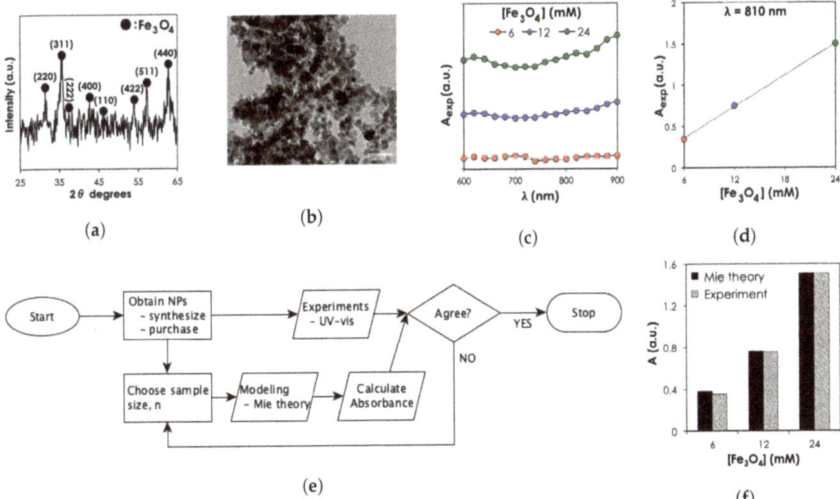

Figure 1. Structural and optical characterization results. (**a**) X-ray diffraction spectra at a power of 45 kV × 40 mA. (**b**) Transmission electron microscopy at magnification of 0.5 mm, Scale bar: 50 nm. Absorbance as a function of (**c**) wavelength (λ) and (**d**) concentration ([Fe_3O_4]) at λ = 810 nm. (**e**) Flow-chart for the comparison of theoretical predictions and experiment measurements. (**f**) Comparison of the A_{pred} and A_{exp} for different [Fe_3O_4] (6, 12, 24 mM).

Photothermal effects of Fe_3O_4 NPs. The influence of laser power (P_0 = 0.5 and 1.0 W) and NP concentration ([Fe_3O_4] = 0–24 mM) on photothermal effects was accessed in aqueous solution (deionized water) to quantify their heat generation capabilities under an irradiation duration of 5 min. Pure deionized water—containing no Fe_3O_4 nanoparticles—was used as a control. The rate of change of the temporal curves increased with concentration at 5 min independent of the laser power that was used (Figure 2a,b). For $P_0 = 0.5$ W, the temperature change, ΔT, increased approximately by 44.4% (from $\approx$9 to 13 °C) when concentration was increased from 0 to 24 mM (Figure 2c). When the power was increased to 1.0 W, ΔT increased by approximately 83.3% (from $\approx$12 to 22 °C) for the same concentration. Photothermal conversion efficiency, η_{exp}, decreased with concentration and laser power (Figure 2d). For instance, η_{exp} for the 6 mM solution decreased from approximately 66% to 51% when P_0 was increased from 0.5 to 1.0 W. Furthermore, when the concentration was increased from 6 to 24 mM, η_{exp} decreased from 46% to 39% using the same power regimes. Generally, the trend of ΔT recorded in this study was in agreement with measured absorbance properties and also consistent with previously reported studies [18,19,26]. For small NPs (<30 nm) and low concentrations, absorption dominates scattering leading to high η_{exp}. On the other hand, scattering dominates the extinction efficiency as nanoparticle size or concentration is increased. As [Fe_3O_4] increases, clusters are formed due to the high surface area to volume ratio of nanoparticles [27]. These clusters act as large particles to enhance scattering leading to the reduction in η_{exp} [28]. Several approaches are available for the prevention of clusters.

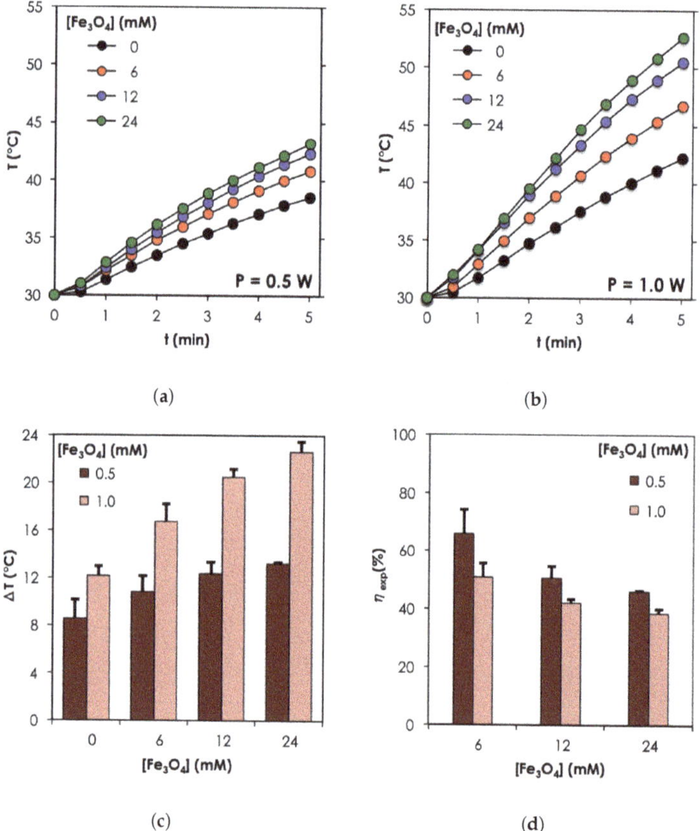

Figure 2. Photothermal characterization results. Temporal response curves for different concentrations after 5 min of irradiation with laser powers: (**a**) $P_0 = 0.5$ W and (**b**) $P_0 = 1.0$ W. Comparison of the corresponding (**c**) temperature change (ΔT) and (**d**) experimental photothermal conversion efficiency (η_{exp}) as a function of laser power. Error bars: s.d.

Computational modeling of NP-mediated photothermal heating of breast tumor. The use of computational model as quantitative frameworks enables assessment and customization of the treatment parameters (NP concentration, treatment duration and irradiation protocols: duration and laser power) to potentially enhance efficacy. Thus, FEM simulations were applied to approximate photothermal heating of a Fe_3O_4-containing tumor embedded within a female breast using the optical diffusion approximation of the transport theory [29] and the Pennes bioheat transfer equation [30].

Figure 3 shows a schematic of 2D representation of the axisymmetric geometry of the computational model. It was configured as a heterogeneously dense [31] multi-layer block of tissue with proportions assigned according to the Breast Imaging Reporting and Data System (BIRADS) developed by American Cancer Research [32]. It consisted of various layers of normal tissue with unequal thickness. The dimensions of the model were chosen to represent a "heterogeneously dense" breast model [31], which consists of 20% muscle layer, 60% glandular layer and 20% fat layer. Also, a tumor is located at 55 mm from the base. The laser source was assumed to be a diode laser 810 nm placed close to the top surface of the breast model. The inset is a fragment of geometry showing control points P1−P4, where temperatures were recorded. The assigned optical, thermal and physical properties of different tissue layers were approximate values obtained from the literature [31,33−36]. Nanoparticles were assumed to be intravenously injected and uniformly distributed throughout the tumor.

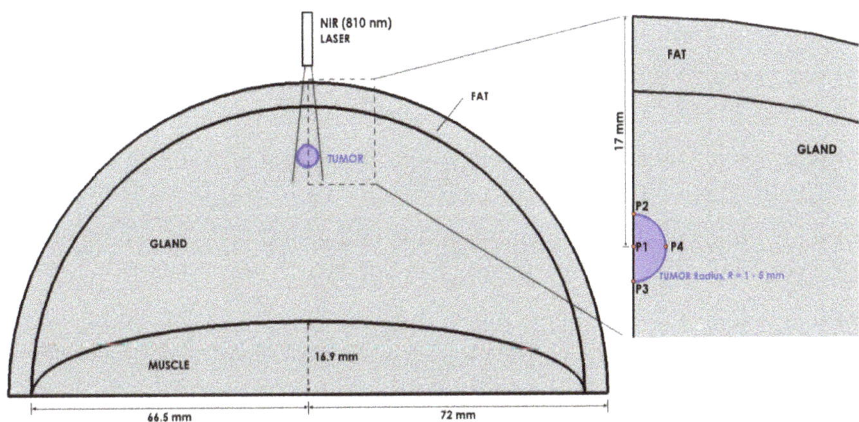

Figure 3. FEM geometry. Schematic of the photothermal therapy consisting of a normal multi-tissue breast domain with an embedded spherical tumor (blue sphere) and NIR (810 nm) laser source. Inset: Fragment of geometry showing controls point P1−P4, where temperature were recorded.

To characterize the temperature and thermal damage profiles, we simulated temperature-controlled heating at a maximum tumor temperature, $T_{max} = 85\ °C$, for $t = 15$ min. The radius of the tumor, R, and P_0, were chosen to be 2.5 mm and 1 W respectively. The predicted temperature distribution (Figure 4a) was revealed to be non-uniform with the maximum temperature occurring within the tumor and decreasing radially outwards into the surrounding tissue. The latter suggests that the heat transfer was predominantly conductive. For the case of the predicted thermal damage shown in Figure 4b,c, it can be seen that the entire tumor area, plus margins of up to 1 mm around it, was completely destroyed ($\Omega = 100\%$). A comparison of temporal response curves for temperatures (Figure 4d) at different control points (Figure 3) within the tumor (P1)and at the tumor-gland boundaries (P2–P4) revealed that the temperature rise as well as the final value was higher at (P1) relative to the boundaries: P2 (top), P3 (bottom) and P4 (side). This phenomenon can be attributed to factors such as relatively low blood perfusion and high metabolic heat of the tumor leading to high retention of heat within the tumor [37]. However, at all the locations, the temperature plateaued after about 2–3 min. The consequence of the high temperature within the tumor is revealed in corresponding predicted temporal curves for the thermal damage (Figure 4e), which shows that 100% thermal damage occurs faster in the innermost part of tumor (P1)—≈3 min—compared to the peripherals, which take up to about ≈10 min (P4). Consistent with the literature [33,38], the model predictions showed the dependence of thermal damage spatial profile on the temperature distribution, which decreased with distance away from center of the tumor (see Figure 4f).

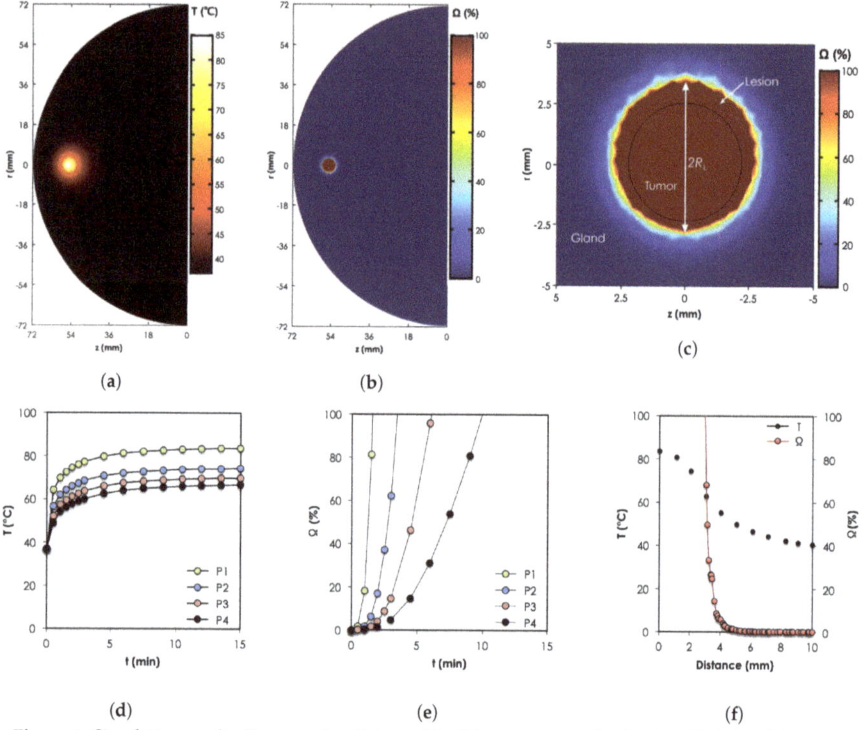

Figure 4. Simulation results. Cross-sectional view of the (**a**) temperature distribution, (**b**) thermal damage, (**c**) thermal damage showing the lesion parameter. Temporal response curves for (**d**) temperature and (**e**) thermal damage at the control points (P1–P4, cf. Figure 3). (**f**) Temperature and thermal damage as a function distance from P1. Simulation settings: $P_0 = 1$ W, $t = 15$ min and $T_{max} = 85$ °C.

Ablative temperatures between 60 and 100 °C cause irreversible damages to key cytosolic and mitochrondrial enzymes [39,40]. For any tumor ablation therapy to be considered successful and thus reduce the chance of recurrence, it is critical to ensure that the entire volume of the tumor reaches therapeutic temperatures that ensures complete thermal damage ($\Omega = 100\%$). Such a goal can be achieved through the use of an appropriate maximum temperature, which takes into consideration the tumor dimensions. For NP assisted photothermal therapies such as the one being proposed in this study, maximum ablative tumor temperatures, T_{max}, can be controlled by varying parameters such as NP number density, N (or volume fraction, ϕ_v), the laser power, and treatment duration, t. To demonstrate this, a parametric study was used to determine N required to achieve a given T_{max} (70, 85, 100 °C) and the corresponding volume of the lesion V_L for different tumor sizes, R (1, 2.5, 5 mm). V_L, was assumed to be spherical [41,42]; its radius, R_L, was calculated as half the axial length of the predicted cross-sectional area where $\Omega = 100\%$ (see Figure 4c). A summary of the results is presented in Table 1. The simulations were run with $P_0 = 1$ W and $t = 15$ min. Generally, it can be observed that T_{max} required to achieve complete thermal damage increased with size of the tumor. For instance, $T_{max} = 70$ °C produced a lesion with $V_L = 2.95$ mm^3, which was insufficient to completely ablate the entire volume of tumor with $R_T = 1$ mm ($V_L = 4.19$ mm^3). On the other hand, $T_{max} = 85$ produced a lesion with $V_L = 2.95$ mm^3, which was big ensure to ensure complete thermal damage. Since P_0 was held constant for all simulation, it meant that N had to be increased to achieve the given T_{max}. The results reveal that N required to achieve $T_{max} = 70\,^\circ$C decreased with tumor size. For instance, N required to achieve $T_{max} = 70\,^\circ$C decreased from 112.37×10^{14} mL^{-1} to 5.54×10^{14} mL^{-1} when R_T was increased from 1 to 5 mm. Lastly, the nanoparticle concentrations that were required to achieve the different values of T_{max} corresponded to volume fractions in the range between 0.004% and 10.6%. A review of the nanoparticle delivery to tumors in the literature between 2006 and 2016 by Wilhelm et al. [43] revealed that only approximately 1% of administered nanoparticle dose reached the tumor. Therefore, it is important that the ϕ_v is kept at the low value for practical applications. This can be achieved by through several means such as increasing the laser power or exploiting the capability of the Fe$_3$O$_4$ NPs to generate synergistic heat during simultaneous exposure to NIR laser and alternating magnetic field as previously reported elsewhere [19].

Table 1. Comparison of volume, V_L, of predicted lesions and the number density, N, of nanoparticles (or volume fraction, ϕ_v) used to achieve maximum tumor temperatures, T_{max} (70, 85, 100 °C) in different tumor sizes, R (1, 2.5, 5 mm). R_L is the radius of the lesion.

T_{max} (°C)	$R = 1$ mm		$R = 2.5$ mm		$R = 5$ mm	
	$N(\phi_v)$ ($\times 10^{14}$/mL)	$V_L(R_L)$ (mm^3)	$N(\phi_v)$ ($\times 10^{14}$/mL)	$V_L(R_L)$ (mm^3)	$N((\phi_v))$ ($\times 10^{14}$/mL)	$V_L(R_L)$ (mm^3)
70	112.37 (5.68%)	2.95 (0.89)	19.36 (0.06%)	20.94 (1.71)	5.54 (0.002%)	44.00 (2.19)
85	166.11 (8.17%)	15.30 (1.54)	27.60 (0.09%)	128.45 (3.13)	7.18 (0.003%)	347.17 (4.36)
100	221.34 (10.6%)	24.43 (1.80)	36.72 (0.13%)	256.20 (3.94)	9.26 (0.004%)	998.31 (6.20)

These predictions are consistent with previously reported experimental and computational results in the literature. Kannadorai et al. [44], developed a treatment planning model for the optimization to parameters such as laser power density, nanoparticle concentration and exposure time in an effort aimed at potential enhancement of treatment outcome. Their predictions showed that any change made to any of the parameters can be compensated by altering the remaining parameters. Using an integrated strategy that combined x-ray computed tomography or ex-vivo with a 4-dimensional FEM model, Maltzahn and co-workers [20] simulated photothermal heating with polyethylene glycol PEGylated gold nanorods (PEG-NR) and used the results to guided pilot therapeutic studies on human xenograft tumors in mice. Their simulations revealed the extension of thermal flux vectors from the region where PEG-NRs were located as well as the expected thermal profile.

3. Discussion

Generally, the efficacy and safety of NP-mediated PPTT depend on several independent factors such as the properties of nanomaterial (e.g., morphology, size distribution, optical absorption coefficient), biological identity (e.g., in-vivo circulation time, stability, tumor-homing) and irradiation protocols (e.g., laser beam power, shape, duration, cross-section, direction). Therefore, it requires an integrated strategy that combines experiments and models to optimally select and customize these parameters towards the realization of a reliable and efficient treatment outcome. Clearly, we acknowledge that the strategy we describe here is not exhaustive; however, our intention was to emphasize the need for a structured procedure that allows a quantitative assessment of the heat generation capabilities and predict critical optical properties of the nanoparticles that can be used in computational modeling.

We show that Fe$_3$O$_4$ NPs exhibit photothermal effects when irradiated with NIR (810 nm) light leading to photothermal generation, which increases with NP concentration and laser power. On the basis of the optical (Figure 1c) and structural (Figure 1b) properties, the absorption coefficient that was used in the computational model was predicted with the Mie scattering theory. It is worth noting here that we used the Mie theory because the NPs were spherical [45], however, the photothermal effect is not unique to only spherical iron-oxide NPs but also cubic [19], hexagonal and wire-like [18]. For such non-spherical geometries, discrete dipole approximation—a discrete solution method of the integral form of Maxwell's equations, should be used [46]. Qin et al. [47] used a combination of the two methods to perform quantitative comparison of photothermal heat generation between gold nanospheres and nanorods. Estimation of η_{exp}, which describes how the NPs dispose (scattering plus absorption) the incident electromagnetic energy, has implications for NP concentration and laser beam power to be used. Although, it was beyond the scope of this work because it has been extensively studied previously [20,48], the biodistribution and effective tumor-homing following intratumoral or i.v. administration is key to the efficacy of treatment. To this end, techniques such as PEGylation and ligand-conjugation of the NPs have been shown to enhance and modulate their performance for biomedical applications and, thus, must be considered as part of efforts to fully characterize the nanoparticles for in-vivo applications.

Due to the complexities of multi-tissue breast tissue and different characteristics of tumors (size, location, shape), coupling of experimental measurements with computational modeling allows for the progressive selection, optimization and customization of parameters including NP concentration, irradiation protocols and treatment duration for in-vivo applications. This approach is essential for mitigation or prevention of collateral damage to healthy tissue surrounding the tumor. Here, we used optical absorption coefficient obtained via Mie theory predictions to develop a FEM model and used a temperature-controlled parametric study to demonstrate that the temperatures of different sized fibroadenomas can reach ablative levels leading to complete thermal damage ($\Omega = 100\%$) during irradiation with different laser powers. Several investigators have shown that the accuracy of FEM models for thermotherapy can be enhanced by using realistic geometries and material properties [20,21,34,44]. Although our model accounted for temperature dependence and blood perfusion effects, the multi layer geometry based on BIRADS [31] is generic and the distribution of the NPs was an assumption. Such simplification can have an adverse effect on integrity of the predicted values. Several reports have shown that using geometries that correlate with real anatomic datasets and include biodistribution data [20,34] have the potential to improve the accuracy of predictions. Elsewhere, such datasets have been obtained via noninvasive techniques such as X-ray computed tomography, sonography and ex-vivo spectrometry [20,34].

Finally, we acknowledge that Au NPs have been the prime candidates for photothermal therapy, however, it still remains an experimental cancer treatment due to issues related to their bio-persistent, which makes them potentially toxic and the use of high irradiation doses to achieve therapeutic temperatures due to the turbidity of biological tissues [28,49]. These issue have led to the recent interest in the photothermal properties of Fe_3O_4 NPs, which have been approved by the food and drugs administration (FDA). Furthermore, recent studies that have explored the simultaneously application (DUAL-mode) of both NIR laser and alternating magnetic field (AMF) to the Fe_3O_4 NPs have shown promising and interesting results. The studies found that the amount of heat generated with DUAL-mode equaled the sum of the heating for NIR laser or AMF only [19,26]. The essence of these results is that the use of the DUAL-mode can be used to overcome the challenges associated with the individual techniques.

4. Materials and Methods

4.1. Materials

The following materials were used in this study: Fe_3O_4 (99.5%, 15–20 nm) NPs (US Research Nanomaterials Inc., Houston, TX, USA).

4.2. Experiments

4.2.1. MNP Characterization

Fe$_3$O$_4$ NPs were characterized by TEM (Philips CM10, Philips Electron Optics, Eindhoven, The Netherlands) and XRD (D8 FOCUS X-ray, Bruker AXS GmbH, Karlsruhe, Germany) for crystal structure and morphology and then UV-vis-NIR spectroscopy (GENESYS 10S UV-vis, Thermo Fisher Scientific, Madison WI, USA) in the wavelength range of 400–900 nm for absorption spectra.

4.2.2. Photothermal Measurement in Water

The sample (Fe$_3$O$_4$ NPs in 0.5 mL of deionized water) contained in a 1.5 mL Eppendorf tube was irradiated by a NIR continuous laser at 810 nm (Photon Soft Tissue Diode Laser, Zolar Technology & MFG, Canada) with an external adjustable power, P_0 (0–3 W). The distance between the sample and the laser was 1–2 cm and the laser spot size was about 1 mm. The laser powers that were used was 0.5 and 1.0 W. Each sample was identically irradiated for 5 min. The resulting temperature rise was recorded by thermocouples (J-type, National Instrument, Austin, TX, USA) connected to a portable data acquisition system (NI USB-9222A, National Instruments, Austin, TX, USA) and recorded every 30 s with NI-DAQmx (National Instruments, Austin, TX, USA) and software (LabVIEW 8.6, National Instruments, Austin, TX, USA). All measurements were obtained in triplicate except stated otherwise.

The experimental photothermal conversion efficiency (η_{exp}) of the NPs was calculated directly from steady-state temperature increase as follows:

$$\eta_{\mathrm{exp}} = \frac{Q_{\mathrm{exp}}}{P(1 - 10^{-A_{exp}})} \tag{1}$$

where Q_{exp} (W) was calculated with previously reported expression [47]: $16.855\Delta T$ (mW), ΔT is the temperature change, P is the incident laser power and A_{exp} is the absorbance of the Fe$_3$O$_4$ nanoparticles at 810 nm.

4.3. Models

4.3.1. Optical Properties Predictions

The experimentally measured absorbance, A_{exp}, of a colloidal solution can be expressed in terms of predicted extinction cross section, σ_{ext} as:

$$A_{\mathrm{pred}} = N\frac{\sigma_{\mathrm{ext}}}{2.303}d_0 \tag{2}$$

where N (m^{-3}) is the number density of the NPs, d_0 (cm) is the path length of the spectrometer. For spherical, homogeneous and isotropic NPs, the "Mie scattering theory" [45,50] can be used to compute the exact values of the Q_{ext}, absorption (Q_{abs}) and scattering (Q_{sca}) efficiency as well as the anisotropy factor (g) as follows [46]:

$$Q_{\mathrm{ext}} = \frac{2}{k^2}\sum_{n=1}^{\infty}(2n+1)\mathrm{Re}(a_n + b_n) \tag{3a}$$

$$Q_{\mathrm{sca}} = \frac{2}{k^2}\sum_{n=1}^{\infty}(2n+1)[(|a_n|^2 + |b_n|^2)] \tag{3b}$$

$$Q_{\mathrm{abs}} = Q_{\mathrm{ext}} - Q_{\mathrm{sca}} \tag{3c}$$

$$g = \frac{4}{k^2 Q_{\mathrm{sca}}}\sum_{n=1}^{\infty}\left[\frac{n(n+2)}{n+1}\mathrm{Re}(a_n a_{n+1}^* + b_n b_{n+1}^*) + \frac{2n+1}{n(n+1)}\mathrm{Re}\left(a_n b_n^*\right)\right] \tag{3d}$$

where k is the NP size parameter ($= 2\pi a/\lambda$). a_n and b_n, the scattering coefficients in terms of the spherical Ricatti-Bessel functions, ψ_n and η_n, respectively, are defined as:

$$a_n = \frac{\psi'_n(mx)\psi_n(x) - m\psi_n(mx)\psi'_n(x)}{\psi'_n(mx)\eta_n(x) - m\psi_n(mx)\eta'_n(x)} \tag{4a}$$

$$b_n = \frac{m\psi'_n(mx)\psi_n(x) - \psi_n(mx)\psi'_n(x)}{m\psi'_n(mx)\eta_n(x) - \psi_n(mx)\eta'_n(x)} \tag{4b}$$

where m is the ratio of complex refractive index ($n_S = \sqrt{\epsilon_S}$) of the sphere to that of the surrounding medium (n_m) asterix ($*$) and prime ($\prime$) indicate complex conjugate and derivative with respect to x and mx, respectively. The numerical calculations were performed with a python code implementation of the original algorithm published by Wiscombe [51]. The wavelength dependent complex refractive index, $n(\lambda)$, was obtained from Ref. [52].

To account for polydispersity, the size range of the nanoparticle was discretized into a varying number of terms (n_t) and then number-averaged to obtain the ensemble optical properties,

$$\overline{\sigma_k} = \frac{1}{n_t} \sum_{r=R_l}^{R_u} \sigma_k(r+i) \quad k = \text{ext, abs, sca,} \quad n_t = 2,3,4,\ldots,N \tag{5}$$

where R_u and R_l are the upper and lower limits of the NP size range, respectively. i is the step size which is calculated as: $i = R_u - R_l/(n-1)$ and $\overline{\sigma_k}$ is the mean k (i.e., extinction, absorption, scattering) cross sections of the NP.

4.3.2. In-Vivo Predictions

The computational model is a multiphysics FEM model, thus, it took into account optical and thermal effects. Light distribution was based on the diffusion approximation of the transport theory [29] and temperature distribution by Pennes bio-heat transfer equation [30], which takes into account the effect of cell death on blood perfusion and the dependence of cell death and properties of the tissue. Cell death was determined by an Arrhenius based integral injury model [53].

Light Distribution. The optical diffusion approximation of the transport theory [29] was used to describe light distribution due to the dominance of scattering over absorption in biological tissues. It is defined by:

$$\frac{1}{c_n}\frac{\partial}{\partial t}\varphi = D\nabla^2\varphi - \mu_a\varphi + S \tag{6}$$

c_n (m s^{-1}) is the speed of light in a medium, φ (W m^{-2}) is the fluence rate, μ_a (m^{-1}) is the absorption coefficient, S (W m^{-3}) is the light source term and $D = \mu_a/\mu_{\text{eff}}^2$ (m) is the diffusion coefficient. $\mu_{\text{eff}} = \sqrt{3\mu_a(\mu_a + \mu'_s)}$ (m^{-1}) is the effective attenuation coefficient and μ'_s (m^{-1}) is the reduced scattering coefficient. Assuming that the light source was a continuous wave Gaussian NIR laser beam that was incident onto the breast model, the φ can be defined by

$$\phi(\vec{r}) = \frac{P_0 \exp(-\mu_{\text{eff}}\vec{r}\cdot\hat{n})}{4\pi D r} \tag{7}$$

where P_0 is the laser power and $\hat{n}$ is the direction of the beam. A summary of the values of the optical properties of the tissue used in the simulation is presented in Table 2.

Table 2. Optical properties of the biological domains that were used in the simulations. The values were obtained from Refs. [34–36].

Tissue	Coefficients, (m^{-1})		Refractive Index, (1)
	Absorption, μ_a	Reduced Scattering, μ'_s	n
Fat [35]	3	950	1.455
Gland [35]	6	1100	1.4
Muscle [34]	23	130	1.37
Tumor [36]	7	1400	1.37

Temperature Distribution. The Pennes bio-heat transfer equation [30] was used to estimate the temperature distribution. An additional term was added to account for the external heat source. The resulting equation is given by:

$$\rho c_\mathrm{p}\frac{\partial T}{\partial t} = \lambda(T)\nabla^2 T + \rho_\mathrm{b}c_\mathrm{b}\omega_\mathrm{b}(\Omega)(T_\mathrm{b} - T) + Q_\mathrm{met} + Q \tag{8}$$

where ρ (kg m^{-3}) is the density, c_p (J kg^{-1} K^{-1}) is the specific heat capacity at constant pressure. $\lambda(T)$ (W m^{-1} K^{-1}) is the temperature dependent thermal conductivity, which is assumed to be a linear function defined by [54]:

$$\lambda(T) = \lambda_{(37\,^\circ\mathrm{C})}[1 + 0.0028(T - 293.15K)] \tag{9}$$

where T (K) and T_b (K) are the normal body and arbitrary temperatures, respectively. ρ_b is the density of blood, c_b, the specific heat capacity of blood and $\omega_\mathrm{b}(\Omega)$ is the coefficient of blood perfusion assumed to be dependent on the cell damage, Ω, and defined by [33,38]:

$$\omega_b(\Omega) = \begin{cases} \omega_b^0 & \text{if } \Omega = 0 \\ (1 + 25\Omega - 260\Omega^2)\omega_b^0, & \text{if } 0 < \Omega \leq 0.1 \\ (1 - \Omega)\omega_b^0, & \text{if } 0.1 < \Omega \leq 1 \\ 0, & \text{if } \Omega > 1 \end{cases} \tag{10}$$

ω_b^0 (s^{-1}) is the baseline coefficient of blood perfusion. Q_met (W m^{-3}) is the metabolic heat. Q accounts for external heat sources, which varies for the different domains of geometry. The heat generated after the absorption of NIR light is defined as $\mu_a\varphi(r)$ (W m^{-3}) and $N\sigma_a\varphi(r)$ (W m^{-3}) for the tissue and tumor domains respectively. N is the number volume of Fe$_3$O$_4$ NPs and σ_a (m^2) is the absorption cross-section of nanoparticles. Table 3 presents a summary of the values of the thermo-physical properties that were used in the simulation.

Table 3. Thermo-physical properties of the biological domains that were used in the simulation. The values were obtained from Refs. [31,33]

Tissue	Specific Capacity Heat c [J (kg K)$^{-1}$]	Thermal Conductivity λ [W (mK)$^{-1}$]	Density ρ [kg m^{-3}]	Metabolic Heat Q_met [W m^{-3}]	Blood Perfusion ω_b [s^{-1}]
Fat	2348	0.21	911	400	0.0002
Gland	2960	0.48	1041	700	0.0005
Muscle	3421	0.48	1090	700	0.0008
Tumor	3770	0.48	1050	8720	0.0001
Blood	3617	-	1050	-	-

Thermal Damage. The Arrhenius injury model was used to estimate tissue destruction. The model, which relates temporal temperature to cell death, is defined by [53]:

$$\Omega(\tau) = A \int_0^\tau \exp\left(\frac{-E_a}{RT(t)}\right) dt \tag{11}$$

where E_a (J mol^{-1}) is the activation energy, A (s^{-1}) is a scaling factor and $R = 8.3$ (J mol^{-1}K^{-1}) is the gas constant. The values for E_a and A were obtained from Ref. [33] as 302 kJ mol^{-1} and 1.18×10^{44} s^{-1} respectively. $\Omega = 1$ corresponds to the 100% irreversible cell damage.

Model Implementation. This FEM model was developed with the COMSOL Multiphysics 5.2 software package (Comsol, Inc. Burlington MA, USA). All properties and dimensions were added explicitly to the FEM model as parameters and variables under the "Global Definition" and "Model" nodes, respectively. Equations (9) and (10) were added as analytic functions under the "Global Definition" node. The 2D axisymmetrical model was used to reduce simulation time.

The light distribution was achieved by implementing Equation (5) as an analytic function. The temperature distribution was achieved using the bio-heat heat transfer application mode. Each tissue was represented with a separate "biological tissue" node. The boundary and initial conditions were specified as follows: a Dirichlet condition, T = 37 °C, at Γ_1; a Neumann condition, $\mathbf{n} \cdot (\lambda \nabla T) = h \cdot (T_{ext} - T)$ at Γ_2 where the heat transfer coefficient, h was equal to 13.5 Wm^{-2}K^{-1} and $T_{ext} = 25$ °C and continuity, $\mathbf{n} \cdot (\lambda_1 \nabla T_1 - \lambda_2 \nabla T_2) = 0$ at all interior boundaries. A temperature of 37 °C (for the normal body) was used as the initial temperatures in all domains of the model. The heat source was added to the bio-heat transfer application mode as a user-defined heat source.

The cell death model was implemented with the "Coefficient Form PDE" application mode. To achieve a time integration, the coefficients d_a and f were set to 1 and Equation (11), respectively. All other coefficients were set to zero. The initial conditions were: $S = \partial S / \partial t = 0$.

In order to enhance the accuracy of results, we resolved the model with successively smaller element sizes and compared results, until an asymptotic behavior of the solution emerged. The comparison was done by analyzing the temperature at the interface between tumor and the tissue. The choice of 2D-axisymmetric model allowed for the use of a physics-controlled mesh with the triangular element with sizes: maximum = 0.24 cm and minimum = 0.0024 cm for the tumor region and element sizes: maximum = 0.42 cm and minimum = 0.018 cm for the other regions. This resulted in 2656 domain elements and 277 boundary elements. The numerical solutions were obtained using the time-dependent solver "GMRES" with its default settings. The simulations were run on a mid-range workstation with Intel(R) Xeon(R) E5-1620 CPU and 8 GB of RAM.

5. Conclusions

Recent developments in imaging techniques have led to early detection of small fibroadenomas. Although observation is recommended for such cases, the agitations by some women due to fear of malignancy [12] coupled with recent report of 41% increase in cancer risk for women diagnosed with fibroadenomas [13] justify the need to develop techniques that can destroy these tumors with minimal or no side effects. We believe our findings demonstrate the potential of NP-mediated photothermal therapy for destroying fibroadenomas. However, we acknowledge the limitations of the study and understand that a future study should incorporate the important aspects discussed earlier so that a proper assessment can be made. Our long term goal is to develop a non-aggressive and noninvasive treatment method for such benign tumors, which is becoming a growing public health concern.

Author Contributions: Conceptualization, A.Y. and K.K.-D.; methodology, I.B.Y., S.W.K.H., Y.K.K.-A., A.Y. and K.K.D.; software, A.Y. and K.K.-D.; formal analysis, I.B.Y., S.W.K.H., Y.K.K.-A., A.Y. and K.K.D.; investigation, I.B.Y., S.W.K.H., Y.K.K.-A., A.Y. and K.K.D.; writing—original draft preparation, A.Y. and K.K.-D; writing—review and editing, I.B.Y., S.W.K.H., Y.K.K.-A., A.Y. and K.K.D.; project administration, K.K.-D.; funding acquisition, K.K.-D. All authors have read and agreed to the published version of the manuscript.

Funding: This research was funded by the Building a New Generation of Academics in Africa (BANGA-Africa) project, which is funded by the Carnegie Corporation of New York.

Conflicts of Interest: The authors declare no conflict of interest.

References

1. Guray, M.; Sahin, A.A. Benign Breast Diseases: Classification, Diagnosis, and Management. *Oncologist* **2006**, *11*, 435–449. [CrossRef] [PubMed]
2. Chinyama, C.N. *Benign Breast Diseases: Radiology-Pathology-Risk Assessment*, 2nd ed.; Springer: Berlin, Germany, 2014; p. 9.
3. Santen R.J.; Mansel R. Benign breast disorders. *N. Engl. J. Med.* **2005**, *353*, 275–285. [CrossRef] [PubMed]
4. El-Wakeel, H.; Umpleby, H.C. Systematic review of fibroadenoma as a risk factor for breast cancer. *Breast (Edinb. Scotl.)* **2003**, *12*, 302–307. [CrossRef]
5. Cerrato, F.; Labow, B.I. Diagnosis and management of fibroadenomas in the adolescent breast. *Semin. Plast. Surg.* **2013**, *27*, 23–25. [CrossRef] [PubMed]
6. Chang, D.S.; McGrath, M.H. Management of benign tumors of the adolescent breast. *Plast. Reconstr. Surg.* **2007**, *120*, 13e–19e. [CrossRef]
7. Derkyi-Kwarteng, L.; Brown, A.A.; Derkyi-Kwarteng, A.; Gyan, E.; Akakpo, K.P.; Abraham, A. Benign Proliferative Breast Disease: A Histopathological Review of Cases at Korle Bu Teaching Hospital. *J. Adv. Med. Med. Res.* **2019**, *31*, 1–5. [CrossRef]
8. Bewtra, C. Fibroadenoma in women in Ghana. *Pan Africa Med. J.* **2009**, *2*, 11. [CrossRef]
9. Oluwole, S.F.; Freeman, H.P. Analysis of benign breast lesions in blacks. *Am. J. Surg.* **1979**, *137*, 786–789. [CrossRef]
10. Newman, L.A. Breast cancer in African-American women. *Oncologist* **2005**, *10*, 1–14. [CrossRef]
11. Zhao, Z.; Wu, F. Minimally-invasive thermal ablation of early-stage breast cancer: A systemic review. *Eur. J. Surg. Oncol.* **2010** *36*, 1149–1155. [CrossRef]
12. Ezer, S.S.; Oguzkurt, P.; Ince, E.; Temiz, A.; Bolat, F.A.; Hicsonmez, A. Surgical Treatment of the Solid Breast Masses in Female Adolescents. *J. Pediatr. Adolesc. Gynecol.* **2013**, *26*, 31–35. [CrossRef]
13. Dyrstad, S.W.; Yan, Y.; Fowler, A.M.; Colditz, G.A. Breast cancer risk associated with benign breast disease: Systematic review and meta-analysis. *Breast Cancer Res. Treat.* **2015**, *149*, 569–575. [CrossRef] [PubMed]
14. Lakoma, A.; Kim, E.S. Minimally invasive surgical management of benign breast lesions. *Gland Surg.* **2014**, *3*, 142–148. [PubMed]
15. Ashikbayeva, Z.; Tosi, D.; Balmassov, D.; Schena, E.; Saccomandi, P.; Inglezakis, V. Application of Nanoparticles and Nanomaterials in Thermal Ablation Therapy of Cancer. *Nanomaterials* **2019**, *9*, 1195. [CrossRef] [PubMed]
16. Maier-Hauff, K.; Ulrich, F.; Nestler, D.; Niehoff, H.; Wust, P.; Thiesen, B.; Orawa, H.; Budach, V.; Jordan, A. Efficacy and Safety of Intratumoral Thermotherapy Using Magnetic Iron-Oxide Nanoparticles Combined with External Beam Radiotherapy on Patients with Recurrent Glioblastoma Multiforme. *J. Neuro-Oncol.* **2011**, *103*, 317–324. [CrossRef]
17. Johannsen, M.; Thiesen, B.; Wust, P.; Jordan, A. Magnetic Nanoparticle Hyperthermia for Prostate Cancer. *Int. J. Hyperthermia* **2010**, *26*, 790–795. [CrossRef]
18. Chu, M.; Shao, Y.; Peng, J.; Dai, X.; Li, H.; Wu, Q.; Shi, D. Near-infrared laser light mediated cancer therapy by photothermal effect of Fe_3O_4 magnetic nanoparticles. *Biomaterials* **2013**, *34*, 4078–4088. [CrossRef]
19. Espinosa, A; Di Corato, R.; Kolosnjaj-Tabi, J.-T.; Flaud, P.; Pellegrino, T.; Wilhelm, C. Duality of Iron Oxide Nanoparticles in Cancer Therapy: Amplification of Heating Efficiency by Magnetic Hyperthermia and Photothermal Bimodal Treatment. *ACS Nano* **2016**, *10*, 2436–2446. [CrossRef]
20. Maltzahn, G.V.; Park, J.-H. Agrawal, A.; Bandaru, N.K.; Das, S.K.; Sailor, M.J.; Bhatia, S.N. Computationally guided photothermal tumor therapy using long-circulating gold nanorod antennas. *Cancer Res.* **2009** *69*, 3892–3900. [CrossRef]
21. Bagley, A.F.; Hill, S.; Rogers, G.S.; Bhatia, S.N. Plasmonic Photothermal Heating of Intraperitoneal Tumors through the Use of an Implanted Near-Infrared Source. *ACS Nano* **2013**, *7*, 8089–8097. [CrossRef]
22. Morozov, K.I.; Lebedev, A.V. The effect of magneto-dipole interactions on the magnetization curves of ferrocolloids. *J. Magn. Magn. Mater.* **1990**, *85*, 51–53. [CrossRef]

23. Pshenichnikov, A.; Lebedev, A.; Ivanov, A.O. Dynamics of Magnetic Fluids in Crossed DC and AC Magnetic Fields. *Nanomaterials* **2019**, *9*, 1711. [CrossRef]

24. Tang, J.; Myers, M.; Bosnick, A.; Brus, L.E. Magnetite Fe_3O_4 Nanocrystals: Spectroscopic Observation of Aqueous Oxidation Kinetics. *J. Phys. Chem. B* **2003**, *107*, 7501–7506. [CrossRef]

25. Shen, S.; Kong, F.; Guo, X.; Wu, L.; Shen, H.; Xie, M.; Wang, X.; Jin, Y.; Ge, Y. Cmcts Stabilized Fe_3O_4 Particles with Extremely Low Toxicity as Highly Efficient near-Infrared Photothermal Agents for in Vivo Tumor Ablation. *Nanoscale* **2013**, *5*, 8056–8066. [CrossRef] [PubMed]

26. Pan, P.; Lin, Y.; Gan, Z.; Luo, X.; Zhou, W.; Zhang, N. Magnetic field enhanced photothermal effect of Fe3O4 nanoparticles. *J. Appl. Phys.* **2018**, *123*, 115115. [CrossRef]

27. Wong, K.V.; Castillo, M.J. Heat transfer mechanisms and clustering in Nanofluids. *Adv. Mech. Eng.* **2010**, *2010*, 795478. [CrossRef]

28. Chen, H.; Shao, L.; Ming, T.; Sun, Z.; Zhao, C.; Yang, B.; Wang, J. Understanding the photothermal conversion efficiency of gold nanocrystals. *Small* **2010**, *6*, 2272–2280. [CrossRef]

29. Kim A.D. Transport theory for light propagation in biological tissue. *J. Opt. Soc. Am. A* **2004**, *21*, 820–827. [CrossRef]

30. Pennes, H.H. Analysis of tissue and arterial blood temperatures in the resting human forearm. *J. Appl. Physiol.* **1948**, *1*, 93–122. [CrossRef]

31. Wahab, A.A.; Salim, M.I.M.; Ahamat, M.A.; Manaf, N.A.; Yunus, J.; Lai, K.W. Thermal distribution analysis of three-dimensional tumor-embedded breast models with different breast density compositions. *Med. Biol. Eng. Comput.* **2016**, *54*, 1363–1373. [CrossRef]

32. Balleyguier, C.; Bidault, F.; Mathieu, M.C.; Ayadi, S.; Couanet, D.; Sigal, R. BIRADS^{TM} mammography: Exercises. *Eur. J. Radiol.* **2007**, *61*, 195–201. [CrossRef] [PubMed]

33. Paruch, M. Mathematical Modeling of Breast Tumor Destruction Using Fast Heating during Radiofrequency Ablation. *Materials* **2020**, *13*, 136. [CrossRef] [PubMed]

34. Mohammed, Y.; Verhey, J.F. A finite element method model to simulate laser interstitial thermo therapy in anatomical inhomogeneous regions. *Biomed. Eng. OnLine* **2005**, *4*, 2. [CrossRef]

35. Sarkar, S.; Gurjarpadhye, A.A.; Rylander, C.G.; Rylander, M.N. Optical properties of breast tumor phantoms containing carbon nanotubes and nanohorns. *J. Biomed. Opt.* **2011**, *16*, 051304. [CrossRef] [PubMed]

36. Dehghani, H.; Brooksby, B.A.; Pogue, B.W.; Paulsen, K.D. Effects of refractive index on near-infrared tomography of the breast. *Appl. Opt.* **2005**, *44*, 1870–1878. [CrossRef] [PubMed]

37. Dewhirst, M.; Stauffer, P.R.; Das, S.; Craciunescu, O.I.; Vujaskovic, Z. *Clinical Radiation Oncology*, 4th ed.; Elsevier: Amsterdam, The Netherlands, 2016; Chapter 21.

38. Sundeep, S.; Repaka, R. Parametric sensitivity analysis of critical factors affecting the thermal damage during RFA of breast tumor. *Int. J. Therm. Sci.* **2018**, *124*, 366–374.

39. Goldberg, S.N.; Gazelle, G.S.; Compton, C.C.; Mueller, P.R.; Tanabe, K.K. Treatment of intrahepatic malignancy with radiofrequency ablation: Radiologic-pathologic correlation. *Cancer* **2000**, *88*, 2452–2463. [CrossRef]

40. Zevas, N.T.; Kuwayama, A. Pathological characteristics of experimental thermal lesions: comparison of induction heating and radiofrequency electrocoagulation. *J. Neurosurg.* **1972**, *37*, 418–422. [CrossRef]

41. Bloom, K.J.; Dowlat, K.; Assad, L. Pathologic changes after interstitial laser therapy of infiltrating breast carcinoma. *Am. J. Surg.* **2001**, *182*, 384–388. [CrossRef]

42. Dowlatshahi, K.; Fan, M.; Gould, V.E.; Bloom, K.J.; Ali, A. Stereotactically guided laser therapy of occult breast tumors: Work-in-progress report. *Arch. Surg.* **2000**, *135*, 1345–1352. [CrossRef]

43. Wilhelm, S.; Tavares, A.J.; Dai, Q.; Ohta, S.; Audet, J.; Dvorak, H.F.; Chan, W.C.W. Analysis of nanoparticle delivery to tumours. *Nat. Rev. Mater.* **2016**, *1*, 1–12. [CrossRef]

44. Kannadorai, R.K.; Liu, Q. Optimization in interstitial plasmonic photothermal therapy for treatment planning. *Med. Phys.* **2013**, *40*, 103301. [CrossRef] [PubMed]

45. Mie, G. Beiträge zur Optik trüber Medien, speziell kolloidaler Metallösungen. *Ann. Phys.* **1908**, *25*, 377–445. [CrossRef]

46. Bohren, C.F.; Huffman, D.R. Absorption and Scattering by a Sphere. In *Absorption and Scattering of Light by Small Particles*; John Wiley & Sons: New York, NY, USA, 2008; pp. 83–129.

47. Qin, Z.; Wanf, Y.; Randrianalisoa, J.; Raesi, V.; Chan, W.C.W.; Lipiński, W.; Bischof, J.C. Quantitative Comparison of Photothermal Heat Generation between Gold Nanospheres ad Nanorods. *Sci. Rep.* **2016**, *6*, 29836. [CrossRef]

48. Popescu, R.C.; Andronescu, E.; Vasile, B.S. Recent Advances in Magnetite Nanoparticle Functionalization for Nanomedicine. *Nanomaterials* **2019**, *9*, 1791. [CrossRef]
49. Soenen, S.J.; Parak, W.J.; Rejman, J.; Manshian, B. Intra Cellular Stability of Inorganic Nanoparticles: Effects on Cytotoxicity, Particle Functionality, and Biomedical Applications. *Chem. Rev.* **2015**, *115*, 2109–2135. [CrossRef]
50. Debye, P. Der Lichtdruck auf Kugeln von beliebigem Material. *Ann. Phys.* **1909**, *335*, 57–136. [CrossRef]
51. Wiscombe, W.J. *Mie Scattering Calculations: Advances in Technique and Fast, Vector-speed Computer Codes (No. NCAR/TN-140+STR);* University Corporation for Atmospheric Research: Boulder, CO, USA, 1979.
52. Complex Refractive Index for Fe_3O_4. Available online: https://refractiveindex.info/?shelf=main&book=Fe3O4&page=Querry (accessed on 15 January 2020)
53. Rylander, M.N.; Feng, Y.; Bass, J.; Diller, K.R. Heat shock protein expression and damage optimization for laser therapy design. *Lasers Surg. Med.* **2007**, *39*, 734–746. [CrossRef]
54. Duck, F.A. *Physical Properties of Tissue: A Comprehensive Reference Book;* Academic Press: Cambridge, MA, USA, 1990.

Article

Numerical 3D Simulation of a Full System Air Core Compulsator-Electromagnetic Rail Launcher

Valentina Consolo, Antonino Musolino *, Rocco Rizzo and Luca Sani

Department of Engineering for Energy System, Territory and Construction (DESTEC), University of Pisa, Largo L. Lazzarino 1, 56122 Pisa, Italy; valentina.consolo@phd.unipi.it (V.C.); rocco.rizzo@unipi.it (R.R.); luca.sani@unipi.it (L.S.)

* Correspondence: antonino.musolino@unipi.it

Received: 5 June 2020; Accepted: 12 August 2020; Published: 26 August 2020

Featured Application: **The electromechanical analysis of full 3D interacting devices is often necessary. This paper presents such an analysis applied to the system constituted by a rail launcher and its feeding generator. The adopted numerical tool has general validity and can be used in other contexts.**

Abstract: Multiphysics problems represent an open issue in numerical modeling. Electromagnetic launchers represent typical examples that require a strongly coupled magnetoquasistatic and mechanical approach. This is mainly due to the high velocities which make comparable the electrical and the mechanical response times. The analysis of interacting devices (e.g., a rail launcher and its feeding generator) adds further complexity, since in this context the substitution of one device with an electric circuit does not guarantee the accuracy of the analysis. A simultaneous full 3D electromechanical analysis of the interacting devices is often required. In this paper a numerical 3D analysis of a full launch system, composed by an air-core compulsator which feeds an electromagnetic rail launcher, is presented. The analysis has been performed by using a dedicated, in-house developed research code, named "EN4EM" (Equivalent Network for Electromagnetic Modeling). This code is able to take into account all the relevant electromechanical quantities and phenomena (i.e., eddy currents, velocity skin effect, sliding contacts) in both the devices. A weakly coupled analysis, based on the use of a zero-dimensional model of the launcher (i.e., a single loop electrical equivalent circuit), has been also performed. Its results, compared with those by the simultaneous 3D analysis of interacting devices, show an over-estimate of about 10–15% of the muzzle speed of the armature.

Keywords: air-core pulsed alternator; electromagnetic rail launcher; coupled analysis; computational electromagnetics; integral formulations

---

1. Introduction

The ElectroMagnetic Launch (EML) technology uses electric propulsion to accelerate objects at high speeds. Because of its superior performance it is substituting several launch systems. Coil and rail launchers are the most used alternative solutions [1,2].

Induction launchers are substantially linear tubular motors, usually air cored. They consist of a barrel formed by an array of (stator) coils and a conductive cylinder moving inside them. Induction launchers are operated as travelling wave induction launchers or as pulsed induction launchers. In the first operation mode, the stator coils are grouped in sections that are energized in a polyphase fashion in order to create a traveling wave of flux density in the region occupied by the sleeve. In the second one the stator coils are fed in sequence by a set of capacitor driven circuit [3–7].

The Electro-Magnetic Aircraft Launch System (EMALS) is another important example of application of the electromagnetic launch technology. It has been introduced in substitution of the steam catapults for the take-off of airplanes from the new class of carriers of the US Navy [8–11]. With respect to the steam catapult EMALS is able to produce a smooth and controllable acceleration profile with a consequent reduction of the stress on the aircrafts. Moreover it is able to accelerate heavier aircrafts with reduced weight, cost and maintenance requirements. EMALS has been recently proposed for civil aircrafts [12]. Ambitious programs for space application of electromagnetic launchers are under investigation [13,14].

A rail launcher is constituted by two conductive rails with a conductive armature (slab or c-shaped) free to slide inside them. At the beginning of the launch the armature is located near the breech of the launcher where a feeding generator is connected between the rails. The current flowing in the rails (and in the armature) produces a flux density distribution in correspondence of the armature, where the interaction with its current produces a thrust force that accelerates the armature. The main drawbacks affecting the rail launchers are a consequence of the Velocity Skin Effect (VSE) which is caused by the limited diffusion rate of the current in the rails as the armature moves; VES produces a concentration of the current in the rear portion of the armature near the rails [15–18]. The importance of VSE increases with the speed and it is one of the causes which may prevent the use of rail launchers at very high speed. The availability of numerical tools for the investigation of VSE and for the design of countermeasures to limit its effects on the launcher performance are of paramount importance [17,19,20].

When considering a solid armature rail launcher, the choice of an air-core compensated pulsed alternator (compulsator) as the feeding device seems to be one of the most promising technology [21]. The absence of ferromagnetic materials allows achieving a very low value of internal inductances. The addition of compensating windings or conductive shields further reduce the internal inductance, so increasing the peak value of the output pulsed current. Moreover, by proper positioning of compensating components, it is possible to shape the current pulse to improve the performance of the launchers, both in terms of muzzle speed and efficiency. As reported in the scientific literature, the maximum speed of an air-core rotor can reach higher values than those in an iron-core one, increasing the stored energy [22,23].

Many papers, based on analytical or numerical models, have been published in the past years to investigate the performance of the air-core compulsator [24–26]. However, the majority of these studies are focused on the performance of the compulsator as a stand-alone device and adopt a simple time varying equivalent circuit to model the rail launcher. Similarly happens for rail launchers, where often the waveforms produced by the feeding devices are assigned, especially when the rotating machines are considered.

Accurate model identification and parameters extraction of the lumped equivalent circuit for these devices may be difficult to achieve since both rail launchers and compulsators are inherently time-varying and nonlinear electromechanical devices and consequently the parameters that identify the equivalent circuit of one device may depend on the operating conditions of the whole system and on the characteristics of the other device. A strong-coupled 3D electromechanical analysis of the interacting devices seems to be the only option able to provide accurate results. This paper discusses the coupled electromechanical analysis of the whole launch package by using the research code EN4EM previously developed by the authors.

In order to avoid confusion, in the remaining of the paper the phrase "strong coupling" will be reserved to the magnetoquasistatic-mechanical problems, arising when analyzing a device with conductors in relative motion. "Strong coupling" is necessary when analyzing high speed devices and it is inherently provided by underlying formulation of EN4EM. The phrase "strong-interaction" is reserved to indicate a simultaneous full 3D "strong coupled" analysis of the rail launcher and its feeding compulsator. Moreover, the phrase "weak-interaction" is reserved for those analyses where one of the devices is substituted with a lumped equivalent circuit and the other is analyzed by a 3D "strong coupled" model. The "equivalent circuit" is reserved for zero-dimensional voltage-current dependencies at the terminals of a device. A lumped "equivalent circuit" is usually unable to provide information about the spatial distribution of the electromechanical quantities inside the device. Finally, the phase "equivalent network" is related to EN4EM and, as it will be shown in the next section, is used to indicate the internal procedure of the code which builds an electric network whose currents are uniquely related to the current density distribution in a device.

To best of the authors' knowledge, scientific literature does not report any "strong interaction" analysis between rail launcher and compulsator capable to consider two mechanical degrees of freedom (one rotation for the compulsator and one translation for the launcher) together with high speed sliding contacts. Considering that the components and the materials used in EML technology are heavily stressed from the electrical, mechanical and thermal point of view, a tools which allows an accurate coupled analysis represents a valuable resource.

The manuscript is organized as follows. Section 2 briefly summarizes the adopted numerical formulation. Section 3 shows two examples of the "weak interaction" analysis and further justifies the motivations of the research by discussing the results of these analyses. In Section 5, the "strong interaction" analysis of the whole system is carried on and its results are compare with those by a "weak interaction" analysis. Finally, some concluding remarks are reported.

2. Numerical Formulation

The 3D numerical analysis of multi degrees of freedom electromechanical devices represents a challenging, still open problem, which poses several critical issues. Several commercial codes, typically based on the Finite Elements Method (FEM), provide packages for coupled electro-mechanical analysis, but they seem to be not very effective with multiple degrees of freedom and with sliding contacts. Moving conductors and sliding contacts usually require remeshing of the domains with consequent increase of computational times and potential numerical instabilities.

Integral Formulations (e.g., the Method of Moments) seem to work quite well with moving domains since they require the discretization of the active regions only (namely conductors and ferromagnetic materials), so avoiding the problem of coupling meshes with different speeds. Integral formulations implicitly enforce the far field boundary conditions, and are able to produce accurate results by using coarse discretization (when compared with those required by FEM).

The numerical investigation of the complete launch system has been performed by the research code EN4EM. It is based on an integral formulation, and it is under continuous development by the authors for investigating electromechanical systems [27–34].

EN4EM applied to the launch package is able to simulate the whole system considering the characteristics of the two devices and taking into account their "strong interaction". Figure 1 shows the steps of a conceptual flow chart of the numerical formulation: discretization in elementary volumes, writing and integration of Ohm's law, arrangement in an electric network and writing of the governing electro-mechanical equations. The usual notations for the electromagnetic quantities in Figure 1 are adopted according to the Table 1.

With reference to the inset in the bottom right of Figure 1, Ohm's law is written in the conductive elementary volumes where a uniform current distribution is assumed (row #1 of the inset). The additive properties of the integrals with respect to the integration domain allows expressing the fields and potentials in the k-th conductive elementary volume as a summation of contribution due to the currents in the other volumes (row #2 of the inset). Finally integration on the k-th elementary volume leads to an equation that can be seen as the voltage-current relationship of a branch that is a series connection of a resistor, an inductor coupled with other inductors, and a voltage generator controlled by the currents in other elementary volumes (row #3). The equivalent network of the device is built by connecting the terminals of the branches so obtained. In case of a multi-device system, the code is able to model separately the different devices by their equivalent networks. Then, all these networks are connected together according to the relative positions of the corresponding elementary volumes and to the presence of electrical contacts between them. The described procedure has been applied to model the launcher and the compulsator, obtaining a whole model composed of thousands of branches.

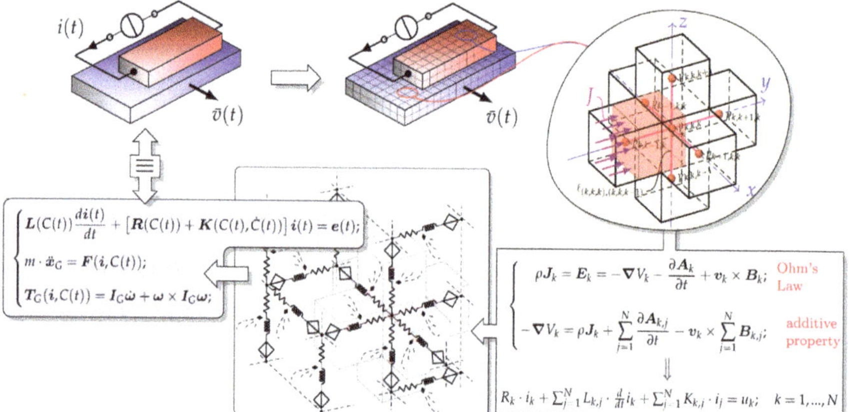

Figure 1. Conceptual flow chart of the numerical formulation. (Reproduced with permission from [30], IEEE, 2017).

Table 1. Notations.

Symbol	Description
J_k	Current density in the k-th conductive element
E_k	Electric field in the k-th conductive element
V_k	Electric potential in the k-th conductive element
A_k	Vector potential in the k-th conductive element
B_k	Flux density in the k-th conductive element
v_k	Velocity of the k-th conductive element
i_k	Current in the k-th conductive element
u_k	Voltage drop across the k-th conductive element
$L_{k,j}$	Mutual induction coeff. between conductive elem.
$K_{k,j}$	Motional voltage coeff. between conductive elem.
R_k	Resistance of the k-th conductive element

Mesh analysis yields to the governing equations written in matrix form:

$$L(C(t))\frac{di}{dt} + \left(R + K\left(C(t), \dot{C}(t)\right)\right)i = e(t) \tag{1}$$

The values of the elements of the matrices in (1) are function of the system configuration $C(t)$ and its derivative $\dot{C}(t)$ (i.e., the relative positions and velocities of the elementary volumes used to discretize the devices respectively). Coupling between electrical and mechanical equations is achieved by the terms $v_k(t) \times B_k(t)$ that, once integrated on the elementary volumes, are assembled to form the matrix of the motional terms K, and by the terms $j_k(t) \times B_k(t)$ which are integrated to provide the forces and the torques on the moving parts of the device. The mechanical equation for the armature of the launcher is:

$$m\ddot{x}_G = F(i, C(t)) \tag{2}$$

while the equation for the compulsator are:

$$I_G \dot{\omega} + \hat{\omega} I_G \omega = M_G(i, C(t)) \tag{3}$$

In the above equations F represents the resultant force on the launcher armature and m is its mass, M_G is the resultant torque on the rotor of the compulsator and I_G is its inertia tensor. The coupled differential equations for electrical and mechanical equilibrium are time varying, and the resulting system is nonlinear; integration is carried out as described in [27,28].

It is worth to mention that the currents in the branches of the equivalent network are not fictitious currents. The current in a branch of the equivalent network is in correspondence with one component of the current density of an elementary volumes as shown by the insets in the upper right and in the lower left corners in Figure 1. By the above described equivalent network the distribution of all the relevant electromechanical quantities can be evaluated everywhere in the whole system. In particular, the expression of the force by the term $j_k(t) \times B_k(t)$ allows to evaluate its distribution in each elementary volumes and to determine the contribution to the total dynamical action in materials that are usually heavily stressed when used in EML technology.

Simultaneous analysis of interacting devices simply requires building a bigger equivalent network constituted by the properly connected equivalent networks of the components. This implements a "strong interaction" between devices for which a "strong coupled" analysis is performed.

If the devices can be assumed magnetically uncoupled (i.e., leakage magnetic fluxes from a device that links the others are negligible) independent networks are built, linked together at the common terminals and simultaneously solved. It is worth to remark that the topology of the resultant network is the same than that of the more general case of magnetically coupled devices. The only difference lies in the filling of the inductance matrix L and of the motional terms matrix K, which are more densely populated in the latter case.

Coupling of the equivalent network so far described with external circuits is straightforward and this will be performed in the next section to investigate the "weak interaction" between the rail launcher and the compulsator.

The presence of sliding contacts has been taken into account by introducing an auxiliary network [20]. Considering that the motion in the rail launcher is characterized by a straight trajectory, all the possible contacts between elementary volumes on the rails and on the armature are known a priori. A new branch is set for each couple of volumes (one on the inner parts of the rails and the other on the faced outer surfaces of the armature) that can have a contact during the motion.

Figure 2 shows an example of a set of auxiliary branches that take into account the sliding contacts. The circuit elements in the auxiliary branches are function of the shared portion of the faces (if any) between the two volumes. When the shared portion is zero, the auxiliary branch opens. The proposed model of the sliding contacts is coherent with the adopted formulation based on an equivalent network of the entire system.

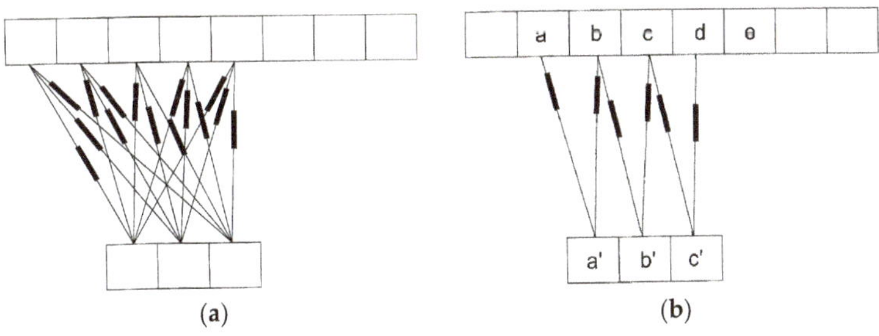

(a)   (b)

Figure 2. Example of the auxiliary network used for taking into account the presence of sliding contacts. (a) All the branches are shown. (b) Only the branches that are not open circuits are evidenced.

Finally the introduction of the equivalent network allows the use of advanced analysis techniques for the evaluation of the sensitivity of the response with respect to parameter variation [35]. This represents an important tool in gradient based optimization processes.

EN4EM was validated by comparison with results produced by other numerical codes and with experimental data [5,6,27,28,31,32].

3. Weak Interaction Analysis

In this section the analysis based on the "weak interaction" between rail launcher and compulsator will be critically reviewed and the difficulties of the equivalent circuit extraction will be discussed.

Figure 3 shows the active parts of a single-phase, two-pole compulsator with selective passive compensation provided by a discontinuous conductive shield. The inner part of the device consists of two stationary field coils which produce a magnetic flux density distribution whose axis is in the vertical direction (Figure 3b). The two series connected armature windings are located on the rotor which is the outer part of the machine. Figure 3c shows the stationary compensating shields that are in the central part of the machine. Figure 3a shows a 3D view of the device. A more detailed description is reported in [29].

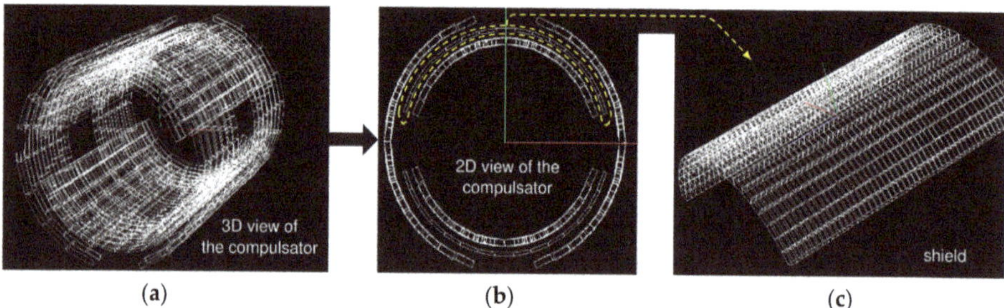

Figure 3. An example of the modeling of a compulsator by EN4EM. Snapshot of (**a**) the 3D view; (**b**) cross section of the machine; (**c**) discretization of the compensating shield. (Reproduced with permission from [29], IEEE, 2017).

When using a compulsator to feed a rail launcher, its rotor has to be preliminarily driven to a proper angular speed at no load conditions; part of the stored kinetic energy of the rotor will be delivered to the railgun armature. At the firing instant, the armature windings of the compulsator are connected to the rails of the launcher. During the launch, the rail armature accelerates and at the same time, the rotating part of the compulsator decreases its angular speed. The rates of change of the speeds of both the moving parts (and consequently their positions) are not known a-priori and substantially depend on the delivered current.

Since the simultaneous 3D analysis of the two electromechanical devices can be very time consuming, a common practice is to substitute one or both the device with equivalent circuits. This procedure may lead to significant error in the current flowing in the devices because of the difficulties in the determination of the topology and of the parameter extraction of the equivalent circuit.

Referring to the compulsator, when looking for a lumped equivalent circuit, we have to distinguish between the two compensation techniques usually adopted. When compensation is achieved by the use of shorted discrete coils, we can consider the self and mutual inductances of all windings (field, armature and compensating ones). Considering that the values of the mutual inductances depend on relative angular positions only, the lumped equivalent circuit of the compulsator is able to give accurate results; its building is a matter of evaluation (experimental or by computations) of self and mutual coefficients and EN4EM can be used as an extraction tool.

Things go in a different way when non-uniform (discontinuous) compensating shields are used. In these devices, the extraction of the values of the equivalent internal inductance L is not as straightforward as with shorted discrete coils, since it is function of the position of the rotor, as well as of the speed of the rotor, which is not known and varies during the system operation.

In fact, the speed of the rotor imposes the frequency of the electrical quantities in the shield, which in turn determines the path and the amplitude of the induced currents and their shielding effects which contribute to the internal equivalent inductance. Figure 4 shows an example of eddy currents distribution on one of the discontinuous compensating shields of the compulsator above described whose operating conditions are reported in [29]. In particular the compulsator was loaded with a simple lumped R-L equivalent circuit representing the rail launcher. The rectangle in Figure 4 represents the flattened cylindrical surface shown Figure 3c.

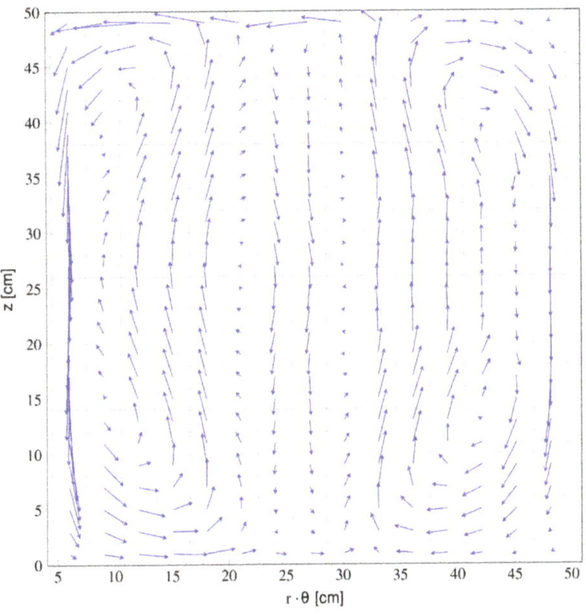

Figure 4. Eddy current distribution on the conductive compensating shield. (Reproduced with permission from [29], IEEE, 2017.)

The equivalent internal inductance of the compulsator, as well as the electromotive force at the terminals of the machine, are functions of the distribution of the currents on the shields. Expressing these functional dependencies represents a challenging problem.

Similar difficulties arise when looking for an equivalent circuit of the rail launcher. Figure 5 shows a rail launcher fed by an ideal voltage source whose waveform is given by $e(t) = 50\left(1 - e^{\frac{-t}{\tau}}\right)V$. Details about the geometry are in Table 2. In this figure, 20 elementary volumes located near the launcher breech are shown. The device was analyzed by EN4EM and the current density waveforms in the evidenced volumes are reported in Figure 6. The solid curves are associated with the labelled sections in reported in the inset of Figure 5; the greatest current density occurs in the section #5 in the inner part of the rail. The peak values became smaller as the outer boundary of the rail is approached. The reported waveforms put into evidence the uneven current density distribution in the rail section for most of the launch time, the ratio between the maximum and the minimum value is greater than two. This behavior is a consequence of the skin and proximity effects; also, the velocity skin effect has an heavy impact on the current distribution, which is a function of the rate of change of the electrical quantities imposed by the generator and the rate of change of the mechanical quantities i.e., the speed of the armature [36,37]. Both these quantities change during the launch.

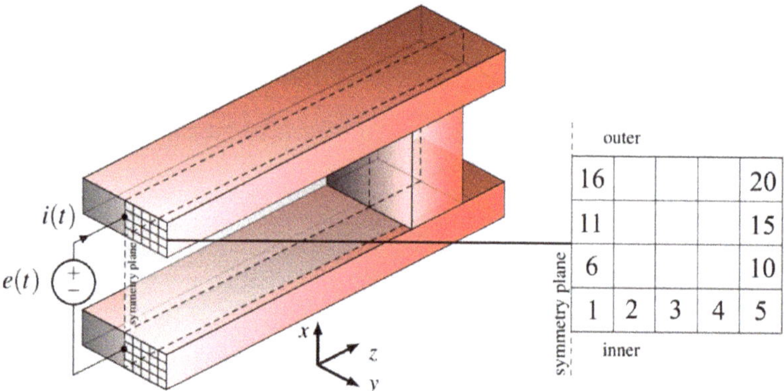

Figure 5. A sketch a rail launcher. The inset put into evidence the cross section of half of the upper rail. The labelled rectangles represent the cross section of the inner layer of the elementary volumes, i.e., those involved in the sliding contacts with the armature.

Table 2. Description of the devices.

Symbol	Description	Value
$r_{in,f.c.}$	internal radius of field coils	10.0 cm
$\Delta r_{f.c.}$	width of the conductors of field coils in radial direction	1.0 cm
$S_{f.c.}$	section of the conductors of the field coils	30.6 mm^2
$r_{in,a.c.}$	internal radius of armature coils	14.0 cm
$\Delta r_{a.c.}$	width of the conductors of armature coils in radial direction	1.0 cm
$S_{a.c.}$	section of the conductors of the armature coils	330 mm^2
$r_{in,shield}$	internal radius of the shield	12.0 cm
Δr_{shield}	width of the shield in radial direction	0.5 cm
J_z	inertia moment of the rotor	1.58 kg·m^2
Ω_0	initial speed of the rotor	12000 rpm
Δz_{rail}	length of the rails of the launcher	40.0 cm
S_{rail}	cross section of the rails	1.5×0.75 cm^2
$\Delta z_{arm.}$	length of the launcher armature	1.4 cm
D_{rail}	distance between the rails	2.0 cm
M_{total}	total launched mass: armature + payload	40 + 130 gr

The basic equivalent circuit of the railgun, here shown in Figure 7, is usually composed of three elements. A resistor and an inductor, both varying with the distance travelled by the armature, and another resistor, that takes into account the motional induced electromagnetic force, is related to the inductance gradient and its resistance depends on the speed of the armature.

In the light of the above discussion, these circuital components should be function of the frequency which influences the effective current density distributions in both the rails and the armature, as well as of the time, because of the speed and of the distance travelled by the armature. Some lumped equivalent circuits of the rail launchers adopt simplified expressions of the form: $R(z, f) = R_a + R_0 + R'z$ and $L(z, f) = L_0 + L'z$, where $R' = \frac{dR}{dz}$ is the rail resistance gradient, $L' = \frac{dL}{dz}$ is the rail inductance gradient, R_a is the armature resistance and R_0 and L_0 are the resistance and the inductance due to the connection wires [1]. All these expressions discard the dependence of these parameters on the frequency and are not able to take into account the velocity skin effect. On the other hand, the formal definition of the parameters as shown in Figure 7 and their estimate pose complex problems.

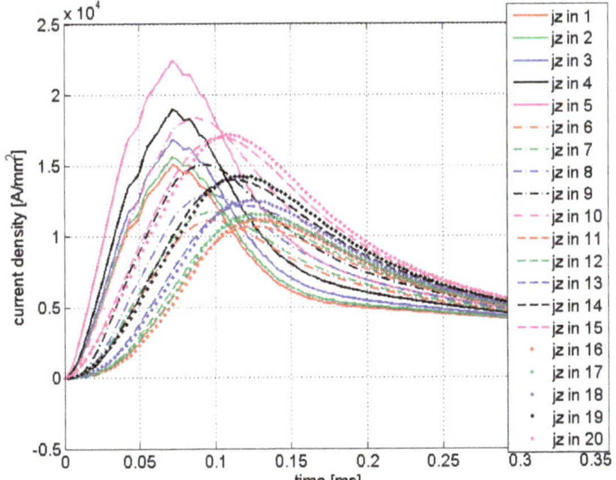

Figure 6. Time waveforms of the z-axis component of current density in the rail at the feeding end. Uneven distribution occurs for about 80% of the launch time.

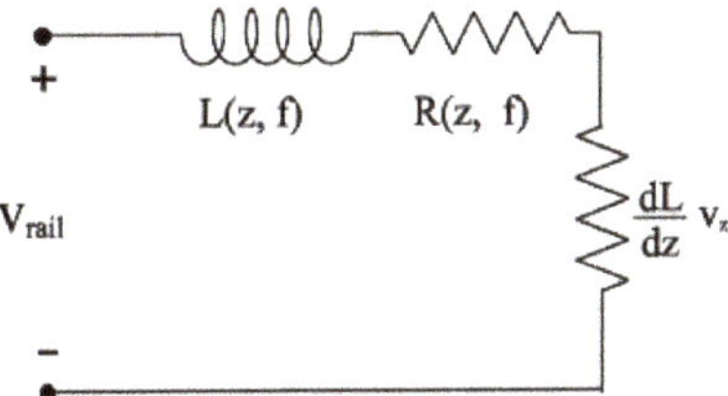

Figure 7. Basic equivalent lumped parameter circuit of a rail launcher. The presence of f (frequency) among the independent variables means that the circuit parameters are function of the rate of change of the electromagnetic phenomena.

4. Strong Interaction Analysis

The considerations developed in the previous section have driven us to consider the strong interaction analysis of a full 3D coupled system constituted by the rail launcher and its feeding compulsator in the time domain. We also considered the weak interaction analysis of the same system where the rail launcher is substituted by its lumped equivalent circuit as in Figure 7. The obtained results have been compared to evaluate the accuracy of the weak interaction analysis.

4.1. Description of the Devices

Figure 8 shows a snapshot of the graphical interface of EN4EM showing a 3D view of the whole system taken at the instant of firing. Figure 9 focuses on the discretization of the rail launchers; for the sake of readability, only a limited number of the branches used to model the sliding contacts (i.e., those related to elements in effective contact) are shown.

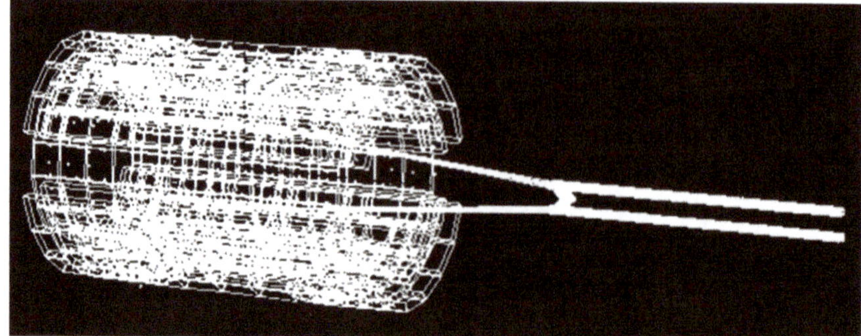

Figure 8. Snapshot of 3D view of the complete launch system: compulsator plus launcher at the instant of firing.

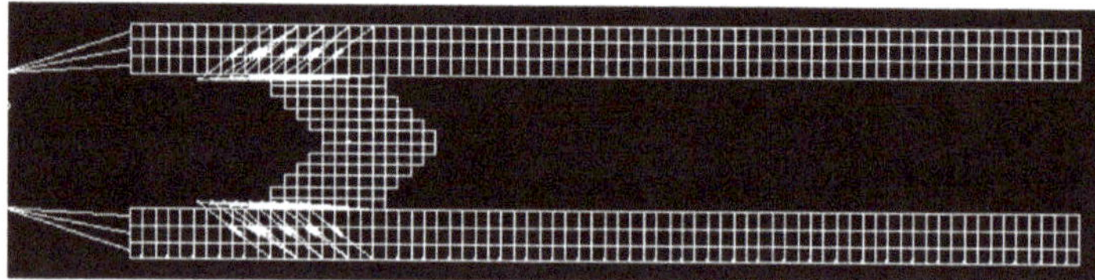

Figure 9. Screenshot of a planar lateral view of the launcher showing the discretization adopted for the strong interaction analysis. Only the active auxiliary branches used to model the sliding contacts are shown. The leftmost segments represent the connections to the compulsator terminals.

The details of the geometries of both the devices are reported in Table 2. The compulsator is a two-poles, single-phase machine. The stationary exciting coils have 25 turns each; they are fed with a direct current of 36 kA, capable to produce a magnetic flux density of 3T in correspondence of the armature coils. The discontinuous stationary shield is aluminum made. The axial length of the compulsator is 50 cm. The two armature coils are copper made, series connected and constituted of four turns each. They are positioned on the rotor. All the active components (field coils, armature coils, and shield) span an angle of 150°.

The snapshots of the 3D views of the active parts of the compulsator are shown in Figure 10. The whole device is built by arranging two items of each of the shown components according to the layout in Figure 3b.

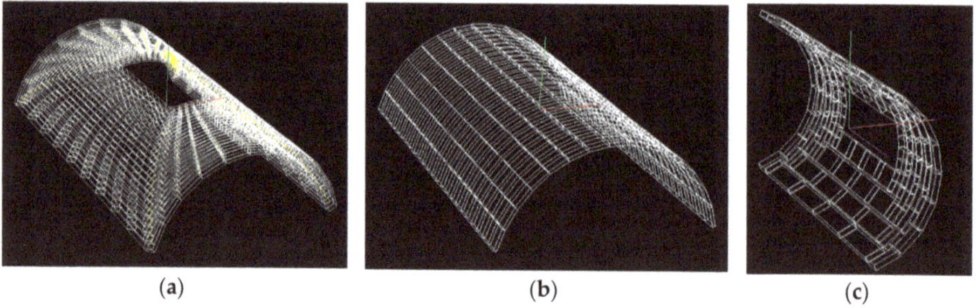

(a)　　　　　　　**(b)**　　　　　　　**(c)**

Figure 10. Screenshots of a the active parts of the compulsator: (**a**) field coil (stationary inner part); (**b**) compensating shield (stationary part just outside the field coils); (**c**) rotating armature (outer part).

At the instant of firing the magnetic axis of the field and of the armature coils are aligned, while the center of the shield is rotated (with respect to the scheme shown in Figure 3 of 60° in the direction of the motion of the rotor [29]. The launcher has copper rectangular rails and a C-shaped armature. The oblique side of the armature forms an angle of 45° with the direction of the rails as shown in Figure 9.

In the light of the discussion in Section 2, the numerical model is able to take into account the relevant components and phenomena in both the compulsator and the rail.

In particular, for the air-core compulsator: (1) the complex armature winding scheme; (2) the presence of excitation/control circuits; (3) the eddy currents in all the conducting parts of the machine (the shield, the shaft, and so forth); (4) the compensating windings of different shapes and arrangements (aluminum sheet, single shorted turns, and so forth.); (5) real winding turns connections; (6) end-turn effects; (7) relative angular velocity between conductors [29]. For the rail launcher: (8) the sliding contacts and the related the velocity skin effect; (9) the current distribution in the solid armature and in the rails [20].

Finally, the numerical formulation can model centrifugal forces and vibrations acting on the shaft of the compulsator due to electric and mechanical unbalances or to misalignments of the shaft from its centered position, as well as the full 3D electromechanical transient behavior of the machine during the real operating conditions.

4.2. Results

By using the proposed numerical formulation we are able to obtain the simultaneous evolution of the both the electrical and mechanical dynamics of the compulsator and of the rail launcher. The results of the strong interaction analysis are compared with those of a weak interaction where the launcher is substituted by a lumped equivalent network whose topology is shown in Figure 7 and parameters are characterized by the simplified expressions discussed in Section 3. These parameters have been evaluated by running EN4EM on a model of the rail launcher characterized by a very coarse discretization which subdivides the rails along the direction of the motion only; this implies that the current is uniformly distributed in the cross section of the rails. As far the discretization of the armature we assumed that the current was concentrated in its most backward quarter (i.e., in the width $\Delta z = 3.5\ mm$). We choose this value analyzing the current distribution (affected by the VES) in the armature of the standing alone rail launcher as described in Section 3 in the range of the speeds obtained by the strong interaction analysis.

Figure 11 shows the current delivered to the railgun. As known, the pulse shape can be adjusted by properly varying the angular position and the extension of the shield. It is important that the zero crossing of the current waveform occurs at the end of the launch, i.e., when the armature exits the launcher. This allows obtaining an increased efficiency in the electromechanical conversion and at the same time to avoid arcing between the armature and the muzzle. In fact, if, at the end of the launch, the current in the system is zero also the magnetic energy stored at the same instant is zero, this means that all the energy delivered from the generators t is converted in kinetic energy of the armature (plus, of course, the energy losses in the resistance). Residual energy stored in the magnetic fields of the system (i.e., in the launcher and in the compulsator) is lost in the arch at the muzzle of the launcher.

The weak interaction analysis predicts a greater current delivered by the compulsator, which in turn produces greater thrust force and speed; the acceleration time is reduced and the zero crossing of the current changes accordingly. As observed, the accurate prediction of the zero crossing allows improving the performance of the system.

The thrust force waveforms on the armature by the two analyses are shown in Figure 12. The ripple superimposed to the thrust force profile predicted by the strong iteration analysis is an artifact due to the commutation at discrete times of the branches of auxiliary network used for the sliding contacts modelling. It is worth noting that the currents do not present this ripple. This is due to the total inductance of the system, i.e., the equivalent inductance of the compulsator and the one of the launchers. A further insight to the cause of the ripple shows that it is due to the discrete variation of the "active" length of the rails, i.e., the portion of the rails behind the armature. Considering the configuration as schematized in Figure 2b, we see that as soon as the contact between element *(a)* in the rail and element *(a')* on the armature is interrupted, the current in *(a)* instantaneously loses its transverse component. Similarly, at some later instant, a contact is set between element *(e)* in the rail and element *(c')* in the armature and the current in element *(d)* of the rail will assume a longitudinal component. The magnitude of the flux density in the armature accordingly changes; also, the terms $j_k(t) \times B_k(t)$, related to the elementary volumes of the armature, suddenly change and produce the discontinuity in the trust force. The ripple is absent in the force profile predicted by the weak interaction analysis since the parameters of the lumped equivalent vary with continuity.

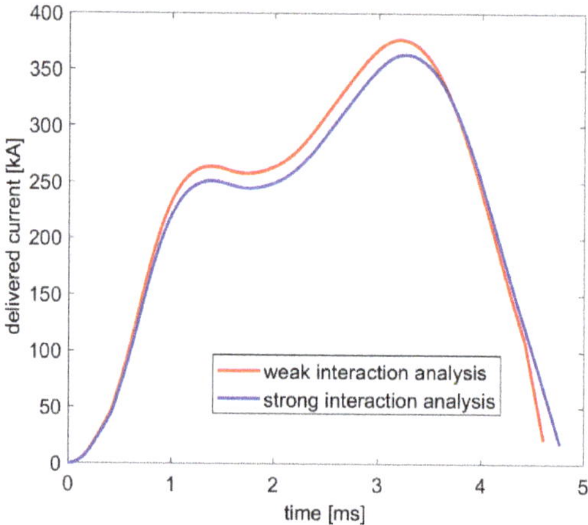

Figure 11. Time waveforms of the current delivered by the compulsator to the launcher in the strong and weak interaction.

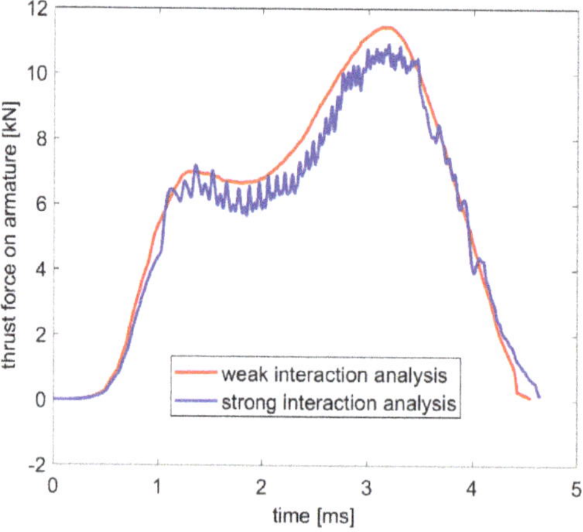

Figure 12. The thrust force on the armature in the strong and weak interaction. The ripple is an artifact due to the commutations that happen in the auxiliary network used to manage the sliding contacts in the strong interaction analysis.

Let us now consider the dynamic quantities on the compulsator. Figure 13 shows the torque acting on the rotor. As expected the torque is mostly negative, producing a decrease of the speed of the rotor. The figure shows that in the last portion of the launch time, the torque assumes positive values so increasing the velocity and the kinetic energy of the rotor, with respect to their minima.

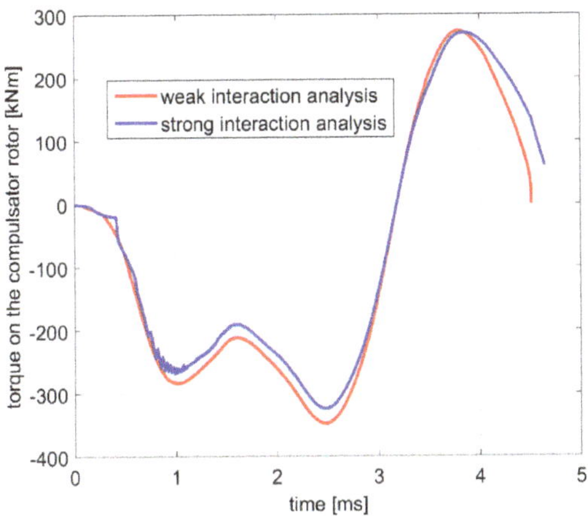

Figure 13. The torque on the rotor of the compulsator. The portion of the curve with positive value corresponds to recovering of magnetic energy as kinetic energy.

The instant when the torque changes its sign is roughly the same as the one when the delivered current reaches its maximum. A decreasing current implies a reduction of the magnetic energy stored in the system. Part of this magnetic energy is converted in mechanical energy by increasing the speeds of the rotor of the compulsator and of the armature of the launcher. The remainder increases the temperature of the conductors. The weak interaction analysis produces a smoother waveform than that of the strong interaction one.

The comparison of the speeds of the armature obtained by the two models is reported in Figure 14. The weak interaction analysis overestimates the speed of about 10%. The ripple in the thrust force is cancelled by the integration and does not affect the speed waveform produced by the strong interaction analysis.

If a lumped equivalent circuit was used for the compulsator, further errors would appear. These errors will be more relevant if components made up of massive conductors are present in the compulsator (e.g., a conductive shield). In this case, the actual distribution of the currents cannot be predicted a priori. Anyhow, the errors are expected to be lower when compensating concentrated windings are used.

The errors in the exit speed, lead to a wrong estimate of the launch time and therefore on the length of the current pulse. If this happens, the exit of the armature from the launcher could occur in correspondence of a non-zero value of the current, with consequent reduction of the system performance (low efficiency and arcing between armature and rail at the muzzle).

Despite the complexity of the problem EN4EM was able to complete the strong interaction analysis in about 150 min on a desktop computer based on an intel i7 6 core and equipped with 20 GB RAM. The maximum allocated memory was about 6 GB. The weak interaction analysis took about 45 min and required about 3 GB.

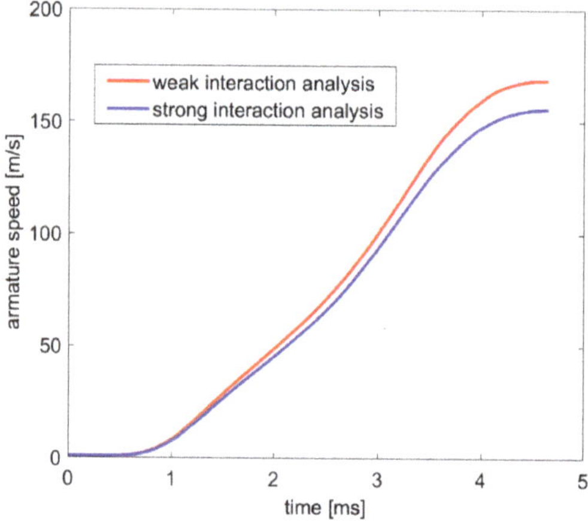

Figure 14. Comparison of the armature speed during the launch as predicted by weak and the strong interaction analysis.

5. Conclusions

The use of lumped equivalent circuits in modeling the coupled electro-mechanical behavior of a rail launcher and its feeding compulsator may produce results whose accuracy is not always satisfactory. The causes are due to the presence of eddy currents in the compensation shield of the compulsator and in the uneven current distribution in the rails and in the armature of the launcher. Coupled 3D electro-mechanical analysis is needed if accurate results are required. The paper has compared the results by the strong and the weak interaction analysis by the research code EN4EM. The availability of such a numerical tool could represent a valuable resource in the design of the launcher and of its feeding compulsator since it allows to determine the more important parameters of the launch.

In particular, it will be possible to prepare a look-up table to arrange the operative parameters of the compulsator (e.g., the excitation current, the initial speed of the rotor, its angular position at the instant of firing) to achieve a designed muzzle speed on a given payload.

Author Contributions: Conceptualization, A.M. and L.S.; methodology A.M., V.C.; software, R.R., A.M.; validation, A.M., and L.S.; formal analysis, V.C., R.R.; investigation, R.R., L.S.; writing—original draft preparation, L.S., R.R.; writing—review and editing, V.C. All authors have read and agreed to the published version of the manuscript.

Funding: This research received no external funding.

Acknowledgments: The authors would like to thank the NVIDIA's Academic Research Team for the donation of two NVIDIA Tesla K20c GPUs that have been extensively exploited for the simulations.

Conflicts of Interest: The authors declare no conflict of interest.

References

1. Ying, W.; Marshall, R.A.; Shukang, C. *Physics of Electric Launch*; Science Press: Beijing, China, 2004.
2. Engel, T.G. Scientific classification method for electromagnetic launchers. *IEEE Trans. Plasma Sci.* **2017**, *45*, 1333–1338. [CrossRef]
3. Gies, V.; Soriano, T. Modeling and Optimization of an Indirect Coil Gun for Launching Non-Magnetic Projectiles. *Actuators* **2019**, *8*, 39. [CrossRef]
4. Gies, V.; Soriano, T.; Marzetti, S.; Barchasz, V.; Barthelemy, H.; Glotin, H.; Hugel, V. Optimisation of Energy Transfer in Reluctance Coil Guns: Application to Soccer Ball Launchers. *Appl. Sci.* **2020**, *10*, 3137. [CrossRef]
5. Musolino, A.; Raugi, M.; Rizzo, R.; Tripodi, E. Modeling of the gyroscopic stabilization in a traveling-wave multipole field electromagnetic launcher via an analytical approach. *IEEE Trans. Plasma Sci.* **2015**, *43*, 1236–1241. [CrossRef]

6. Musolino, A.; Rizzo, R.; Tripodi, E. Travelling wave multipole field electromagnetic launcher: An SOVP analytical model. *IEEE Trans. Plasma Sci.* **2013**, *41*, 1201–1208. [CrossRef]

7. Niu, X.; Liu, K.; Zhang, Y.; Xiao, G.; Gong, Y. Multiobjective optimization of multistage synchronous induction coilgun based on NSGA-II. *IEEE Trans. Plasma Sci.* **2017**, *45*, 1622–1628. [CrossRef]

8. Doyle, M.R.; Samuel, D.J.; Conway, T.; Klimowski, R.R. Electromagnetic aircraft launch system-EMALS. *IEEE Trans. Magn.* **1995**, *31*, 528–533. [CrossRef]

9. Musolino, A.; Raugi, M.; Rizzo, R.; Tucci, M. Optimal design of EMALS based on a double-sided tubular linear induction motor. *IEEE Trans. Plasma Sci.* **2015**, *43*, 1326–1331. [CrossRef]

10. Musolino, A.; Raugi, M.; Rizzo, R.; Tucci, M. Force optimization of a double-sided tubular linear induction motor. *IEEE Trans. Magn.* **2014**, *50*, 1–11. [CrossRef]

11. Bushway, R.R. Electromagnetic aircraft launch system development considerations. *IEEE Trans. Magn.* **2001**, *37*, 52–54. [CrossRef]

12. Bertola, L.; Cox, T.; Wheeler, P.; Garvey, S.; Morvan, H. Electromagnetic launch systems for civil aircraft assisted take-off. *Arch. Electr. Eng.* **2015**, *66*, 535–546. [CrossRef]

13. Roesler, G. Mass Estimate for a Lunar Resource Launcher Based on Existing Terrestrial Electromagnetic Launchers. *Machines* **2013**, *1*, 50–62. [CrossRef]

14. McNab, I.R. Launch to space with an electromagnetic railgun. *IEEE Trans. Magn.* **2003**, *39*, 295–304. [CrossRef]

15. Hsieh, K.T.; Stefani, F.; Levinson, S. Numerical modeling of the velocity skin effects: An investigation of issues affecting accuracy. *IEEE Trans. Magn.* **2001**, *37*, 416–420. [CrossRef]

16. Benton, T.; Stefani, F.; Satapathy, S.; Hsieh, K.T. Numerical modeling of melt-wave erosion in conductors. *IEEE Trans. Magn.* **2003**, *39*, 129–133. [CrossRef]

17. Liebfried, O.; Schneider, M.; Stankevič, T.; Balevičius, S.; Žurauskienė, N. Velocity-induced current profiles inside the rails of an electric launcher. *IEEE Trans. Plasma Sci.* **2013**, *41*, 1520–1525. [CrossRef]

18. Wang, Z.; Li, B. A High Reliability 3D Scanning Measurement of the Complex Shape Rail Surface of the Electromagnetic Launcher. *Sensors* **2020**, *20*, 1485. [CrossRef]

19. Kim, H.K.; Kang, B.S.; Moon, Y.H.; Kim, J. Analytical Performance Prediction of an Electromagnetic Launcher and Its Validation by Numerical Analyses and Experiments. *Appl. Sci.* **2019**, *9*, 4063. [CrossRef]

20. Musolino, A. Numerical analysis of a rail launcher with a multilayered armature. *IEEE Trans. Plasma Sci.* **2011**, *39*, 788–793. [CrossRef]

21. Wu, S.; Zhao, W.; Wang, S.; Cui, S. Overview of pulsed alternators. *IEEE Trans. Plasma Sci.* **2017**, *45*, 1078–1085. [CrossRef]

22. Zhao, W.; Wu, S.; Cui, S.; Wang, X. Electromagnetic shields of the air-core compulsator. *IEEE Trans. Plasma Sci.* **2015**, *43*, 1497–1502.

23. Wu, S.; Cui, S.; Song, L.; Zhao, W.; Zhang, J. Design, simulation, and testing of a dual stator-winding all-air-core compulsator. *IEEE Trans. Plasma Sci.* **2010**, *39*, 328–334. [CrossRef]

24. Crawford, M.; Mallick, J.A.; Pappas, J. Use of SABER circuit simulation software for the modeling of compensated pulsed alternators driving a railgun load. *IEEE Trans. Magn.* **2003**, *39*, 337–342. [CrossRef]

25. Cui, S.; Zhao, W.; Wang, S.; Wang, T. Investigation of multiphase compulsator systems using a co-simulation method of FEM-circuit analysis. *IEEE Trans. Plasma Sci.* **2013**, *41*, 1247–1253. [CrossRef]

26. Xie, X.; Yu, K.; Zhang, F.; Tang, P.; Yao, J. Simulation of a seven-phase air-core pulsed alternator driving the electromagnetic rail gun. *IEEE Trans. Plasma Sci.* **2017**, *45*, 1251–1256. [CrossRef]

27. Tripodi, E.; Musolino, A.; Rizzo, R.; Raugi, M. A new predictor–corrector approach for the numerical integration of coupled electromechanical equations. *Int. J. Num. Meth. Eng.* **2016**, *105*, 261–285. [CrossRef]

28. Tripodi, E.; Musolino, A.; Rizzo, R.; Raugi, M. Numerical integration of coupled equations for high-speed electromechanical devices. *IEEE Trans. Magn.* **2015**, *51*, 1–4. [CrossRef]

29. Musolino, A.; Raugi, M.; Rizzo, R.; Sani, L.; Di Dio, V. Electromechanical numerical analysis of an air-core pulsed alternator via equivalent network formulation. *IEEE Trans. Plasma Sci.* **2017**, *45*, 1429–1435. [CrossRef]

30. Musolino, A.; Rizzo, R.; Sani, L.; Tripodi, E. Null-flux coils in permanent magnets bearings. *IEEE Trans. Plasma Sci.* **2017**, *45*, 1545–1552. [CrossRef]

31. Tucci, M.; Sani, L.; Di Dio, V. Optimization of a Novel Magneto-Rheological Device with Permanent Magnets. *Prog. Electromagn. Res.* **2017**, *62*, 175–188. [CrossRef]

32. Musolino, A. Finite-Element Method/Method of Moments Formulation for the Analysis of Current Distribution in Rail Launchers. IEEE Trans. Mag. **2005**, *41*, 387–392.

33. Iachininoto, M.G.; Camisa, V.; Leone, L.; Pinto, R.; Lopresto, V.; Merla, C.; Giorda, E.; Carsetti, R.; Zaffina, S.; Podda, M.V.; et al. Effects of exposure to gradient magnetic fields emitted by nuclear magnetic resonance devices on clonogenic potential and proliferation of human hematopoietic stem cells. *Bioelectromagnetics* **2016**, *37*, 201–211. [CrossRef] [PubMed]

34. Di Dio, V.; Sani, L. Coupled Electromechanical Analysis of a Permanent-Magnet Bearing. *Applied Comp. Electromagnetic Soc. J.* **2017**, *32*, 736–741.

35. Barmada, S.; Musolino, A.; Rizzo, R.; Tucci, M. Multi-resolution based sensitivity analysis of complex non-linear circuits. *IET Circuits Devices Syst.* **2012**, *6*, 176–186. [CrossRef]

36. Liebfried, O.; Schneider, M.; Balevicius, S. Current distribution and contact mechanisms in static railgun experiments with brush armatures. *IEEE Trans. Plasma Sci.* **2010**, *39*, 431–436. [CrossRef]

37. Wild, B.; Alouahabi, F.; Simicic, D.; Schneider, M.; Hoffman, R. A comparison of C-shaped and brush armature performance. *IEEE Trans. Plasma Sci.* **2017**, *45*, 1227–1233. [CrossRef]

MDPI

St. Alban-Anlage 66

4052 Basel

Switzerland

Tel. +41 61 683 77 34

Fax +41 61 302 89 18

www.mdpi.com

Applied Sciences Editorial Office

E-mail: applsci@mdpi.com

www.mdpi.com/journal/applsci